# Lecture Notes in Mathematics 1983

Habib Ammari (Ed.)

# Mathematical Modeling in Biomedical Imaging I

## Electrical and Ultrasound Tomographies, Anomaly Detection, and Brain Imaging

Springer

*Editor*
Habib Ammari
Laboratoire Ondes et Acoustique
ESPCI, 10 Rue Vauquelin
75231 Paris, Cedex 05
France
habib.ammari@espci.fr

and

Centre de Mathématiques Appliquées
Ecole Polytechnique
91128, Palaiseau, Cedex
France
habib.ammari@polytechnique.fr

ISBN: 978-3-642-03443-5 e-ISBN: 978-3-642-03444-2
DOI: 10.1007/978-3-642-03444-2
Springer Heidelberg Dordrecht London NewYork

Lecture Notes in Mathematics ISSN print edition: 0075-8434
ISSN electronic edition: 1617-9692

Library of Congress Control Number: 2009932673

Mathematics Subject Classification (2000): 92C55, 31A25, 31A10, 35B25, 35J55, 35J65, 34A55



*Cover design*: SPi Publisher Services

Printed on acid-free paper

springer.com

# Preface

Mathematical sciences are contributing more and more to advances in life science research, a trend that will grow in the future.

Realizing that the mathematical sciences can be critical to many areas of biomedical imaging, we organized a three-day minicourse on mathematical modelling in biomedical imaging at the Institute Henri Poincaré in Paris in March 2007. Prominent mathematicians and biomedical researchers were paired to review the state-of-the-art in the subject area and to share mathematical insights regarding future research directions in this growing discipline.

The speakers gave presentations on hot topics including electromagnetic brain activity, time-reversal techniques, elasticity imaging, infrared thermal tomography, acoustic radiation force imaging, electrical impedance and magnetic resonance electrical impedance tomographies. Indeed, they contributed to this volume with original chapters to give a wider audience the benefit of their talks and their thoughts on the field.

This volume is devoted to providing an exposition of the promising analytical and numerical techniques for solving important biomedical imaging problems and to piquing interest in some of the most challenging issues. We hope that it will stimulate much needed progress in the directions that were described during the course. The biomedical imaging problems addressed in this volume trigger the investigation of interesting and difficult problems in various branches of mathematics including partial differential equations, harmonic analysis, complex analysis, numerical analysis, optimization, image analysis, and signal theory.

The partial support offered by the ANR project EchoScan (AN-06-Blan-0089) is acknowledged. We also thank the staff at the Institute Henri Poincaré.

Paris
March 2009

*Habib Ammari*

# List of Contributors

**Habib Ammari** Laboratoire Ondes et Acoustique, 10 rue Vauquelin, CNRS, E.S.P.C.I, Université Paris 7, 75005, Paris, France, habib.ammari@espci.fr

**Sylvain Baillet** Cognitive Neuroscience and Brain Imaging Laboratory, Université Pierre and Marie Curie-Paris 6, CNRS UPR 640 – LENA, Paris, France, sylvain.baillet@chups.jussieu.fr

**George Dassios** Department of Applied Mathematics and Theoretical Physics, University of Cambridge, Cambridge CB3 0WA, UK, G.Dassios@damtp.cam.ac.uk

**Mathias Fink** Laboratoire Ondes et Acoustique, 10 rue Vauquelin, CNRS, E.S.P.C.I, Université Paris 7, 75005, Paris, France, mathias.fink@espci.fr

**Hyeonbae Kang** Department of Mathematics, Inha University, Incheon 402-751, Korea, hbkang@inha.ac.kr

**Julien Lefèvre** Cognitive Neuroscience and Brain Imaging Laboratory, Université Pierre and Marie Curie-Paris 6, CNRS UPR 640 – LENA, Paris, France, julien.lefevre@chups.jussieu.fr

**Jin Keun Seo** Department of Mathematics, Yonsei University, 262 Seongsanno, Seodaemun, Seoul 120-749, Korea, seoj@yonsei.ac.kr

**Mickael Tanter** Laboratoire Ondes et Acoustique, 10 rue Vauquelin, CNRS, E.S.P.C.I, Université Paris 7, 75005, Paris, France, mickael.tanter@espci.fr

**Eung Je Woo** College of Electronics and Information, Kyung Hee University, Seoul 130-701, Korea, ejwoo@khu.ac.kr

# Contents

# Introduction

Medical imaging modalities such as computerized tomography using X-ray and magnetic resonance imaging have been well established providing three-dimensional high-resolution images of anatomical structures inside the human body. Computer-based mathematical methods have played an essential role for their image reconstructions. However, since each imaging modality has its own limitations, there have been much research efforts to expand our ability to see through the human body in different ways. Lately, biomedical imaging research has been dealing with new imaging techniques to provide knowledge of physiologic functions and pathological conditions in addition to structural information.

Electrical impedance tomography, ultrasound imaging, and electrical and magnetic source imaging are three of such attempts for functional imaging and monitoring of physiological events.

The aim of this book is to review the most recent advances in the mathematical and numerical modelling of these three emerging modalities. Although they use different physical principles for signal generation and detection, the underlying mathematics are quite similar. We put a specific emphasis on the mathematical concepts and tools for image reconstruction. Other promising modalities such as photo-acoustic imaging and fluorescence microscopy as well as those in nuclear medicine will be discussed in a forthcoming volume.

Electrical impedance tomography uses low-frequency electrical current to probe a body; the method is sensitive to changes in electrical conductivity. By injecting known amounts of current and measuring the resulting electrical potential field at points on the boundary of the body, it is possible to invert such data to determine the conductivity or resistivity of the region of the body probed by the currents. This method can also be used in principle to image changes in dielectric constant at higher frequencies. However, the aspect of the method that is most fully developed to date is the imaging of conductivity. Potential applications of electrical impedance tomography include determination of cardiac output, monitoring for pulmonary edema, and screening for breast cancer.

Electrical source imaging is an emerging technique for reconstructing brain electrical activity from electrical potentials measured away from the brain. The concept of electrical source imaging is to improve on electroencephalography by determining the locations of sources of current in the body from measurements of voltages.

Ion currents arising in the neurons of the brain produce magnetic fields outside the body that can be measured by arrays of superconducting quantum interference device detectors placed near the chest; the recording of these magnetic fields is known as magnetoencephalography. Magnetic source imaging is the reconstruction of the current sources in the brain from these recorded magnetic fields. These fields result from the synchronous activity of tens or hundreds of thousands of neurons.

Both magnetic source imaging and electrical source imaging seek to determine the location, orientation, and magnitude of current sources within the body.

Ultrasound imaging is a noninvasive, easily portable, and relatively inexpensive diagnostic modality which finds extensive use in the clinic. The major clinical applications of ultrasound include many aspects of obstetrics and gynecology involving the assessment of fetal health, intra-abdominal imaging of the liver, kidney, and the detection of compromised blood flow in veins and arteries.

Operating typically at frequencies between 1 and 10 MHz, ultrasound imaging produces images via the backscattering of mechanical energy from interfaces between tissues and small structures within tissue. It has high spatial resolution, particularly at high frequencies, and involves no ionizing radiation. The weakness of the technique include the relatively poor soft-tissue contrast and the fact that gas and bone impede the passage of ultrasound waves, meaning that certain organs can not easily be imaged.

As we said before, in this book not only the basic mathematical principles of these three emerging modalities are reviewed but also the most recent developments to improve them are reported. We emphasize the mathematical concepts and tools for image reconstruction. Our main focuses are, on one side, on promising anomaly detection techniques in electrical impedance tomography and in elastic imaging using the method of small-volume expansions and in ultrasound imaging using time-reversal techniques, and on the other side, on emerging multi-physics or hybrid imaging approaches such as the magnetic resonance electrical impedance, impediography, and magnetic resonance elastography.

The book is organized as follows. Chapter 1 is devoted to electrical impedance tomography and magnetic resonance electrical impedance tomography. It focuses on robust reconstructions of conductivity images under practical environments having various technical limitations of data collection equipments and fundamental limitations originating from its inherent ill-posed nature. The mathematical formulation of the magnetic resonance electrical impedance tomography and multi-frequencies electrical impedance tomography are rigorously described. Efficient image reconstruction algorithms are provided and their limitations are discussed.

Chapter 2 outlines the basic physical principles of time-reversal techniques and their applications in ultrasound imaging. It gives a good introduction to this very interesting subject.

Chapter 3 covers the method of small-volume expansions. A remarkable feature of the method of small-volume expansions is that it allows a stable and accurate reconstruction of the location and of geometric features of the anomalies, even for moderately noisy data. Based on this method robust and efficient algorithms for imaging small thermal conductivity, electromagnetic, and elastic anomalies are provided. Emerging multi-physics or hybrid imaging approaches, namely impediography, magneto-acoustic imaging, and magnetic resonance elastography are also discussed. In these techniques, different physical types of radiation are combined into one tomographic process to alleviate deficiencies of each separate type of waves, while combining their strengths. Finally, a mathematical formulation of the concept of time reversing waves is provided and its use in imaging is described.

Chapter 4 deals with electrical and magnetic source imaging reconstruction methods for focal brain activity. Mathematical formulations and uniqueness and non-uniqueness results for the inversion source problems are given. The basic mathematical model is described by the Biot–Savart law of magnetism, which makes the mathematical difficulties for solving the inverse source problem very similar to those in magnetic resonance electrical impedance tomography discussed in Chap. 1.

Chapter 5 considers time-resolved imaging of brain activity. It discusses optical flow techniques in order to infer on the stability of brain activity.

# Chapter 1
# Multi-frequency Electrical Impedance Tomography and Magnetic Resonance Electrical Impedance Tomography

**Jin Keun Seo and Eung Je Woo**

## 1.1 Introduction

Medical imaging modalities such as computerized tomography (CT) using X-ray and magnetic resonance imaging (MRI) have been well established providing three-dimensional high-resolution images of anatomical structures inside the human body and computer-based mathematical methods have played an essential role for their image reconstructions. However, since each imaging modality has its own limitations, there have been much research efforts to expand our ability to see through the human body in different ways. Lately, biomedical imaging research has been dealing with new imaging techniques to provide knowledge of physiologic functions and pathological conditions in addition to structural information. Electrical impedance tomography (EIT) is one of such attempts for functional imaging and monitoring of physiological events.

EIT is based on numerous experimental findings that different biological tissues inside the human body have different electrical properties of conductivity and permittivity. Viewing the human body as a mixture of distributed resistors and capacitors, we can evaluate its internal electrical properties by injecting a sinusoidal current between a pair of surface electrodes and measuring voltage drops at different positions on the surface. EIT is based on this bioimpedance measurement technique using multiple surface electrodes as many as 8 to 256. See Figs. 1.1a and 1.2. In EIT, we inject linearly independent patterns of sinusoidal currents through all or chosen pairs of electrodes and measure induced boundary voltages on all or selected electrodes.

J.K. Seo (✉)
Department of Mathematics, Yonsei University, 262 Seongsanno, Seodaemun, Seoul 120-749, Korea
e-mail: seoj@yonsei.ac.kr

E. Je Woo
College of Electronics and Information, Kyung Hee University, Seoul 130-701, Korea
e-mail: ejwoo@khu.ac.kr

H. Ammari, *Mathematical Modeling in Biomedical Imaging I*,
Lecture Notes in Mathematics 1983, DOI 10.1007/978-3-642-03444-2_1,

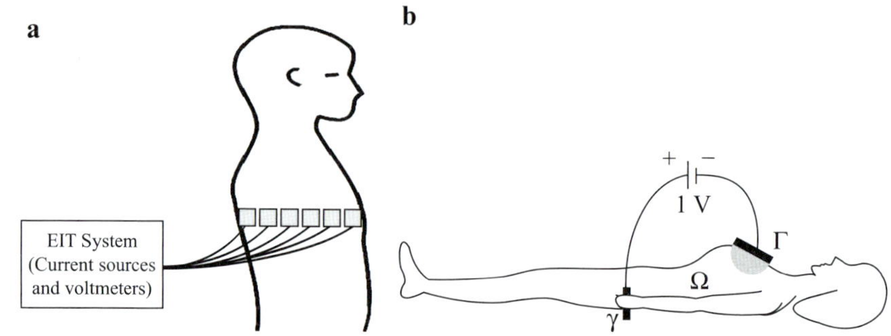


**Fig. 1.1** (**a**) EIT system and (**b**) TAS system

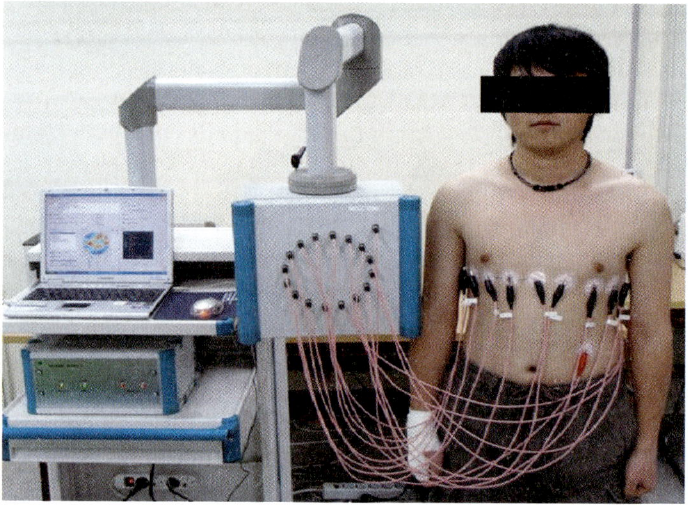

**Fig. 1.2** EIT system at Impedance Imaging Research Center (IIRC) in Korea

The measured boundary current–voltage data set is used to reconstruct cross-sectional images of the internal conductivity and/or permittivity distribution. The basic idea of the impedance imaging was introduced by Henderson and Webster in 1978 [13], and the first clinical application of a medical EIT system was described by Barber and Brown [7]. Since then, EIT has received considerable attention and several review papers described numerous aspects of the EIT technique [8, 10, 14, 36, 49, 62]. To support the theoretical basis of the EIT system, mathematical theories such as uniqueness and stability were developed [2, 6, 16, 19, 25, 29, 38, 39, 48, 52, 57–59, 61] since Calderón's pioneering contribution in 1980 [9].

Most EIT imaging methods use a forward model of an imaging object with a presumed conductivity and permittivity distributions. Injecting the same currents into the model, boundary voltages are computed to numerically simulate measured data. Using differences between measured and computed (or referenced) current-to-voltage data, we produce EIT images through a misfit minimization process. However, the inverse problem in EIT has suffered from its ill-posed characteristic due to the inherent insensitivity of boundary measurements to any changes of interior conductivity and permittivity values.

In practice, it is very difficult to construct an accurate forward model of the imaging object due to technical difficulties in capturing the boundary shape and electrode positions with a reasonable accuracy and cost. Therefore, there always exist uncertainties in these geometrical data needed for the model and this causes systematic errors between measured and computed voltages without considering mismatch in the true and model conductivity and permittivity distributions. The ill-posedness of EIT together with these systematic artifacts related with inaccurate boundary geometry and electrode positions make it difficult to reconstruct accurate images with a high spatial resolution in clinical environments. Primarily due to the poor spatial resolution and accuracy of EIT images, its practical applicability has been limited in clinical applications. Taking account of these restrictions, it is desirable for EIT to find clinical applications where its portability and high temporal resolution to monitor changes in electrical properties are significant merits.

Magnetic resonance electrical impedance tomography (MREIT) was motivated to deal with the well-known severe ill-posedness of the image reconstruction problem in EIT. In MREIT, we inject current $I$ (Neumann data) into an object $\Omega$ through a pair of surface electrodes to produce internal current density $\mathbf{J} = (J_x, J_y, J_z)$ and magnetic flux density $\mathbf{B} = (B_x, B_y, B_z)$ in $\Omega$. The distribution of the induced magnetic flux density $\mathbf{B}$ is governed by the Ampere law $\mathbf{J} = \frac{1}{\mu_0}\nabla \times \mathbf{B}$ where $\mu_0$ is the magnetic permeability of the free space. Let $z$ be the direction of the main magnetic field of an MRI scanner. Then, the $B_z$ data can be measured by using an MRI scanner as illustrated in Fig. 1.3. MREIT takes advantage of the MRI scanner as a tool to capture the $z$-component $B_z$ of the induced magnetic flux density in $\Omega$. Conductivity imaging in MREIT is based on the relationship between the injection current $I$ and the measured $B_z$ data which conveys the information about any local change of the conductivity $\sigma$ via the Biot–Savart law:

$$B_z(x,y,z) = \frac{\mu_0}{4\pi}\int_{\Omega} \frac{\sigma(\mathbf{r})\left[(x-x')\frac{\partial u}{\partial y}(\mathbf{r}') - (y-y')\frac{\partial u}{\partial x}(\mathbf{r}')\right]}{|\mathbf{r}-\mathbf{r}'|^3}\, d\mathbf{r}',\ \mathbf{r} = (x,y,z) \in \Omega$$

where $u$ is the induced electrical potential due to the injection current. This supplementary use of the internal $B_z$ data enables MREIT to bypass the ill-posedness problem in EIT.

The technique to measure the internal magnetic flux density $\mathbf{B}$ using an MRI scanner was originally developed for magnetic resonance current density imaging (MRCDI) in late 1980s [17]. In MRCDI, we have to rotate the

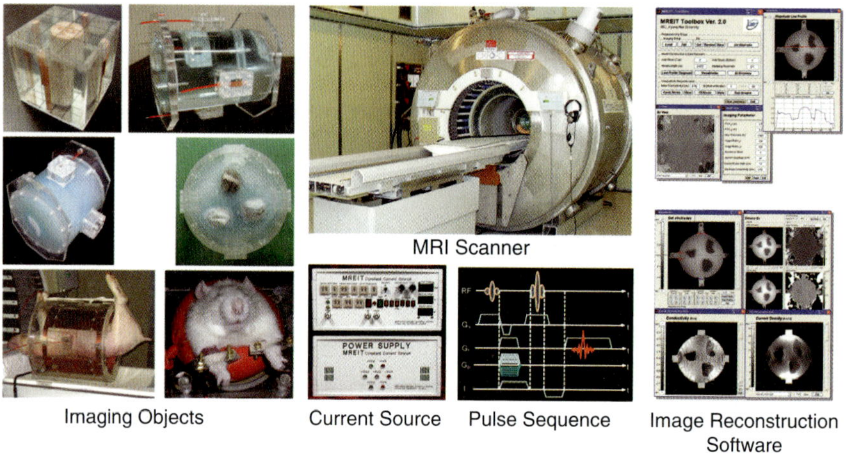


**Fig. 1.3** MREIT system at Impedance Imaging Research Center (IIRC) in Korea and its image reconstruction software

object in three orthogonal directions to obtain all three components of **B** since the MRI system can measure only the component of **B** that is parallel to the direction of its main magnetic field (usually denoted as the $z$-direction). Once we get all three components of **B**, we can visualize the internal current density distribution **J** via the Ampere law $\mathbf{J} = \frac{1}{\mu_0}\nabla \times \mathbf{B}$ [50, 51]. This MRCDI provided a strong motivation of MREIT which combines EIT and MRI techniques [15, 30, 63, 65].

In order for MREIT to be practical, it is obvious that we must be able to produce images of the internal conductivity distribution without rotating the object. This means that the $z$-component $B_z$ of **B** is the only available data. The first constructive $B_z$-based MREIT algorithm called the harmonic $B_z$ algorithm was proposed by Seo et al. in 2001 [56]. It is based on the following curl of the Ampere law:

$$\frac{1}{\mu_0}\Delta B_z = \langle \nabla\sigma \, , \begin{pmatrix} 0 & 1 & 0 \\ -1 & 0 & 0 \\ 0 & 0 & 0 \end{pmatrix} \nabla u \rangle \qquad (1.1)$$

where $\langle \cdot, \cdot \rangle$ denotes the inner product and $\Delta$ is the Laplacian. After the harmonic $B_z$-algorithm, various image reconstruction algorithms based on the $B_z$-based MREIT model have been developed [41, 42, 44, 45, 54, 55]. Recent published numerical simulations and phantom experiments show that conductivity images with a high spatial resolution are achievable as long as the measured $B_z$ data has an enough signal-to-noise ratio (SNR).

Although imaging techniques in MREIT have been advanced rapidly, rigorous mathematical theories of the numerical algorithms such as stability and convergence have not been supported yet. Theoretical as well as experimental studies in MREIT are essential for the progress of the technique. In this lecture note, we explains recent results on the convergence behavior and numerical stability of the harmonic $B_z$ algorithm based on a mathematical model replicating the actual MREIT system [34, 35]. Before clinical applications of MREIT, it is necessary to study how errors in the raw data are propagated to the final result of the conductivity imaging. Hence, it is highly necessary to set up an actual mathematical model for MREIT describing the accurate relationship among input current, $B_z$ data and conductivity distribution. For the real MREIT model, boundary conditions are different from conventional styles in PDE and great care is required in using non-standard boundary conditions. The disadvantages of MREIT over EIT may include the lack of portability, potentially long imaging time and requirement of an expensive MRI scanner. Hence, EIT still has various advantages over MREIT although we should not expect EIT to compete with MREIT in terms of spatial resolution.

Lately, a frequency-difference electrical impedance tomography (fdEIT) has been proposed to deal with technical difficulties of the conventional static EIT imaging caused by unknown boundary geometry, uncertainty in electrode position and other systematic measurement artifacts [21, 33, 43]. Conductivity ($\sigma$) and permittivity ($\epsilon$) spectra of numerous biological tissues show frequency-dependent changes indicating that we can view a complex conductivity ($\sigma + i\epsilon$) distribution inside an imaging object as a function of frequency. In fdEIT, we inject currents with at least two different frequencies and use the difference between induced boundary voltages at different frequencies to eliminate unknown common modelling errors. To test its feasibility, we consider anomaly detection problems where an explicit representation formula for the potential is available. The formula provides a clear connection between its Cauchy data and the anomaly [3, 29]. As an example of such an anomaly detection problem, let us consider the breast cancer detection problem. In this case, the inverse problem is reduced to detect a suspicious abnormality (instead of imaging) underneath the breast skin from measured boundary data. Figure 1.1b depicts trans-admittance scanner (TAS) which is a device for breast cancer diagnosis. Most of anomaly detection methods used a difference between measured data and reference data in the absence of anomaly. However, in practice, the reference data is not available and its computation is not possible since the inhomogeneous complex conductivity of the background is unknown. To deal with this problem, multi-frequency TAS system has been proposed where a frequency difference of measured data sets at a certain moment is used for anomaly detection [43].

This lecture note focuses on robust reconstructions of conductivity images under practical environments having various technical limitations of data collection equipments and fundamental limitations originating from its inherent nature. We describe the mathematical formulation of MREIT and multi-frequency EIT in clinical environments, image reconstruction algorithms, measurement techniques and examples of images.

## 1.2 Electrical Impedance Tomography

### 1.2.1 Inverse Problem in RC-Circuit

The human body can be viewed as a mixture of distributed resistors and capacitors and a circuit model containing resistors and capacitors can be used to explain the one-dimensional EIT problem. Let us begin with considering a simple RC-circuit. Electrical impedance, denoted by $Z$, is a measure of the total opposition of a circuit to a time-varying electrical current flow. It comprises resistance and reactance taking account of the effects from resistors and capacitors, respectively.

Consider a linear circuit containing a resistor, capacitor and sinusoidally time-varying current source connected in series. If the current source in the circuit is given by $I(t) = I_0 \cos(\omega t)$ where $I_0$ is the amplitude and $\omega$ is the angular frequency, then the resulting voltage $V(t)$ is also sinusoidal with the same angular frequency $\omega$. The relation between $I(t)$ and $V(t)$ is governed by

$$RI(t) + \frac{1}{C}\int I(t)dt = V(t) \tag{1.2}$$

where $R$ is the resistance and $C$ is the capacitance. The voltage can be expressed as

$$V(t) := RI_0 \cos(\omega t) + \frac{I_0}{\omega C}\sin(\omega t) = V_0 \cos(\omega t - \phi)$$

where $V_0 = \sqrt{(RI_0)^2 + \left(\frac{I_0}{\omega C}\right)^2}$ is the amplitude and $\phi$ is the phase angle such that $\tan\phi = \frac{1}{\omega RC}$ and $0 \le \phi \le \frac{\pi}{2}$. In order to see the interrelation among the impedance $Z := R + \frac{1}{i\omega C}$, voltage and current, it is convenient to express sinusoidally time-varying functions $I(t)$ and $V(t)$ in terms of time-independent phasors $\tilde{I}$ and $\tilde{V}$ such as

$$I(t) = \Re\{\tilde{I}e^{i\omega t}\} \quad \text{and} \quad V(t) = \Re\{\tilde{V}e^{i\omega t}\}$$

where $\tilde{I} = I_0$ and $\tilde{V} = V_0 e^{-i\phi}$. The phasor $\tilde{V}$, corresponding to the time function $V(t)$, contains the amplitude $|\tilde{V}| = V_0$ and phase $\arg(\tilde{V}) = \phi$. With the use of phasors $\tilde{I}$ and $\tilde{V}$, (1.2) can be expressed as

$$\left[R + \frac{1}{i\omega C}\right]\tilde{I}e^{i\omega t} = \tilde{V}e^{i\omega t}$$

or simply

$$\left[R + \frac{1}{i\omega C}\right] I_0 = V_0 e^{-i\phi}. \tag{1.3}$$

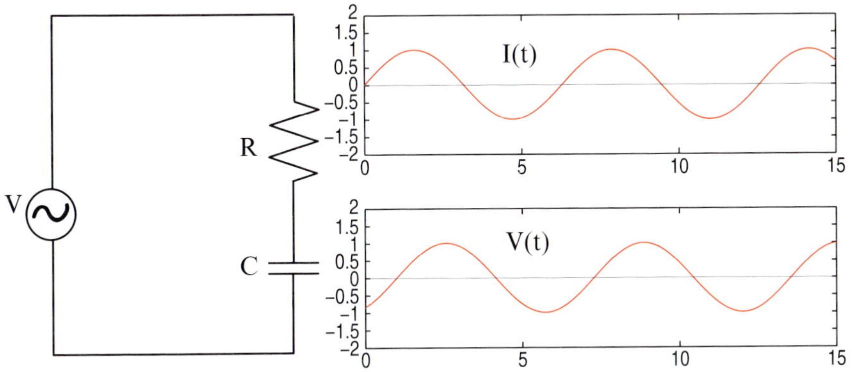


**Fig. 1.4** RC-circuit and current–voltage data

Now, let us consider a very elementary inverse problem in the simple RC-circuit (Fig. 1.4).

*Example 1.1 (One-dimensional inverse problem of a simple RC-circuit).* Consider the simple RC-circuit and current–voltage data shown below. Find the electrical impedance $Z := R + \frac{1}{i\omega C}$ from the relation between $V(t)$ and $I(t)$.

The impedance $Z := R + \frac{1}{i\omega C}$ can be obtained from the ratio of the phasor voltage $\tilde{V}$ to the phasor current $\tilde{I}$:

$$Z = \frac{\tilde{V}}{\tilde{I}} = \frac{V_0}{I_0} e^{-i\phi}.$$

Hence, $R = \Re\{Z\}$ and $C = \frac{-1}{\omega \Im\{Z\}}$. □

*Example 1.2 (Non-uniqueness in one-dimensional inverse problem).* Consider the circuit containing two resistors $R_1$ and $R_2$ and two capacitors $C_1$ and $C_2$ (Fig. 1.5). Find $R_1$, $R_2$, $C_1$ and $C_2$ from $I(t)$ and $V(t)$.

From (1.3), we have $(R_1 + R_2) + \frac{1}{i\omega}\left(\frac{1}{C_1} + \frac{1}{C_2}\right) = \frac{\tilde{V}}{\tilde{I}}$ Hence, the relation between $\tilde{V}$ and $\tilde{I}$ determines $R = R_1 + R_2$ and $\frac{1}{C} = \frac{1}{C_1} + \frac{1}{C_2}$ uniquely. On the other hand, the inverse problem of finding $R_1, R_2, C_1$ and $C_2$ from $V(t)$ and $I(t)$ has infinitely many solutions; all positive real numbers $R_1, R_2, C_1, C_2$ satisfying $R = R_1 + R_2$ and $\frac{1}{C} = \frac{1}{C_1} + \frac{1}{C_2}$ are solutions of the inverse problem. □

*Example 1.3 (One-dimensional inverse problem with many uniform Rs and Cs).* Consider the circuit containing many uniform resistors and capacitors shown in Fig. 1.6. Find $R$ and $C$ from $I(t)$ and $V(t)$.

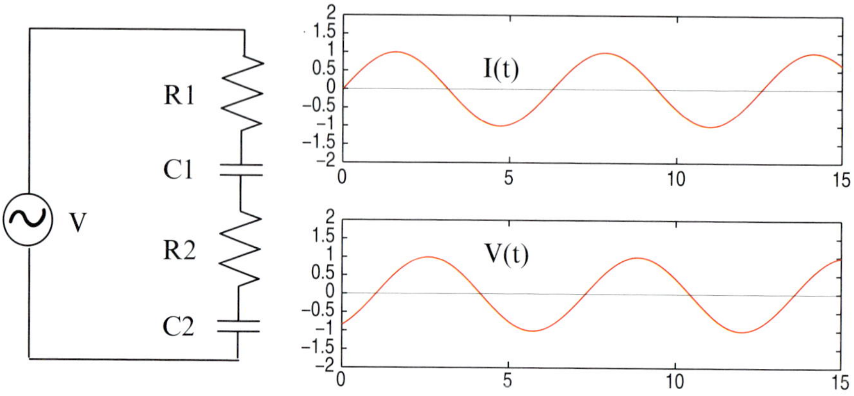


**Fig. 1.5** Circuit containing two resistors and two capacitors

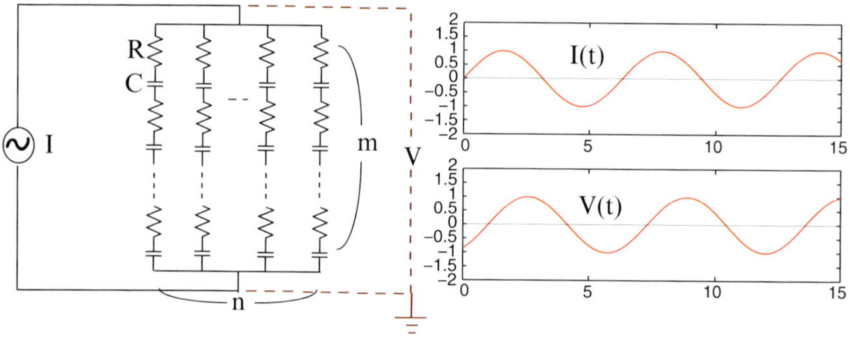


**Fig. 1.6** Circuit containing many uniform resistors and capacitors

Since $\frac{m}{n}\left(R+\frac{1}{i\omega C}\right) = \frac{\tilde{V}}{\tilde{I}}$ from (1.3), the ratio $\frac{m}{n}$ would be the most important factor in the reconstruction. Indeed, $R = \frac{m}{n}\Re\{\frac{\tilde{V}}{\tilde{I}}\}$ and $C = \left[-\omega\frac{n}{m}\Im\{\frac{\tilde{V}}{\tilde{I}}\}\right]^{-1}$. □

### 1.2.2 Governing Equation in EIT

Let the imaging object occupy a three-dimensional region $\Omega$ bounded by its surface $\partial\Omega$. Assume that the boundary $\partial\Omega$ is connected and smooth. We denote the conductivity by $\sigma = \sigma(\mathbf{r},\omega)$ and the permittivity by $\epsilon = \epsilon(\mathbf{r},\omega)$ which depend on the position $\mathbf{r} = (x,y,z)$ and the angular frequency $\omega$. We assume that the magnetic permeability $\mu$ of the entire object is $\mu_0$, the magnetic permeability of the free space.

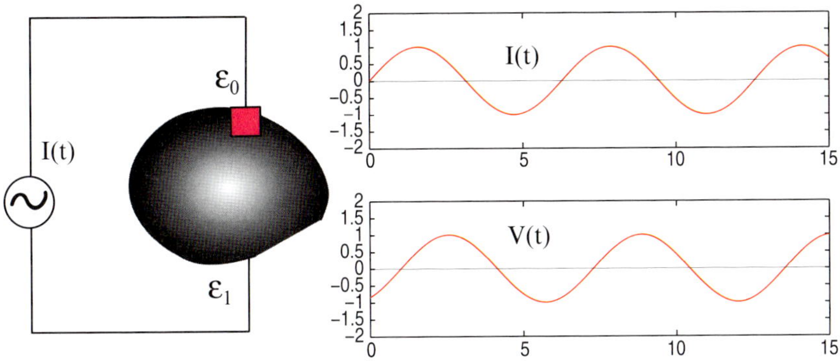


**Fig. 1.7** Principles of EIT

To understand the relation among $(\sigma, \epsilon)$, electrical field and magnetic field, let us review the electromagnetic phenomena that can be explained by Maxwell's equations. When we inject current $I(t) = I \cos \omega t$ into $\Omega$ through a pair of electrodes $\mathcal{E}_0$ and $\mathcal{E}_1$, it produces a sinusoidal electric field $\tilde{\mathbf{E}}(\mathbf{r}, t)$ with the same angular frequency $\omega$. Throughout this section, we assume that

$$\frac{\omega}{2\pi} \leq 10^5\, Hz \quad \text{and} \quad \text{diam}(\Omega) \leq 3 \text{ m}.$$

*Example 1.4.* As in Fig. 1.7, we attached two electrodes $\mathcal{E}_0$ and $\mathcal{E}_1$ on $\partial\Omega$. We inject $I(t) = I \cos \omega t$ through $\mathcal{E}_0$ and $\mathcal{E}_1$ to generate the electrical potential $V(\mathbf{r}, t)$ which is also a sinusoidal function with the same angular frequency $\omega$. Let us derive the governing equation for $V(\mathbf{r}, t)$. For the reference potential, we assume $V|_{\mathcal{E}_0} = 0$.

To derive the governing equation for $V$, we use the time-varying electromagnetic phenomena that is based on the following four relations known as Maxwell's equations.

| Law | Time-varying field | Time-harmonic field |
|---|---|---|
| Faraday's law | $\nabla \times \tilde{\mathbf{E}} = -\frac{\partial \tilde{\mathbf{B}}}{\partial t}$ | $\nabla \times \mathbf{E} = -i\omega \mathbf{B}$ |
| Ampére's law | $\nabla \times \tilde{\mathbf{H}} = \tilde{\mathbf{J}} + \frac{\partial \tilde{\mathbf{D}}}{\partial t}$ | $\nabla \times \mathbf{H} = \mathbf{J} + i\omega \mathbf{E}$ |
| Gauss's law | $\nabla \cdot \tilde{\mathbf{D}} = \rho$ | $\nabla \cdot \mathbf{D} = \rho$ |
| Gauss's law for magnetism | $\nabla \cdot \tilde{\mathbf{B}} = 0$ | $\nabla \cdot \mathbf{B} = 0$ |

Here, we summarize notations:

- Time-varying fields: $\mathbf{E}(\mathbf{r}, t) :=$ electric field ($\text{V m}^{-1}$), $\tilde{\mathbf{J}}(\mathbf{r}, t) :=$ current density ($\text{A m}^{-2}$), $\tilde{\mathbf{D}} :=$ electric flux density ($\text{C m}^{-2}$) and $\mathbf{B} :=$ magnetic flux density (T).
- $\tilde{\mathbf{J}} = \tilde{\mathbf{J}}_c + \tilde{\mathbf{J}}_s$, sum of conduction and source currents. Here, $\tilde{\mathbf{J}}_s = 0$ in $\Omega$ for EIT.

- Time-harmonic fields: $\mathbf{E}(\mathbf{r})$, $\mathbf{J}(\mathbf{r})$, $\mathbf{D}(\mathbf{r})$ and $\mathbf{B}(\mathbf{r})$ where $\tilde{\mathbf{J}}(\mathbf{r},t) = \Re\{\mathbf{J}(\mathbf{r})e^{i\omega t}\}$, $\tilde{\mathbf{D}}(\mathbf{r},t) = \Re\{\mathbf{D}(\mathbf{r})e^{i\omega t}\}$ and $\tilde{\mathbf{B}}(\mathbf{r},t) = \Re\{\mathbf{B}(\mathbf{r})e^{i\omega t}\}$.
- $\rho :=$ electrical charge density per unit volume ($\mathrm{C\,m^{-3}}$).
- $\mathbf{J} = \sigma\mathbf{E}$, $\mathbf{D} = \epsilon\mathbf{E}$, $\mathbf{B} = \mu\mathbf{H}$ where $\mu :=$ magnetic permeability.
- In the free space, $\mu_0 = 4\pi \times 10^{-7}$ ($\mathrm{H\,m^{-1}}$), $\epsilon_0 = 8.85 \times 10^{-12}\,\mathrm{F\,m^{-1}}$ and $c = \frac{1}{\sqrt{\mu_0\epsilon_0}} = 3 \times 10^8\,\mathrm{m\,s^{-1}}$, speed of light.

From the assumption that $\mu = \mu_0$, $\frac{\omega}{2\pi} \le 10^5$ and $\mathrm{diam}(\Omega) \le 3\,\mathrm{m}$, we may approximate

$$\nabla \times \mathbf{E} = \mu_0\omega\mathbf{H} \approx 0.$$

Since $\mathbf{E}$ is approximately irrotational, it follows from Stokes's theorem that we define a complex potential $u$ based on the line integral:

$$u(\mathbf{r}_2) - u(\mathbf{r}_1) = -\int_{C_{\mathbf{r}_1\to\mathbf{r}_2}} \mathbf{E}\cdot dl \quad \text{is approximately path independent}$$

where $C_{\mathbf{r}_1\to\mathbf{r}_2}$ is a curve lying in $\Omega$ with the starting point $\mathbf{r}_1$ and the ending point $\mathbf{r}_2$. Hence, the complex potential $u$ satisfies

$$-\nabla u(\mathbf{r}) \approx \mathbf{E}(x) \quad \text{in } \Omega.$$

For the uniqueness of $u$, we set the reference voltage $u|_{\mathcal{E}_0} = 0$.

From $\nabla \times \mathbf{H} = \mathbf{J} + i\omega\mathbf{D} = (\sigma + i\omega\epsilon)\mathbf{E}$, we have the following relation:

$$\nabla \times \mathbf{H}(\mathbf{r}) = \underbrace{(\sigma(\mathbf{r},\omega) + i\omega\epsilon(\mathbf{r},\omega))}_{:=\gamma(\mathbf{r},\omega)}\mathbf{E}(\mathbf{r}) = -\gamma\nabla u.$$

Since $\nabla\cdot(\nabla\times\mathbf{H}) = 0$, the complex potential $u$ satisfies

$$-\nabla\cdot(\gamma\nabla u) = \nabla\cdot\nabla\times\mathbf{H} = 0 \quad \text{in } \Omega.$$

Neglecting the contact impedance at the surface $\mathcal{E}_0 \cup \mathcal{E}_1 \subset \partial\Omega$, the injection current $I\cos(\omega t)$ into $\Omega$ through the electrodes $\mathcal{E}_0$ and $\mathcal{E}_1$ provides the boundary condition for $\tilde{\mathbf{J}}$ and $\tilde{\mathbf{D}}$:

- $\int_{\mathcal{E}_1} \mathbf{n}\cdot\tilde{\mathbf{J}}ds = -\int_{\mathcal{E}_0} \mathbf{n}\cdot\tilde{\mathbf{J}}ds = I\cos(\omega t)$ where $\mathbf{n}$ is the unit outward normal vector, $ds$ is the surface element and $\mathcal{E}_k$ is the portion of the surface $\partial\Omega$ contacting the corresponding electrode.
- $\tilde{\mathbf{J}}\cdot\mathbf{n} = 0 = \tilde{\mathbf{D}}\cdot\mathbf{n}$ on $\partial\Omega\setminus(\mathcal{E}_0\cup\mathcal{E}_1)$.
- Since voltages on electrodes $\mathcal{E}_0$ and $\mathcal{E}_1$ are constants,

$$\nabla\times\tilde{\mathbf{J}} = 0 = \nabla\times\tilde{\mathbf{D}} \quad \text{on } \mathcal{E}_0\cup\mathcal{E}_1.$$

The above boundary condition for the time-varying electric field leads to that of the time-independent complex potential $u$:

- $-\int_{\mathcal{E}_1} \mathbf{n}\cdot(\gamma\nabla u)ds = \int_{\mathcal{E}_0} \mathbf{n}\cdot(\gamma\nabla u) = I$ (approximation).

- $(\gamma\nabla u)\cdot\mathbf{n}=0 \quad \text{on } \partial\Omega\setminus(\mathcal{E}_0\cup\mathcal{E}_1)$.
- $\mathbf{n}\times\nabla u=0 \quad \text{on } \mathcal{E}_0\cup\mathcal{E}_1$ (neglecting contact impedance).

Hence, the complex potential approximately satisfies the following mixed boundary value problem:

$$\begin{cases} -\nabla\cdot(\gamma\nabla u)=0 \quad \text{in } \Omega \\ I=-\int_{\mathcal{E}_1}\gamma\frac{\partial u}{\partial\mathbf{n}}ds=\int_{\mathcal{E}_0}\gamma\frac{\partial u}{\partial\mathbf{n}}ds \\ \gamma\frac{\partial u}{\partial\mathbf{n}}=0 \quad \text{on } \partial\Omega\setminus\overline{\mathcal{E}_0\cup\mathcal{E}_1} \\ \nabla u\times\mathbf{n}|_{\mathcal{E}_1}=0, \quad u|_{\mathcal{E}_0}=0. \end{cases} \tag{1.4}$$

Denoting the real and imaginary parts of the potential $u$ by $v_\omega=\Re u$ and $h_\omega=\Im u$, the boundary value problem (1.4) can be expressed as the following coupled system:

$$\begin{cases} \nabla\cdot(\sigma\nabla v_\omega)-\nabla\cdot(\omega\epsilon\nabla h_\omega)=0 & \text{in } \Omega \\ \nabla\cdot(\omega\epsilon\nabla v_\omega)+\nabla\cdot(\sigma\nabla h_\omega)=0 & \text{in } \Omega \\ \int_{\mathcal{E}_j}\mathbf{n}\cdot(-\sigma\nabla v_\omega(x)+\omega\epsilon\nabla h_\omega(x))ds=(-1)^j I & \text{for } j=0,1 \\ \quad \nabla v_\omega\times\mathbf{n}|_{\mathcal{E}_1}=0, \quad v_\omega|_{\mathcal{E}_0}=0 & \\ \mathbf{n}\cdot(-\sigma\nabla v_\omega+\omega\epsilon\nabla h_\omega)=0 & \text{on } \partial\Omega\setminus\overline{\mathcal{E}_0\cup\mathcal{E}_1} \\ \mathbf{n}\cdot(-\sigma\nabla h_\omega-\omega\epsilon\nabla v_\omega)=0 & \text{on } \partial\Omega. \end{cases} \tag{1.5}$$

*Remark 1.1.* The mixed boundary value problem (1.4) is well-posed and has a unique solution in the Sobolev space $H^1(\Omega)$. It is easy to see the relation between $u$ and $v$ where $v$ is the solution of the following mixed boundary value problem:

$$\begin{cases} -\nabla\cdot(\gamma\nabla v)-0 \quad \text{in } \Omega \\ \gamma\frac{\partial v}{\partial\mathbf{n}}=0 \quad \text{on } \partial\Omega\setminus\overline{\mathcal{E}_0\cup\mathcal{E}_1} \\ v|_{\mathcal{E}_1}=1, \quad v|_{\mathcal{E}_0}=0. \end{cases} \tag{1.6}$$

Due to the edge effect of the electrodes, the Neumann data $g=\nabla u\cdot\mathbf{n}|_{\partial\Omega}$ lies on $g\in H^{-1/2}(\partial\Omega)\setminus L^2(\partial\Omega)$.

*Example 1.5 (Two-channel EIT).* Assume that the conductivity $\sigma$ and the permittivity $\epsilon$ are constants in the subject $\Omega$. Let us reconstruct $\sigma$ and $\epsilon$ from the applied $I(t)=I_0\cos\omega t$ and the measured $V(t)$ at $\mathcal{E}_1$. Throughout this note, we assume $I_0=1$ for simplicity.

Denote $u_0=\gamma u$. Then, $u_0$ is the solution of (1.4) with $\gamma$ replaced by 1. The complex conductivity $\gamma$ can be obtained by

$$\frac{\gamma}{|\gamma|^2}=\frac{\int_\Omega\gamma|\nabla u|^2}{\int_\Omega|\nabla u_0|^2}=\frac{\int_{\mathcal{E}_1}\gamma\frac{\partial u}{\partial\mathbf{n}}\bar{u}}{\int_\Omega|\nabla u_0|^2}=\frac{\bar{u}|_{\mathcal{E}_1}}{\int_\Omega|\nabla u_0|^2}$$

where $\bar{u}$ is the complex conjugate of $u$. It should be noted that without detailed information of the boundary shape $\partial\Omega$ and electrode positions, the reconstructed $\gamma$ would be a little bit different from the true one.

*Example 1.6.* This example comes from a TAS system for breast cancer detection [4,53]. Let $\Omega = \mathbb{R}^3_- := \{\mathbf{r} : z < 0\}$ and $\Gamma = \{(x, y, 0) : \sqrt{x^2 + y^2} < 1\}$. Let $v$ be the $H^1(\Omega)$-solution of the following mixed boundary value problem:

$$\begin{cases} -\nabla^2 v = 0 & \text{in } \Omega \\ \frac{\partial v}{\partial \mathbf{n}} = 0 & \text{on } \partial\Omega \setminus \Gamma \\ v|_\Gamma = 1. \end{cases} \tag{1.7}$$

Then $g = \nabla v \cdot \mathbf{n}|_{\partial\Omega}$ satisfies the integral equation:

$$1 = \frac{1}{2\pi} \int_\Gamma \frac{g(x', y')}{\sqrt{(x - x')^2 + (y - y')^2}} dx'dy' \quad \text{if } \sqrt{x^2 + y^2} < 1.$$

From this, we can find the formula for $v$ and the behavior of $g$ near the boundary circle of the disk $\Gamma$ which helps us to design a TAS probe.

### *1.2.3 EIT System and Measured Data Set*

In EIT, we attach multiple electrodes $\mathcal{E}_0, \mathcal{E}_1, \cdots, \mathcal{E}_n$ on $\partial\Omega$. With these $n$ surface electrodes, we can apply $n$ linearly independent time-harmonic electrical currents. Assume that we use the $n$ pairs of electrodes $(\mathcal{E}_0, \mathcal{E}_j), j = 1, \cdots, n$ to inject the sinusoidal current $I_j \cos(\omega t)$. (*In practice, we may use the adjacent pair of electrodes* $\mathcal{E}_j$ *and* $\mathcal{E}_{j+1}$ *to inject current. Only for reader's convenience, we fix one electrodes* $\mathcal{E}_0$.) Assume that $I_j = 1\,\text{mA}$, $\frac{\omega}{2\pi} \leq 500\,\text{kHz}$ and the hight of the subject is less than $2\,\text{m}$ (Fig. 1.8).

Let $V_j(\mathbf{r}, t)$ be the corresponding electric potential with $V_j|_{\mathcal{E}_0} = 0$ subject to the injection current $I_j \cos(\omega t)$. We measure the resulting electric potential $\tilde{f}_{jk}(t) = V_j|_{\mathcal{E}_k}$ at all electrodes $\mathcal{E}_k, k = 1, 2, \cdots, n$. Here, we neglect contact impedances. *The inverse problem in EIT* is to determine $\sigma$ and $\epsilon$ from the measured data $\mathbb{F}_{\sigma,\epsilon}$:

$$\widetilde{\mathbb{F}}_{\sigma,\epsilon} := \begin{pmatrix} \tilde{f}_{11}(t) & \tilde{f}_{11}(t) & \cdots\cdots & \tilde{f}_{1n}(t) \\ \tilde{f}_{21}(t) & \tilde{f}_{22}(t) & \cdots\cdots & \tilde{f}_{2n}(t) \\ \vdots & \vdots & & \vdots \\ \vdots & \vdots & & \vdots \\ \tilde{f}_{n1}(t) & \tilde{f}_{n2}(t) & \cdots\cdots & \tilde{f}_{nn}(t) \end{pmatrix} \begin{matrix} \leftarrow I_1 \sin(\omega t) \\ \leftarrow I_2 \sin(\omega t) \\ \vdots \\ \vdots \\ \leftarrow I_n \sin(\omega t). \end{matrix} \tag{1.8}$$

Hence, the inverse problem is to invert the map

$$\underbrace{(\sigma, \epsilon)}_{\text{within some admissible class}} \longrightarrow \underbrace{\widetilde{\mathbb{F}}_{\sigma,\epsilon}(t)}_{n\times n \text{ matrix}} .$$

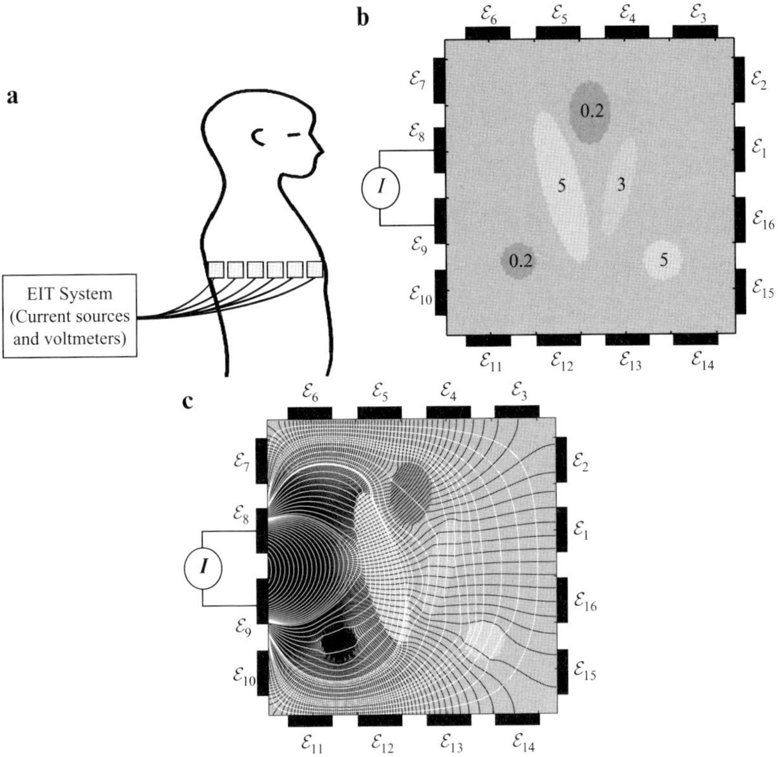


**Fig. 1.8** (**a**) EIT system. (**b**) Surface electrodes $\mathcal{E}_j, j = 1, \cdots, 16$ are attached on the boundary of a simplified rectangular model with a given conductivity distribution $\sigma$. We inject current $I$ using the electrodes $\mathcal{E}_8$ and $\mathcal{E}_9$. (**c**) *Black and white lines* are equipotential lines and electric field streamlines, respectively. In this case where $\omega = 0$, the data set in (1.8) is time-independent

This requires us to investigate the relationship between $(\sigma, \epsilon)$ and the data set $\mathbb{F}_{\sigma,\epsilon}$ in (1.8).

## (n+1)-Channel EIT System

Let $u_j$ be the complex potential such that $V_j = \Re\{u_j e^{i\omega t}\}$ where $V_j$ is the potential in the previous section. From (1.4), the complex potential $u_j$ satisfies the following mixed boundary value problem:

$$\begin{cases} \nabla \cdot (\gamma \nabla u_j) = 0 \quad \text{in } \Omega \\ I = \int_{\mathcal{E}_1} \gamma \frac{\partial u_j}{\partial \mathbf{n}} \, ds = -\int_{\mathcal{E}_j} \gamma \frac{\partial u_j}{\partial \mathbf{n}} ds \\ u_j|_{\mathcal{E}_0} = 0, \quad \nabla u_j \times \mathbf{n}|_{\mathcal{E}_j} = 0 \\ \gamma \frac{\partial u_j}{\partial \mathbf{n}} = 0 \quad \text{on } \partial\Omega \setminus \overline{\mathcal{E}_0 \cup \mathcal{E}_j} \end{cases} \tag{1.9}$$

where $\frac{\partial u}{\partial \mathbf{n}} = \nabla u \cdot \mathbf{n}$. For the image reconstruction, we use the measured potential on each electrodes $\mathcal{E}_k$. For each $j, k = 1, \cdots, n$, the measured data is

$$f_j(k) := u_j|_{\mathcal{E}_k} \qquad \text{(neglecting the contact impedance over } \mathcal{E}_j\text{)}.$$

Note that $\tilde{f}_{jk}(t) = \Re\{f_j(k)\, e^{i\omega t}\}$. The inverse problem of EIT is to reconstruct $\gamma$ from the $n^2$ data set $\{f_j(k) \;:\; j, k = 1, 2, \cdots, n\}$.

*Remark 1.2.* We have the reciprocity relation:

$$f_j(k) = f_k(j) \qquad j, k = 1, \cdots, n.$$

The symmetry follows from the identity:

$$f_k(j) = \frac{1}{I}\int_{\partial\Omega} \gamma \frac{\partial u_j}{\partial \mathbf{n}}\, u_k ds = \frac{1}{I}\int \gamma \nabla u_j \cdot \nabla u_k d\mathbf{r} = \frac{1}{I}\int_{\partial\Omega} \gamma \frac{\partial u_k}{\partial \mathbf{n}}\, u_j ds = f_j(k).$$

The reciprocity relation tells that the following matrix of the data set is symmetric:

$$\mathbb{F}_\gamma := \begin{pmatrix} f_1(1) & f_1(2) & \cdots\cdots & f_1(n) \\ f_2(1) & f_2(2) & \cdots\cdots & f_2(n) \\ \vdots & \vdots & & \vdots \\ \vdots & \vdots & & \vdots \\ f_n(1) & f_n(2) & \cdots\cdots & f_n(n) \end{pmatrix} \qquad \begin{matrix} \leftarrow I_1 \text{ current} \\ \leftarrow I_2 \text{ current} \\ \vdots \\ \vdots \\ \leftarrow I_n \text{ current} \end{matrix} \tag{1.10}$$

Hence, the number of the independent data is $\frac{n(n+1)}{2}$ and the number $\frac{n(n+1)}{2}$ is the maximum number of the unknown parameters of $\gamma$ which can be reconstructed from the above mentioned EIT data set $\mathbb{F}_\gamma$.

*Remark 1.3 (16-channel EIT system at IIRC).* Figure 1.9 shows a 16-channel EIT system at IIRC. We inject *15 linearly independent currents* using the adjacent pair of electrodes $\mathcal{E}_j$ and $\mathcal{E}_{j+1}$ for $j = 1, \cdots, 15$. Here, $u_1$ is the complex potential due to injection current using $\mathcal{E}_1$ and $\mathcal{E}_2$.

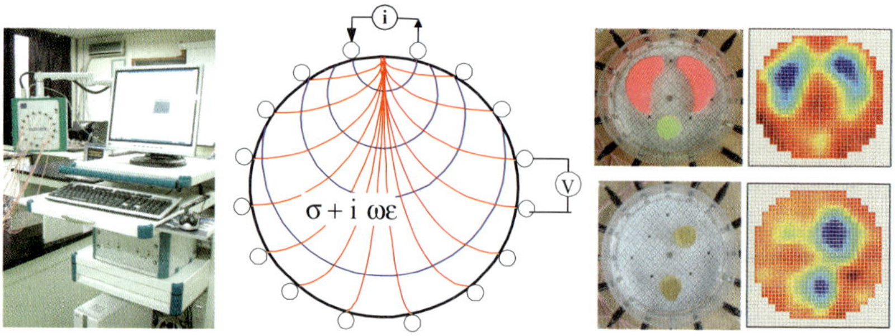


**Fig. 1.9** A 16-channel EIT system at IIRC

### 1.2.4 The Standard Reconstruction Method: 4-Channel EIT System

To understand the data set $\mathbb{F}_\gamma$ precisely, we consider a 4-channel EIT system (Fig. 1.10). In this case, we can apply *at most three linearly independent currents.*

**Problem 1.1 (Reconstruction of $\gamma$ in a 4-channel EIT system).** For a simple numerical simulation, we assume that $\Omega$ is a rectangular region. We assume that $\omega = 0$ and $\gamma = \sigma$ is real and isotropic. Find a rough structure of $\gamma = \sigma$ from the following baby Neumann-to-Dirichlet (NtD) map.

| | $k=1$ | $k=2$ | $k=3$ |
|---|---|---|---|
| $u_1\|_{\mathcal{E}_k}$ | −2.0285 | −1.3025 | −1.0962 |
| $u_2\|_{\mathcal{E}_k}$ | −1.3068 | −2.3413 | −1.3633 |
| $u_3\|_{\mathcal{E}_k}$ | −1.1053 | −1.3724 | −2.5987 |

It should be noticed that, with this limited data set $\mathbb{F}_\gamma$ having at most six data, it is impossible to reconstruct $\gamma$ having more than six parameters. So, we try to find a very rough distribution of $\gamma$. Most of the conventional reconstruction methods are some variations of the following least square method:

- For a given $\gamma$, let the complex potential $u_j^\gamma$ be the $H^1(\Omega)$-solution of $\mathcal{P}_j[\gamma]$:

$$\mathcal{P}_j[\gamma]: \begin{cases} \nabla \cdot (\gamma \nabla u_j^\gamma) = 0 \text{in } \Omega \\ \mathbf{n} \times \nabla u_j^\gamma|_{\mathcal{E}_j} = 0, \quad u_j^\gamma|_{\mathcal{E}_0} = 0 \\ \gamma \frac{\partial u_j^\gamma}{\partial n}|_{\partial\Omega \setminus (\mathcal{E}_i \cup \mathcal{E}_0)} = 0 \\ \int_{\mathcal{E}_0} \gamma \frac{\partial u_j^\gamma}{\partial n} = I = -\int_{\mathcal{E}_i} \gamma \frac{\partial u_j^\gamma}{\partial \mathbf{n}}. \end{cases}$$

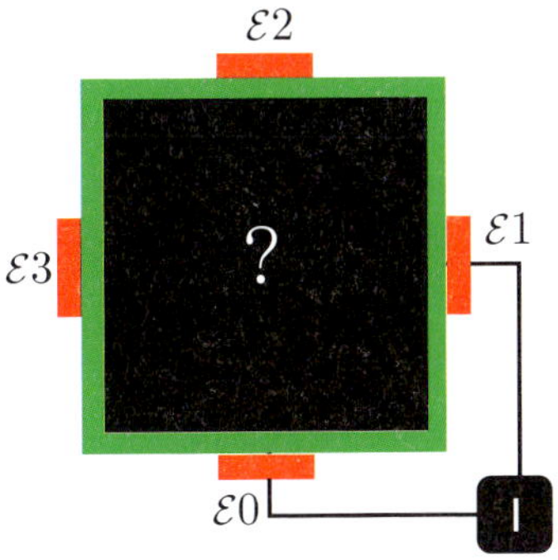


**Fig. 1.10** A 4-channel EIT system

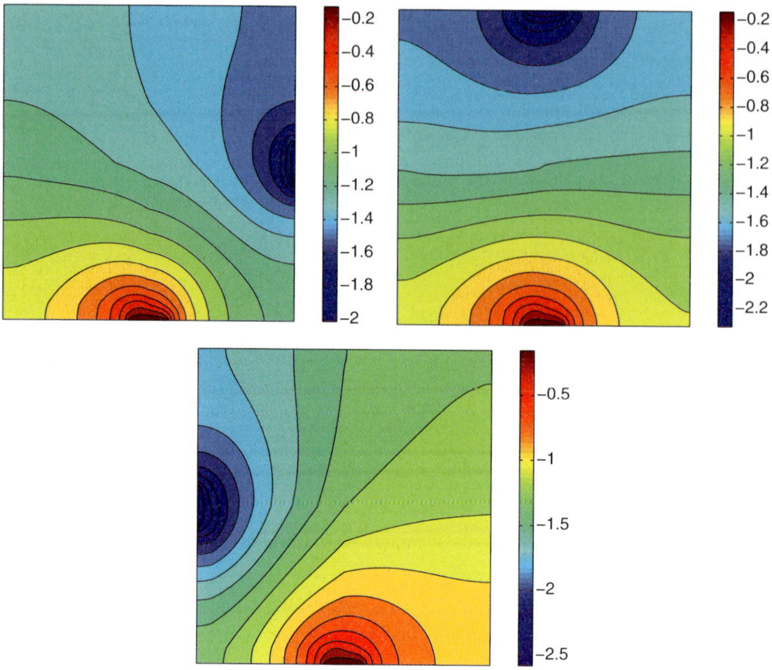


**Fig. 1.11** Examples of complex potentials

- By solving $\mathcal{P}_j[\gamma]$, we produce the following simulated data $\mathbf{u}^\gamma$:

$$\mathbf{u}^\gamma := \underbrace{\begin{bmatrix} u_1^\gamma|_{\mathcal{E}_1} & u_1^\gamma|_{\mathcal{E}_2} & u_1^\gamma|_{\mathcal{E}_3} \\ u_2^\gamma|_{\mathcal{E}_1} & u_2^\gamma|_{\mathcal{E}_2} & u_2^\gamma|_{\mathcal{E}_3} \\ u_3^\gamma|_{\mathcal{E}_1} & u_3^\gamma|_{\mathcal{E}_2} & u_3^\gamma|_{\mathcal{E}_3} \end{bmatrix}}_{\text{computed voltage set with the guessed } \gamma.}$$

Figure 1.11 shows examples of the complex potentials $u_1^\gamma, u_2^\gamma$ and $u_3^\gamma$, respectively.

This simulated data $\mathbf{u}^\gamma$ will be compared with the measure data

$$\mathbf{f} := \underbrace{\begin{bmatrix} f_1|_{\mathcal{E}_1} & f_1|_{\mathcal{E}_2} & f_1|_{\mathcal{E}_3} \\ f_2|_{\mathcal{E}_1} & f_2|_{\mathcal{E}_2} & f_2|_{\mathcal{E}_3} \\ f_3|_{\mathcal{E}_1} & f_3|_{\mathcal{E}_2} & f_3|_{\mathcal{E}_3} \end{bmatrix}}_{\text{measured voltage set}}$$

where $f_j|_{\mathcal{E}_k}$ is the measured voltage at the electrode $\mathcal{E}_k$ subject to the $j$-th current.

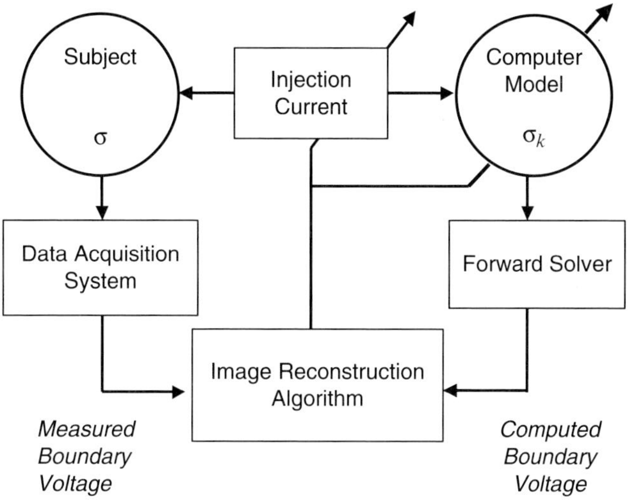


**Fig. 1.12** Minimization procedure

- *Minimization problem.* We try to find $\gamma$ which minimizes the misfit between simulated and measured data (Fig. 1.12):

$$\Phi(\gamma) = \|\mathbf{f} - \mathbf{u}^\gamma\|^2 + \eta(\gamma) = \sum_{k=1}^{3}\sum_{j=1}^{3} \left| f_{ij} - u_j^\gamma|_{\mathcal{E}_k} \right|^2 + \underbrace{\eta(\gamma)}_{\text{regularization term}}$$

## Standard Approach for Solving Minimization Problem of $\Phi(\gamma)$

- The goal is to find a minimizer $\gamma$ for $\Phi(\gamma)$. For a moment, let us assume that $\eta(\gamma) = 0$. Then,

$$\Phi(\gamma) = \sum_{j,k=1}^{3} \left| f_{jk} - u_j^\gamma|_{\mathcal{E}_k} \right|^2 = \sum_{j,k=1}^{3} \left[ \int_{\partial\Omega} [f_j - u_j^\gamma]\, \gamma \frac{\partial u_k^\gamma}{\partial \mathbf{n}} ds \right]^2 .$$

  - Computation of the Frechét derivative of the functional $\Phi(\gamma)$ requires us to investigate a linear change $\delta u := u^{\gamma+\delta\gamma} - u^\gamma$ due to a small conductivity perturbation $\delta\gamma$.
  - Note that $\Phi(\gamma + \delta\gamma) \approx \Phi(\gamma) + D\Phi(\gamma)(\delta\gamma) + \frac{1}{2}D^2\Phi(\gamma)(\delta\gamma, \delta\gamma)$.
  - For simplicity, assume that $\delta\gamma = 0$ near $\partial\Omega$.

- The relationship between $\delta\gamma$ and the linear change $\delta u$ can be explained by

$$\begin{cases} \nabla \cdot (\delta\gamma \nabla u) \approx -\nabla \cdot (\gamma \nabla \delta u) \text{ in } \Omega \\ \frac{\partial(\delta u)}{\partial \mathbf{n}}|_{\partial\Omega} = 0, \qquad \delta u|_{\mathcal{E}_0} = 0. \end{cases}$$

The above relation is based on the following linear approximation:

- Linear approximation:

$$\begin{aligned} 0 &= \nabla\cdot((\gamma+\delta\gamma)\nabla(u+\delta u)) \\ &= \nabla\cdot(\delta\gamma\nabla u)+\nabla\cdot(\gamma\nabla\delta u)+\underbrace{\nabla\cdot(\delta\gamma\nabla\delta u)}_{\text{negligible term}} \\ \Rightarrow \quad \nabla\cdot(\delta\gamma\nabla u) &\approx -\nabla\cdot(\gamma\nabla\delta u). \end{aligned}$$

- Note that $u+\delta u$ satisfies

$$\begin{cases} \nabla\cdot((\gamma+\delta\gamma)\nabla(u+\delta u))=0 \quad \text{in} \quad \Omega \\ \gamma\frac{\partial(u+\delta u)}{\partial\mathbf{n}}|_{\partial\Omega}=\gamma\frac{\partial u}{\partial\mathbf{n}}|_{\partial\Omega}, \qquad \delta(u+\delta u)|_{\mathcal{E}_0}=0. \end{cases}$$

- Next, we investigate the relation between $\delta\gamma$ and $\delta u_j^\gamma|_{\mathcal{E}_k}$. For $j,k=1,2,3$, we have the following approximation:

$$\int_\Omega \delta\gamma\nabla u_j^\gamma\cdot\nabla u_k^\gamma dx \approx \delta u_j^\gamma|_{\mathcal{E}_k}.$$

  - Using the identity $\nabla\cdot(\delta\gamma\nabla u_j)=-\nabla\cdot(\gamma\nabla\delta u_j)$,

$$\begin{aligned} &\Rightarrow \quad \nabla\cdot(\delta\gamma\nabla u_j)\, u_k = -\nabla\cdot(\gamma\nabla\delta u_j)\, u_k \\ &\Rightarrow \quad \int_\Omega \delta\gamma\nabla u_j\cdot\nabla u_k = -\int_\Omega \gamma\nabla\delta u_j\cdot\nabla u_k \\ &\Rightarrow \quad \int_\Omega \delta\gamma\nabla u_j\cdot\nabla u_k = -\int_{\partial\Omega} \delta u_j\, \gamma\frac{\partial u_k}{\partial\mathbf{n}}. \end{aligned}$$

  - Neumann boundary condition for $u_k$ yields

$$-\int_{\partial\Omega}(\delta u_j)\,\gamma\frac{\partial u_k}{\partial n}ds = \delta u_j|_{\mathcal{E}_k}.$$

- To get a rough insight for finding the steepest descent direction, consider

$$\int_\Omega \delta\gamma\nabla u_j^\gamma\cdot\nabla u_k^\gamma dx = u_j^\gamma|_{\mathcal{E}_k}-f_{jk} \quad j,k=1,2,3.$$

  - Recall $\Phi(\gamma)=\sum_{j,k=1}^3\left|f_{jk}-u_j^\gamma|_{\mathcal{E}_k}\right|^2$.
  - $\Phi(\gamma+\delta\gamma)-\Phi(\gamma)\approx 2\sum_{j,k=1}^3\left(f_{jk}-u_j^\gamma|_{\mathcal{E}_k}\right)\,\delta u_j^\gamma|_{\mathcal{E}_k}$.
  - As a direction $\delta\gamma$ which makes $\Phi(\gamma+\delta\gamma)-\Phi(\gamma)$ smallest with a given norm $\|\delta\gamma\|$, we choose $\delta\gamma$ such that $\delta u_j^\gamma|_{\mathcal{E}_k}=u_j^\gamma|_{\mathcal{E}_k}-f_{jk}$.
  - To get $\delta\gamma$, use the approximation $\int_\Omega\delta\gamma\nabla u_j^\gamma\cdot\nabla u_k^\gamma dx\approx\delta u_j^\gamma|_{\mathcal{E}_k}$ and obtain

$$\int_\Omega \delta\gamma\nabla u_j^\gamma\cdot\nabla u_k^\gamma dx \approx u_j^\gamma|_{\mathcal{E}_k}-f_{jk}.$$

- To compute the derivatives $D\Phi(\gamma)$ more precisely, we define

$$U_j(\gamma) := \begin{pmatrix} u_j^\gamma|_{\mathcal{E}_1} \\ u_j^\gamma|_{\mathcal{E}_2} \\ u_j^\gamma|_{\mathcal{E}_3} \end{pmatrix}, \quad \mathbf{f}_j := \begin{pmatrix} f_{j1} \\ f_{j2} \\ f_{j3} \end{pmatrix}, \quad \Phi(\gamma) = \sum_j |U_j(\gamma) - \mathbf{f}_j|^2.$$

  - For $j = 1, 2, 3$, let $DU(\gamma)(\delta\gamma) := \begin{pmatrix} \delta u_j^\gamma|_{\mathcal{E}_1} \\ \delta u_j^\gamma|_{\mathcal{E}_2} \\ \delta u_j^\gamma|_{\mathcal{E}_3} \end{pmatrix}, \quad \mathbf{e}_j := \begin{pmatrix} -\int_{\mathcal{E}_1} \gamma \frac{\partial u_j^\gamma}{\partial \mathbf{n}} \\ -\int_{\mathcal{E}_2} \gamma \frac{\partial u_j^\gamma}{\partial \mathbf{n}} \\ -\int_{\mathcal{E}_3} \gamma \frac{\partial u_j^\gamma}{\partial \mathbf{n}} \end{pmatrix}.$
  - Note that $\mathbf{e}_1 = (1,0,0)^T, \mathbf{e}_2 = (0,1,0)^T$ and $\mathbf{e}_3 = (0,0,1)^T$ where $A^T$ is the transpose of $A$.

- For the computation of $D\Phi(\gamma)$, consider the map $DU(\gamma)(\delta\gamma) : L^2(\Omega) \to \mathbb{R}^3$. Then, its adjoint is the map $[DU(\gamma)]^* : \mathbb{R}^2 \to L^2(\Omega)$ given by

$$\langle DU_j(\gamma)(\delta\gamma), \mathbf{e}_k \rangle = \int_\Omega \delta\gamma \; [DU_j(\gamma)]^*(\mathbf{e}_k) d\mathbf{r}.$$

Integrating by parts yields $[DU_j(\gamma)]^*$:

$$\int_\Omega \delta\gamma \; [DU_j(\gamma)]^*(\mathbf{e}_k) d\mathbf{r} = -\int_\Omega \delta\gamma \nabla u_j^\gamma \nabla u_k^\gamma$$

because

$$\langle DU_j(\gamma)(\delta\gamma), \mathbf{e}_k \rangle = \delta u_j^\gamma|_{\mathcal{E}_k} = -\int_\Omega \delta\gamma \nabla u_j^\gamma \nabla u_k^\gamma.$$

- Now, we are ready to compute $D\Phi(\gamma)(\delta\gamma)$.

$$\begin{aligned} D\Phi(\gamma)(\delta\gamma) &\approx -2 \sum_{j=1}^{3} \langle [DU_j(\gamma)](\delta\gamma), (U_j(\gamma) - \mathbf{f}_j) \rangle \\ &= -2 \sum_{j,k=1}^{3} \langle [DU_j(\gamma)](\delta\gamma), \mathbf{e}_k \rangle \;\; \langle \mathbf{e}_k, (U_j(\gamma) - \mathbf{f}_j) \rangle \\ &= -2 \sum_{j,k=1}^{3} \langle (\delta\gamma), [DU_j(\gamma)]^* \mathbf{e}_k \rangle_{L^2(\Omega)} \;\; \langle \mathbf{e}_k, (U_j(\gamma) - \mathbf{f}_j) \rangle \\ &= -2 \sum_{j,k=1}^{3} \left( \int_\Omega \delta\gamma \nabla u_j^\gamma \nabla u_k^\gamma \right) \;\; \langle \mathbf{e}_k, (U_j(\gamma) - \mathbf{f}_j) \rangle. \end{aligned}$$

Note that $[DU_j(\gamma)]^*(\mathbf{e}_k) = \nabla u_j^\gamma \nabla u_k^\gamma$.

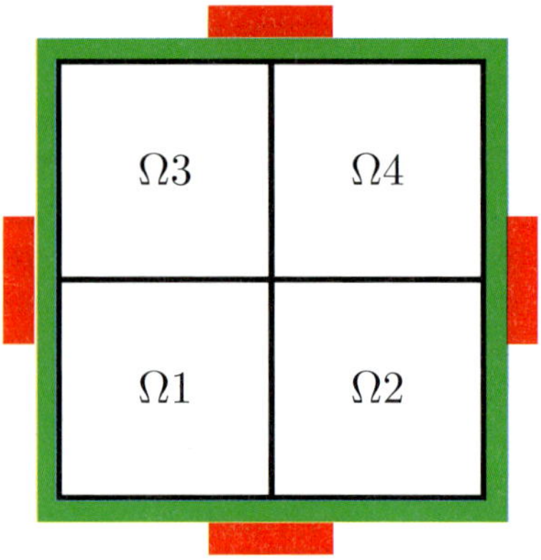


**Fig. 1.13** Iteration process

- The steepest descent direction $\delta\gamma$ satisfies the relation

$$\int_{\Omega} \delta\gamma \nabla u_j^\gamma \cdot \nabla u_k^\gamma dx = u_j^\gamma|_{\mathcal{E}_k} - f_{jk} \quad j,k = 1,2,3.$$

  - $D\Phi(\gamma)(\delta\gamma) \approx -2\sum_{j,k=1}^{3} \left(\int_{\Omega} \delta\gamma \nabla u_j^\gamma \nabla u_k^\gamma\right) \ \langle \mathbf{e}_k, (U_j(\gamma) - \mathbf{f}_j)\rangle$.
  - For the steepest descent direction, we choose $\delta\gamma$ satisfying

$$\left(\int_{\Omega} \delta\gamma \nabla u_j^\gamma \nabla u_k^\gamma\right) = \langle \mathbf{e}_k, (U_j(\gamma) - \mathbf{f}_j)\rangle, \quad j,k = 1,2,3.$$

- Iteration process for the reconstruction of $\gamma$ (Fig. 1.13).
  - Divide $\Omega$ into $\Omega = \Omega_0 \cup \Omega_1 \cup \Omega_2 \cup \cdots \cup \Omega_N$.
  - Assume $\gamma|_{\Omega_0} = 1$ and $\gamma$ is a constant on each $\Omega_m, m = 1, 2, \cdots, L$. This $L$ should be $L \leq 6$.
  - Write it as a matrix form $\mathbb{A}\mathbf{x} = \mathbf{b}$.

$$\begin{pmatrix} \int_{\Omega_1} \nabla u_1 \cdot \nabla u_1 \ldots \int_{\Omega_N} \nabla u_1 \cdot \nabla u_1 \\ \int_{\Omega_1} \nabla u_1 \cdot \nabla u_2 \quad\quad \int_{\Omega_N} \nabla u_1 \cdot \nabla u_2 \\ \int_{\Omega_1} \nabla u_1 \cdot \nabla u_3 \quad\quad \int_{\Omega_N} \nabla u_1 \cdot \nabla u_3 \\ \int_{\Omega_1} \nabla u_2 \cdot \nabla u_2 \quad\quad \int_{\Omega_N} \nabla u_2 \cdot \nabla u_2 \\ \int_{\Omega_1} \nabla u_2 \cdot \nabla u_3 \quad\quad \int_{\Omega_N} \nabla u_2 \cdot \nabla u_3 \\ \int_{\Omega_1} \nabla u_3 \cdot \nabla u_3 \ldots \int_{\Omega_L} \nabla u_3 \cdot \nabla u_3 \end{pmatrix} \begin{pmatrix} \delta\gamma_1 \\ \delta\gamma_2 \\ \vdots \\ \vdots \\ \delta\gamma_L \end{pmatrix} = \begin{pmatrix} u_1^\gamma|_{\mathcal{E}_1} - f_{11} \\ u_1^\gamma|_{\mathcal{E}_2} - f_{12} \\ u_1^\gamma|_{\mathcal{E}_3} - f_{13} \\ u_2^\gamma|_{\mathcal{E}_2} - f_{22} \\ u_2^\gamma|_{\mathcal{E}_3} - f_{23} \\ u_3^\gamma|_{\mathcal{E}_3} - f_{33} \end{pmatrix}.$$

  - From the singular value decomposition (SVD), we use only four singular values, that is, reconstruct an image of $L = 4$ independent pixels.
  - The natural solution would be the least squares solution of $\delta\gamma = \arg\min_x \|Ax - b\|$.
    1. Start with the initial guess $\gamma = 1$.
    2. Calculate $\mathbb{A}$ and $\mathbf{b}$ by solving the direct problem $\mathcal{P}(\gamma)$.

3. Calculate
$$\delta\gamma = \arg\min_x \|Ax - b\|.$$
4. Update $\gamma + \delta\gamma$.
5. Repeat the process 2, 3 and 4.

### *1.2.5 Boundary Geometry Effect and Rough Analysis of Relation Between Conductivity and Potential*

In this section, we investigate how the conductivity $\sigma$ can be perceived from a potential $u$ due to a low frequency injection current. Since $\nabla \cdot (\sigma \nabla u) = 0$, we can write $\nabla u \cdot \nabla \log \sigma = -\Delta u$. Hence, the electrical field $\mathbf{E} = -\nabla u$ probes the $\nabla u-$directional change of $\log \sigma$ in such a way that

$$\frac{\nabla u}{|\nabla u|} \cdot \nabla \log \sigma = -\frac{1}{|\nabla u|} \Delta u \quad \text{in } \Omega.$$

This leads to the following properties:

- If $u$ is convex, $\log \sigma$ is decreasing in the direction $\nabla u$.
- If $u$ is concave, $\log \sigma$ is increasing in the direction $\nabla u$.
- If $u$ is harmonic, $\log \sigma$ is constant in the direction $\nabla u$.

To get some quantitative insight, assume that $\sigma$ is a small perturbation of a known reference conductivity $\sigma_0$. We know that $\Delta u$ dictates the change of $\log \sigma$ along the current flow line for the vector field $\frac{\nabla u}{|\nabla u|}$. Under the low contrast assumption, the direction vector $\frac{\nabla u}{|\nabla u|}$ is mostly dictated by the structure of the boundary geometry $\partial\Omega$ and the Neumann data $g$ instead of the distribution of $\log \frac{\sigma}{\sigma_0}$. Since the Neumann data $g$ is determined by electrode positions, accurate knowledge of geometry of $\partial\Omega$ and electrode positions are essential for a reliable estimate of the vector field $\frac{\nabla u}{|\nabla u|}$ which would be close to $\frac{\nabla u_0}{|\nabla u_0|}$ where $u_0$ is the corresponding solution with the conductivity $\sigma_0$. Using the knowledge of the vector field $\frac{\nabla u_0}{|\nabla u_0|}$, we can use a linearization technique for the reconstruction of $\log \frac{\sigma}{\sigma_0}$.

Let $u_j$ be the potential due to the $j$-th current pattern $g_j$ in an $N$-channel EIT system. Viewing $u_j$ as a highly nonlinear function of $\sigma$ and $g_j$, we seek a best possible solution $\sigma$ within a ceratin admissible class which fits the following highly nonlinear system in a sense of a sum of squared errors:

$$\begin{cases} \nabla u_j \cdot \nabla \log \sigma \approx -\Delta u_j \ \text{ in } \Omega \\ \sigma \frac{\partial u_j}{\partial \mathbf{n}}|_{\partial\Omega} \approx g_j, \quad u_j|_{\partial\Omega} \approx f_j \end{cases} \quad j = 1, \cdots, N-1.$$

Roughly, we can define a nonlinear operator $\Psi$ which is a map from the conductivity to the set of boundary voltages:

$$\Psi(\sigma)(\mathbf{r}) := \mathbf{f}^{\sigma,\partial\Omega}(\mathbf{r}) = \int_{\partial\Omega} \mathcal{N}_{\sigma,\partial\Omega}(\mathbf{r},\mathbf{r}') \begin{bmatrix} g_1(\mathbf{r}') \\ g_2(\mathbf{r}') \\ \vdots \\ g_{N-1}(\mathbf{r}') \end{bmatrix} dS \quad \mathbf{r} \in \partial\Omega$$

where $\mathcal{N}_{\sigma,\partial\Omega}$ is the Neumann function and $\mathbf{f}^{\sigma,\partial\Omega} = [f_1, \cdots, f_{N-1}]^T$ is the corresponding Dirichlet data set. We should note that the Neumann function depends heavily on the geometry of $\partial\Omega$. Under the assumption that $\sigma$ is a small perturbation of $\sigma_0$, most reconstruction algorithms use a sensitivity matrix $\mathbb{A}_{\sigma_0,\partial\Omega}$ at $\sigma_0$ with the subject boundary $\partial\Omega$. The matrix expresses changes of boundary voltages due to changes in conductivity from $\sigma_0$. With this linearization, we usually seek an approximate solution $\sigma$ minimizing the misfit functional:

$$F(\sigma) := \|\mathbb{A}_{\sigma_0,\partial\Omega}(\sigma - \sigma_0) - [\mathbf{f}_{meas} - \mathbf{f}_{\sigma_0,\partial\Omega}]\|^2 + \underbrace{\|\log\sigma\|_{TV}}_{\text{regularization}} \tag{1.11}$$

where $\mathbf{f}_{meas}$ is the measured voltage data set. Hence, the corresponding reconstruction algorithm requires to compute $\mathbb{A}_{\sigma_0,\partial\Omega}(\sigma-\sigma_0)$ and $\mathbf{f}_{\sigma_0,\partial\Omega}$ and we need an exact forward model including the accurate boundary geometry and electrodes positions. We should note that the computed voltage $\mathbf{f}_{\sigma_0,\partial\Omega}$ is very sensitive to $\partial\Omega$ and electrodes positions determining $g_j$, while it is insensitive to local changes of internal conductivity distribution $\sigma_0$. Hence, without having a reasonably accurate computer model of the boundary geometry and electrode positions, we should not expect to reconstruct reliable static conductivity images no matter how good the algorithm is. Unfortunately, in practice, there are serious technical difficulties obtaining these geometrical data with a reasonable accuracy and cost. Without eliminating these technical difficulties, any reconstructed image of the static EIT can not be trusted. Figure 1.14 shows an example of artifacts due to unknown geometry errors.

These technical difficulties in the static EIT could be eliminated in the time-difference EIT (tdEIT) where we try to produce difference images between time $t = t_0 + \delta t$ and $t_0$ from $\mathbf{f}^{t_0+\delta t}_{meas} - \mathbf{f}^{t_0}_{meas}$. Taking advantage of the difference which cancels out common errors, it was demonstrated that time-difference conductivity images can be successfully produced [10, 37]. Here, note that if the measured data set $\mathbf{f}^{t_0}_{meas}$ from a homogeneous phantom is used as a reference Dirichlet data set, we regard such a case as the time-difference EIT. The Sheffield group described numerous in-vivo three-dimensional EIT images using the Mk3.5 EIT system [36,37]. Figures 1.15a–c are examples of time-difference images showing conductivity changes during ventilation. Figure 1.15a show eight images at eight different cross-sectional

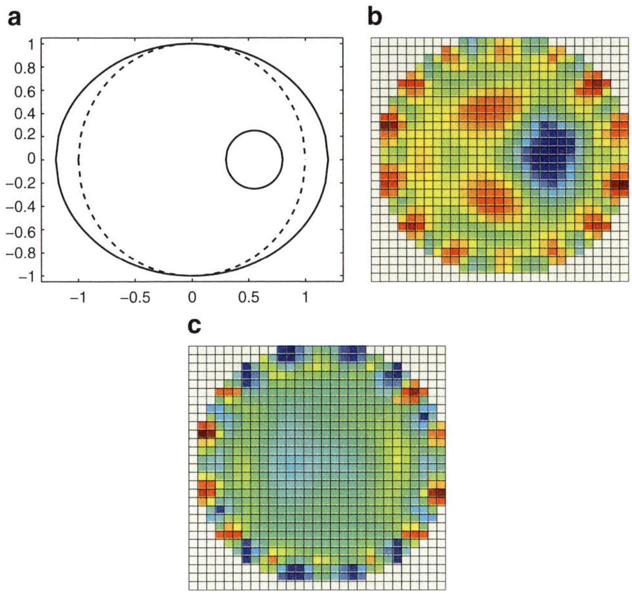


**Fig. 1.14** Effect of boundary geometry errors in reconstructed image. (**a**) The true domain is an *ellipse in solid line* and the computational domain of the model is a *circle in dashed line*. The *small disk inside the ellipse* is an anomaly. The complex conductivity of the background is $0.137\,\mathrm{S\,m^{-1}}$ and the anomaly has $0.007 + i2.28 \times 10^{-6}\,\mathrm{S\,m^{-1}}$ at 100 Hz and $0.025 + i0.033\,\mathrm{S\,m^{-1}}$ at 50 kHz. (**b**) and (**c**) show the real parts of the reconstructed images by using the reference boundary voltage data at 100 Hz and 50 kHz, respectively, computed by using the circular model

planes of a human thorax with both lungs at residual volume. The images are difference images with respect to a set of reference data at total lung capacity. Figures 1.15b,c are corresponding images at functional residual capacity and peak tidal volume, respectively. Figure 1.15d shows a two-dimensional conductivity phantom with 32 electrodes constructed by the RPI group [10]. The phantom is circular with 30 cm diameter and saline-filled. On its inside surface, there are 32 stainless steel electrodes with the size of $2.54 \times 2.54\,\mathrm{cm}^2$. Two lung-shaped structures and a heart-shaped structure made of agar are immersed in saline. Figure 1.15e is an EIT image of the phantom reconstructed from a set of boundary data.

Tidswell et al. [60] applied the EIT technique to image functional activity in the brain. Figure 1.16 shows a time series of EIT images during visual stimulation. The goggles were used to produce a bright flash which stimulated the infant's vision. Each column represents averaged images over 4 s at six slices of the baby's head. From the EIT images, we can see that visual stimulation produced conductivity changes at the front and the back of the head. The conductivity increases over the back of the head corresponding to the position of the visual cortex probably due to an increased blood volume

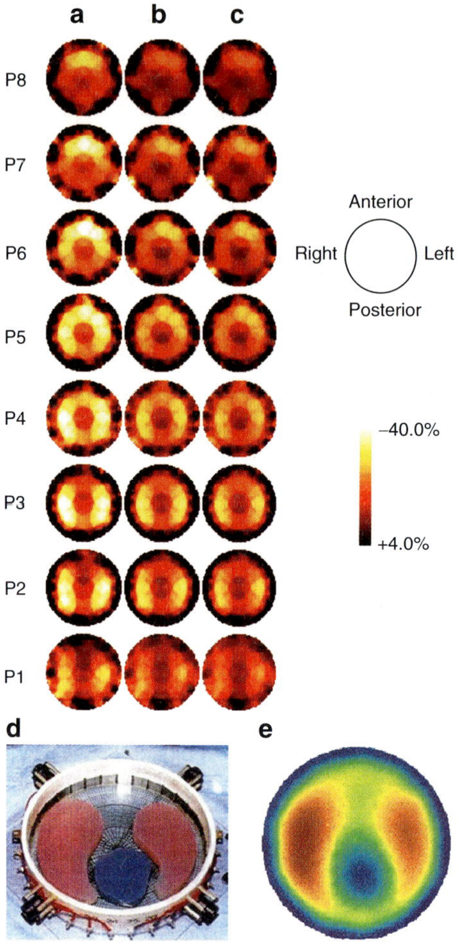


**Fig. 1.15** (**a**)–(**c**) Three-dimensional in-vivo time-difference EIT images of a human thorax using the Sheffiel Mk3.5 system. See [37] for details and also http://www.shef.ac.uk/uni/academic/I-M/mpce/rsch/funimg.html. (**d**) Two-dimensional conductivity phantom and (**e**) reconstructed time-difference conductivity image using the ACT-3 EIT system from the RPI group. See [10] for details and also http://www.rpi.edu/newelj/eit.html

there. The larger change at the front of the head was interpreted as artifacts due to blinking or muscle movement during the bright visual stimulus. More information on the EIT technique and numerous applications can be found at http://www.eit.org.uk/index.html.

For the time-difference imaging, Griffiths et al. [12] used a modified sensitivity matrix $\tilde{\mathbb{A}}_{\sigma_0,\partial\Omega}$ which expresses changes in the logarithm of the Dirichlet data with respect to changes in the logarithm of the conductivity at $\sigma_0$. As

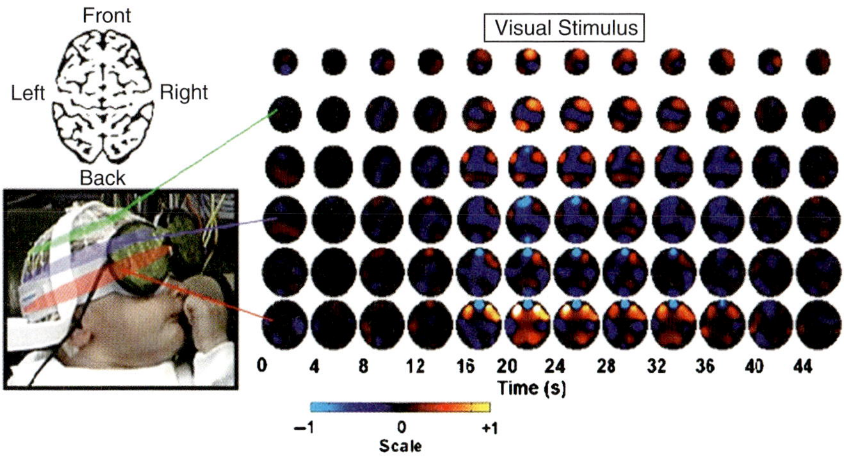


**Fig. 1.16** Time series EIT images of a neonatal brain during visual stimulation from the UCL group. See text for details and also http://madeira.physiol.ucl.ac.uk/midx-group/

in (1.11), in this case, we try to find an approximate solution $\sigma$ minimizing the following modified misfit functional:

$$\tilde{F}(\sigma_t) := \left\| \tilde{\mathbb{A}}_{\sigma_0,\partial\Omega}\left(\log\frac{\sigma_t}{\sigma_0}\right) - \log\frac{\mathbf{f}^t_{meas}}{\mathbf{f}^{t_0}_{meas}} \right\|^2 + \|\log\sigma_t\|_{TV} \tag{1.12}$$

where $\log \frac{\mathbf{f}^t_{meas}}{\mathbf{f}^{t_0}_{meas}]}$ denotes the logarithm of the voltage ratio in each component. This approach would be effective to eliminate electrode spacing errors.

## *1.2.6 Frequency-Difference EIT*

We may consider a frequency-difference EIT (fdEIT) technique to deal with the drawbacks of the static EIT caused by modelling errors described in the previous section [33]. In fdEIT, viewing the complex conductivity distribution inside of an object as a function of the frequency, we apply several difference frequencies $\omega_1, \omega_2, \cdots, \omega_m$ ranging over $0 \le \frac{\omega_1}{2\pi} \le \frac{\omega_2}{2\pi} \le \cdots \le \frac{\omega_m}{2\pi} \le 500\,\text{kHz}$. For the image reconstruction, we use frequency-difference voltage at each electrode. We hope that the use of the frequency difference somehow eliminates the drawback of the conventional single-frequency EIT method since the difference may cancel out various common errors related with inaccuracy of background conductivity, unknown boundary geometry, measurement artifacts and so on.

For a feasibility study of fdEIT, it is important to understand the sensitivity of a frequency-difference voltage data to a frequency-dependent change of a complex conductivity distribution. In order to test its feasibility, it would be best to use the simplest model of a 16-channel EIT system with a circular phantom. We assume that there is an anomaly inside $\Omega$ which occupies a region $D$ and the complex conductivity $\gamma = \sigma + i\omega\epsilon$ changes abruptly across $\partial D$. In order to distinguish between them, we set

$$\sigma(\mathbf{r}, \omega) = \begin{cases} \sigma_b(\omega) & \text{if } \mathbf{r} \in \Omega \setminus \overline{D} \\ \sigma_a(\omega) & \text{if } \mathbf{r} \in D \end{cases}$$

and

$$\epsilon(\mathbf{r}, \omega) = \begin{cases} \epsilon_b(\omega) & \text{if } \mathbf{r} \in \Omega \setminus \overline{D} \\ \epsilon_a(\omega) & \text{if } \mathbf{r} \in D. \end{cases}$$

Complex conductivities $\sigma_a + i\omega\epsilon_a$ and $\sigma_b + i\omega\epsilon_b$ are constants at each frequency $\omega$ but they are changing with the frequency $\omega$. Figures 1.17a,b show measured

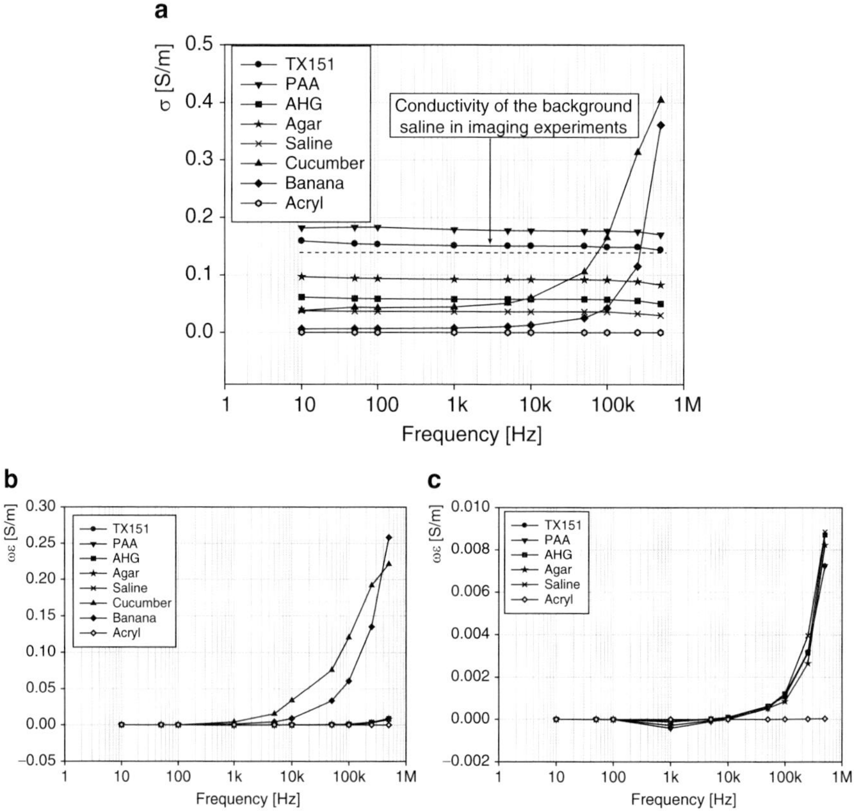


**Fig. 1.17** (**a**) $\sigma$ and (**b**) $\omega\epsilon$ spectra. (**c**) Magnified $\omega\epsilon$ spectra of non-biological materials only

$\sigma$ and $\omega\epsilon$ spectra of seven different materials. Conductivities of non-biological materials including saline, agar, polyacrylamide (PAA), animal hide gelatin (AHG) and TX151 can be adjusted by changing the amount of NaCl. Since there exist significant differences in $\omega\epsilon$ spectra between biological and non-biological materials, we plotted $\omega\epsilon$ spectra of non-biological materials only in Fig. 1.17c using a different scale.

In fdEIT, we inject currents at two different frequencies $\omega_1, \omega_2$ and measure induced boundary voltages $f^j_{\omega_1}(k) = u^j_{\omega_1}|_{\mathcal{E}_k}$ and $f^j_{\omega_2}(k) = u^j_{\omega_2}|_{\mathcal{E}_k}$ at each electrode. Here, the corresponding time-harmonic potential $u^j_\omega$ satisfies the following boundary value problem:

$$\begin{cases} \nabla \cdot \left((\sigma(\mathbf{r},\omega) + i\omega\epsilon(\mathbf{r},\omega))\nabla u^j_\omega\right) = 0 & \text{in } \Omega \\ -(\sigma(\mathbf{r},\omega) + i\omega\epsilon(\mathbf{r},\omega))\nabla u^j_\omega \cdot \mathbf{n} = g_j & \text{on } \partial\Omega. \end{cases} \tag{1.13}$$

| | Conductivity($\sigma$[S m$^{-1}$]) | | | | | | |
|---|---|---|---|---|---|---|---|
| Frequency (Hz) | TX151 | PAA | AHG | AGAR | CUCUMBER | BANANA | SALINE |
| 100 | 1.53$E$−1 | 1.83$E$−1 | 5.80$E$−2 | 9.30$E$−2 | 4.30$E$−2 | 7.00$E$−3 | 3.60$E$−2 |
| 10,000 | 1.51$E$−1 | 1.77$E$−1 | 5.80$E$−2 | 9.20$E$−2 | 1.05$E$−1 | 2.50$E$−2 | 3.60$E$−2 |
| 100,000 | 1.49$E$−1 | 1.77$E$−1 | 5.70$E$−2 | 9.10$E$−2 | 1.65$E$−1 | 4.20$E$−2 | 3.60$E$−2 |
| 250,000 | 1.49$E$−1 | 1.76$E$−1 | 5.50$E$−2 | 8.80$E$−2 | 3.13$E$−1 | 1.15$E$−1 | 3.30$E$−2 |
| 500,000 | 1.44$E$−1 | 1.70$E$−1 | 5.00$E$−2 | 8.30$E$−2 | 4.05$E$−1 | 3.61$E$−1 | 3.00$E$−2 |

| | Permittivity($\omega\epsilon$[S m$^{-1}$]) | | | | | | |
|---|---|---|---|---|---|---|---|
| Frequency (Hz) | TX151 | PAA | AHG | AGAR | CUCUMBER | BANANA | SALINE |
| 100 | 0 | 0 | 0 | 0 | 1.46$E$−6 | 2.28$E$−6 | 0 |
| 10,000 | 5.17$E$−4 | 5.17$E$−4 | 6.26$E$−4 | 5.23$E$−4 | 7.59$E$−2 | 3.32$E$−2 | 5.94$E$−4 |
| 100,000 | 1.19$E$−3 | 1.19$E$−3 | 1.08$E$−3 | 8.47$E$−4 | 1.21$E$−1 | 6.07$E$−2 | 1.21$E$−3 |
| 250,000 | 3.18$E$−3 | 3.18$E$−3 | 3.15$E$−3 | 2.65$E$−3 | 1.92$E$−1 | 1.35$E$−1 | 3.96$E$−3 |
| 500,000 | 7.24$E$−3 | 7.24$E$−4 | 8.72$E$−3 | 8.24$E$−3 | 2.21$E$−1 | 2.58$E$−1 | 8.86$E$−3 |

The goal is to provide an image of any frequency-dependent changes of complex conductivity using measured data $f^j_{\omega_1}(k)$ and $f^j_{\omega_2}(k)$. For the image reconstruction, we try to use a weighted frequency-difference voltage $u^j_{\omega_k} - \alpha_b u^j_{\omega_l}$ at each electrode.

A careful analysis shows that the weighted frequency-difference data $f_{\omega_2} - \alpha_b f_{\omega_1}$ is connected with the anomaly $D$ through the following representation formula:

$$f_{\omega_2}(\mathbf{r}) - \alpha_b f_{\omega_1}(\mathbf{r}) = \int_D \frac{\mathbf{r}-\mathbf{r}'}{\pi|\mathbf{r}-\mathbf{r}'|^2} \cdot \left[\tau_2 \nabla u_{\omega_2}(\mathbf{r}') - \tau_1 \nabla u_{\omega_1}(\mathbf{r}')\right] d\mathbf{r}', \quad \mathbf{r} \in \partial\Omega \tag{1.14}$$

where

$$\alpha_b = \frac{\sigma_b(\omega_1) + i\omega_1\epsilon_b(\omega_1)}{\sigma_b(\omega_2) + i\omega_2\epsilon_b(\omega_2)}$$

and

$$\tau_j = \frac{(\sigma_b(\omega_j) - \sigma_a(\omega_j)) + i\omega_j(\epsilon_b(\omega_j) - \epsilon_a(\omega_j))}{\sigma_b(\omega_2) + i\omega_2\epsilon_b(\omega_2)}, \ j = 1, 2.$$

Hence, the real and imaginary parts of $f_{\omega_2} - \alpha_b f_{\omega_1}$ correspond to those of $\tau_2 \nabla u_{\omega_2} - \tau_1 \nabla u_{\omega_1}$, respectively.

**Observation 1.1.** *Average directions of $\nabla v_1, \nabla v_2, \nabla h_1, \nabla h_2$ on $D$ are approximately parallel or anti-parallel:*

$$\left| \int_D \nabla v_1(\mathbf{r}) d\mathbf{r} \times \int_D \nabla v_2(\mathbf{r}) d\mathbf{r} \right| \approx 0 \quad \text{and} \quad \left| \int_D \nabla v_j(\mathbf{r}) d\mathbf{r} \times \int_D \nabla h_j(\mathbf{r}) d\mathbf{r} \right| \approx 0$$

*for $j = 1, 2$. Under these approximations, real and imaginary parts of (1.14) can be approximated by*

$$\begin{array}{ll} \Re\{f_{\omega_2}(\mathbf{r}) - \alpha_b f_{\omega_1}(\mathbf{r})\} \approx C_1(\omega_1, \omega_2, \frac{\sigma_a}{\sigma_b}, \frac{\epsilon_a}{\epsilon_b}) \int_D \frac{\mathbf{r}-\mathbf{r}'}{\pi|\mathbf{r}-\mathbf{r}'|^2} \cdot \nabla v_1(\mathbf{r}') d\mathbf{r}', & \mathbf{r} \in \partial\Omega \\ \Im\{f_{\omega_2}(\mathbf{r}) - \alpha_b f_{\omega_1}(\mathbf{r})\} \approx C_2(\omega_1, \omega_2, \frac{\sigma_a}{\sigma_b}, \frac{\epsilon_a}{\epsilon_b}) \int_D \frac{\mathbf{r}-\mathbf{r}'}{\pi|\mathbf{r}-\mathbf{r}'|^2} \cdot \nabla v_1(\mathbf{r}') d\mathbf{r}', & \mathbf{r} \in \partial\Omega \end{array} \tag{1.15}$$

*where $C_1$ and $C_2$ are constants depending on $\omega_1, \omega_2, \frac{\sigma_a}{\sigma_b}, \frac{\epsilon_a}{\epsilon_b}$.*

Now, let us construct an image reconstruction algorithm for an $N$-channel EIT system. Assume $\gamma_{\omega_1}$ and $\gamma_{\omega_2}$ have homogeneous background with the complex conductivities $\hat{\gamma}_1$ and $\hat{\gamma}_2$, respectively, and $\gamma_{\omega_1} = \hat{\gamma}_1$ and $\gamma_{\omega_2} = \hat{\gamma}_2$ near $\partial\Omega$. Let $\hat{u}_l^j, l = 1, 2$, be the solutions of (1.13) with $\gamma_\omega = \sigma(\mathbf{r}, \omega) + i\omega\epsilon(\mathbf{r}, \omega)$ replaced by $\hat{\gamma}_l$. In this case,

$$\alpha_b = \frac{\hat{\gamma}_1}{\hat{\gamma}_2} \quad \text{and} \quad \gamma_{\omega_1} - \alpha_b \gamma_{\omega_2} = 0 \text{ on } \partial\Omega.$$

Writing $\delta\gamma = \alpha_b \gamma_{\omega_2} - \gamma_{\omega_1}$, the normalized change of the complex conductivity due to the frequency change from $\omega_1$ to $\omega_2$ is computed from the relation

$$\int_\Omega \delta\gamma \nabla \left( \frac{\hat{u}_0^j}{\hat{\gamma}_1} \right) \cdot \nabla \left( \frac{\hat{u}_0^k}{\hat{\gamma}_2} \right) d\mathbf{r} = \left( f_{\omega_2}^j(k) - \alpha_b f_{\omega_1}^j(k) \right) I \tag{1.16}$$

for $j, k = 1, 2, 3, \cdots, 15$. Numerical simulations are conducted on a circular phantom with 16 electrodes around its circumference. The circular phantom can be regarded as a unit disk $\Omega = \{(x, y) : x^2 + y^2 \le 1\}$ normalizing the length scale. The phantom was filled with saline and contained one anomaly occupying the region $D = \{(x, y) : (x - 0.45)^2 + y^2 \le 0.25^2\}$.

In fdEIT algorithm, the weighted difference of voltage data at two frequencies is used to reconstruct the frequency-dependent change of complex conductivity distribution. To show the effect of the weight $\alpha_b$, reconstructed

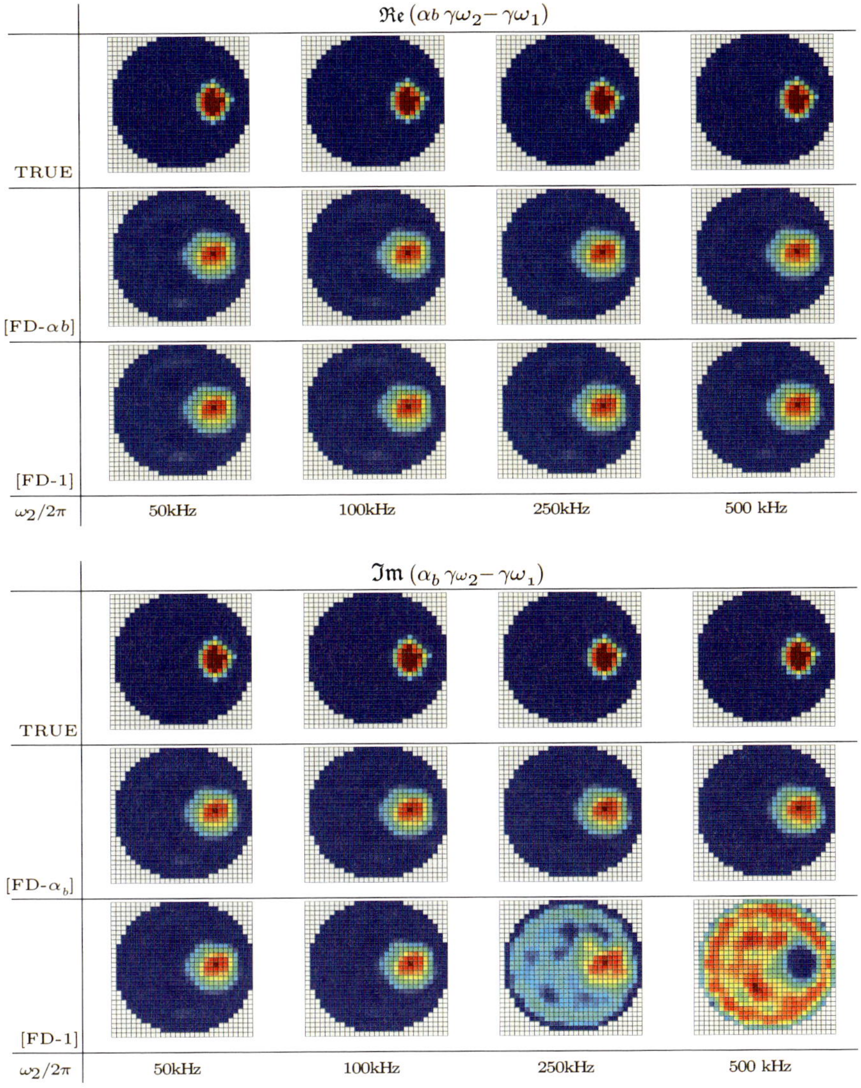


**Fig. 1.18** Frequency-difference images of a banana object inside a circular saline phantom. *The first and fourth rows* are true images of real and imaginary parts of $\alpha_b \gamma_{\omega_2} - \gamma_{\omega_1}$, *the second and fifth rows* are the reconstructed images using the weighted difference and *the third and sixth rows* are the reconstructed images without using the weighted difference

images using a weighted difference and a simple difference are compared. Figure 1.18 shows reconstructed frequency-difference images of a banana object using four different high frequencies, $\omega_2/2\pi = 50, 100, 250$ and $500\,\mathrm{kHz}$ having a fixed low frequency $\omega_1/2\pi = 100\,\mathrm{Hz}$.

## 1.3 Trans-Admittance Scanner for Breast Cancer Detection

### 1.3.1 Review on Lesion Estimation Using EIT

This section handles the problem of estimating or detecting lesions or anomalies inside an electrically conducting object using boundary measurements of current–voltage data. We assume that there exists a high contrast between conductivity or permittivity values of a lesion and the surrounding medium. Here, the major difficulty basically comes from the followings. First, its reconstruction map from the current–voltage data to the geometry of an anomaly is highly nonlinear. Second, the sensitivity of the current–voltage data to the inhomogeneity due to the anomaly is very low. Therefore, as already discussed in the previous section on EIT, the cross-sectional conductivity and/or permittivity imaging of the subject may not be able to provide enough spatial resolution needed to localize the anomaly. Without appropriately managing this difficulty, any static or absolute EIT image would be suspicious in terms of its accuracy. This section focuses on the feature extraction of anomalies inside the object instead of its cross-sectional imaging. We should mention that the major drawback of this feature detection approach is the requirement of a homogeneous background and the use of voltage data in the absence of any anomaly.

In 1999, Kwon et al. [28] looked at the inverse problem in a different way and developed the idea of the location search method. In 2000, they extended this idea to lesion detection in EIT [29] and showed that the estimation of locations and size is a well-posed problem [3, 5, 27]. Suppose anomalies $D_1, \cdots, D_M$ occupy a region $D = D_1 \cup \cdots \cup D_M$ inside a background medium $\Omega$. Since the conductivity $\sigma$ changes abruptly across the interface $\partial D$, a clear contrast exists between the anomalies and the surrounding medium. To distinguish them, we denote

$$\sigma = \begin{cases} \sigma_0 & \text{in } \Omega \setminus \overline{\cup_j D_j} \\ \sigma_j & \text{in each } D_j \end{cases} \tag{1.17}$$

for $j = 1, \cdots, M$. Along the interface $\partial D$, the tangential component of the electric field is continuous while the normal component changes abruptly. If $u$ is the voltage in (1.13), it satisfies the transmission conditions of

$$\begin{aligned} \sigma_j\, \mathbf{n}(\mathbf{r}) \cdot \nabla u^{int}(\mathbf{r}) &= \sigma_0\, \mathbf{n}(\mathbf{r}) \cdot \nabla u^{int}(\mathbf{r}) \quad \text{for each } \mathbf{r} \in \partial D_j, \\ u^{ext}(\mathbf{r}) = u^{int}(\mathbf{r}) &\quad \text{for } \mathbf{r} \in \partial D \end{aligned} \tag{1.18}$$

where $u^{int} := u|_D$ and $u^{ext} := u|_{\Omega \setminus \bar{D}}$ are voltages inside and outside of $D$, respectively. The inverse problem is to recover anomalies $D$ from the relationship between current $g = -\sigma\nabla u \cdot \mathbf{n}|_{\partial\Omega}$ and boundary voltage $u|_{\partial\Omega}$. The goal of this problem is to develop an algorithm for extracting a quantitative core information of $D$ with a few measured data in such a way that the core information of $D$ is reasonably stable against measurement errors. This section considers non-iterative anomaly estimation algorithms for searching location of anomaly and estimating its size.

To provide a feasible representation formula for detecting anomaly, we begin with considering the simplest case where $\sigma_0, \sigma_1, \cdots, \sigma_M$ are constants. We assume that the domains $\{D_j\}_{j=1}^M$ are small relative to $\Omega$, separated apart from each other, and away from the boundary. As in Fig. 1.8, we place surface electrodes $\mathcal{E}_j$ for $j = 1, \cdots, E$ on the boundary $\partial\Omega$. Let $u$ be the solution of the Neumann boundary value problem in (1.13) and $f$ be its boundary voltage.

In order to extract the information of the collection of anomalies $D$, it is desirable to express $u$ in terms of $D$. In this section, we derive a representation formula of $u$ involving $D$ [18, 27, 29]. Let $\Phi$ be the fundamental solution of the Laplace equation:

$$\Phi(\mathbf{r}, \mathbf{r}') = -\frac{1}{4\pi|\mathbf{r} - \mathbf{r}'|} = \frac{-1}{4\pi\sqrt{(x-x')^2 + (y-y')^2 + (z-z')^2}}. \tag{1.19}$$

Carefully using the transmission condition in (1.18), we have the following identities:

$$\begin{aligned} u(\mathbf{r}) &= \sum_{j=1}^M \left(\frac{\sigma_j}{\sigma_0} - 1\right) \int_{D_j} \nabla u(\mathbf{r}) \cdot \nabla\Phi(\mathbf{r}, \mathbf{r}')\, d\mathbf{r}' + H(\mathbf{r}; g, f) \quad \text{for } \mathbf{r} \in \Omega, \\ 0 &= \sum_{j=1}^M \left(\frac{\sigma_j}{\sigma_0} - 1\right) \int_{D_j} \nabla u(\mathbf{r}) \cdot \nabla\Phi(\mathbf{r}, \mathbf{r}')\, d\mathbf{r}' + H(\mathbf{r}; g, f) \quad \text{for } \mathbf{r} \in \mathbb{R}^3 \setminus \overline{\Omega}. \end{aligned} \tag{1.20}$$

Here, $H(\mathbf{r}; g, f)$ is a harmonic function that is computed directly from the data $g$ and $f$:

$$H(\mathbf{r}; g, f) := \frac{1}{\sigma_0} \int_{\partial\Omega} \Phi(\mathbf{r}, \mathbf{r}')\, g(\mathbf{r}') dS_{\mathbf{r}'} + \int_{\partial\Omega} \mathbf{n}(\mathbf{r}') \cdot \nabla_{\mathbf{r}'}\Phi(\mathbf{r}, \mathbf{r}') f(\mathbf{r}')\, dS_{\mathbf{r}'} \tag{1.21}$$

for $\mathbf{r} \in \mathbb{R}^3 \setminus \partial\Omega$. The derivation of the representation formula (1.20) can be done using the classical layer potential techniques. In order to extract information of $D$, we need to find a direct interplay between the unknown $D$ and known $H(\mathbf{r}; g, f)$ from (1.20). Since (1.20) involves the unknown $u$ that depends on $D$ in a highly nonlinear way, it is desirable to approximate $u$ in terms of $H(\mathbf{r}; g, f)$ and $D$.

We express the integral term involving $D_j$ in (1.20) as a single layer potential with the weight $\varphi_j$:

$$\left(\frac{\sigma_j}{\sigma_0}-1\right)\int_{D_j}\nabla_y\Phi(\mathbf{r}-\mathbf{r}')\cdot\nabla u(\mathbf{r}')\,d\mathbf{r}'=\int_{\partial D_j}\Phi(\mathbf{r}-\mathbf{r}')\varphi_j(\mathbf{r}')dS_{\mathbf{r}'} \quad (1.22)$$

where $\varphi_j,, j=1,\ldots,M$ is the normal component of $\nabla u$ on the interface $\partial D$ multiplied by the constant $\frac{\sigma_j}{\sigma_0}-1$:

$$\varphi_j=\left(\frac{\sigma_j}{\sigma_0}-1\right)\nabla u\cdot\mathbf{n}|_{\partial D_j}.$$

The main advantage of introducing $\varphi_j$ is that we can represent $\varphi_j$ as a function depending only on $D_j$ and the known function $H$. To be precise, $\varphi_j$ is the solution of the following integral identity:

$$\frac{\sigma_0+\sigma_j}{2(\sigma_j-\sigma_0)}\varphi_j(\mathbf{r})-\mathcal{K}^*_{D_j}\varphi_j(\mathbf{r})=\mathbf{n}(\mathbf{r})\cdot\nabla H_j(\mathbf{r}) \quad \text{for } \mathbf{r}\in\partial D_j \quad (1.23)$$

where

$$\mathcal{K}^*_{D_j}\varphi_j(\mathbf{r})=\frac{1}{4\pi}\int_{\partial D_j}\frac{(\mathbf{r}-\mathbf{r}')\cdot\mathbf{n}(\mathbf{r})}{|\mathbf{r}-\mathbf{r}'|^3}\varphi_j(\mathbf{r}')dS_{\mathbf{r}'} \quad \text{for } \mathbf{r}\in\partial D_j,$$

$$H_j(\mathbf{r}):=H(\mathbf{r};g,f)+\sum_{k\neq j}\int_{\partial D_j}\Phi(\mathbf{r}-\mathbf{r}')\varphi_k(\mathbf{r}')dS_{\mathbf{r}'} \quad \text{for } \mathbf{r}\in\mathbb{R}^3.$$

This enable us to provide the following approximations:

$$\begin{aligned}u(\mathbf{r})&\approx H(\mathbf{r};g,f)+\sum_{j=1}^{m}\frac{1}{\lambda_j}\int_{D_j}\nabla_{\mathbf{r}'}\Phi(\mathbf{r}-\mathbf{r}')\cdot\nabla H_j(\mathbf{r}')d\mathbf{r}' \quad \text{for} \mathbf{r}\in\Omega,\\ H(\mathbf{r};g,f)&\approx-\sum_{j=1}^{m}\frac{1}{\lambda_j}\int_{D_j}\nabla_{\mathbf{r}'}\Phi(\mathbf{r}-\mathbf{r}')\cdot\nabla H_j(\mathbf{r}')d\mathbf{r}' \quad \text{for } \mathbf{r}\in\mathbb{R}^3\setminus\overline{\Omega},\end{aligned} \quad (1.24)$$

where $\lambda_j=\frac{\sigma_0+\sigma_j}{2(\sigma_j-\sigma_0)}$. In the next section, we use the above identities for extracting features of $D$.

### Location Search Method

The location of an anomaly can be determined by the pattern of a simple weighted combination of injection current and boundary voltage [29]. We assume that the object contains a single anomaly $D=D_1$ that is small compared with the object itself and separated away from the boundary $\partial\Omega$. We also assume that $\sigma_0$ and $\sigma_1$ are constants. According to (1.24), $D$ and $H(\mathbf{r}, g, f)$ satisfy the following approximate identity:

$$\mathbf{H}(\mathbf{r}; g, f) \approx -\frac{1}{\lambda_1} \int_D \nabla_{\mathbf{r}'} \Phi(\mathbf{r} - \mathbf{r}') \cdot \nabla H(\mathbf{r}'; g, f) d\mathbf{r}', \quad \mathbf{r} \in \mathbb{R}^3 \setminus \bar{\Omega}. \quad (1.25)$$

The location search algorithm is based on simple aspects of the function $H(\mathbf{r}; g, f)$ outside the domain $\Omega$ which can be computed directly from the data $g$ and $f$.

For the injection current pattern, we choose $g(\mathbf{r}) = \mathbf{a} \cdot \mathbf{n}(\mathbf{r})$ for some fixed constant vector $\mathbf{a}$. Now, we are ready to explain the location search method using only the boundary current–voltage data:

*(1) Take two observation regions $\Sigma_1$ and $\Sigma_2$ contained in $\mathbb{R}^3 \setminus \Omega$ given by*

$$\Sigma_1 := \textit{a line parallel to } \mathbf{a},$$
$$\Sigma_2 := \textit{a plane (or a line if } n = 2\textit{) normal to } \mathbf{a}.$$

*(2) Find two points $P_i \in \Sigma_i (i = 1, 2)$ so that*

$$H(P_1; g, f) = 0$$

*and*

$$H(P_2; g, f) = \begin{cases} \min_{\mathbf{r} \in \Sigma_2} H(\mathbf{r}; g, f) & \textit{if } \sigma_j - \sigma_0 > 0, \\ \max_{\mathbf{r} \in \Sigma_2} H(\mathbf{r}; g, f) & \textit{if } \sigma_j - \sigma_0 < 0. \end{cases}$$

*(3) Draw the corresponding plane $\Pi_1(P_1)$ and the line $\Pi_2(P_2)$ given by*

$$\Pi_1(P_1) := \{\mathbf{r} \,;\, \mathbf{a} \cdot (\mathbf{r} - P_1) = 0\},$$
$$\Pi_2(P_2) := \{\mathbf{r} \,;\, (\mathbf{r} - P_2) \textit{ is parallel to } \mathbf{a}\}.$$

*(4) Find the intersecting point $P$ of the plane $\Pi_1(P_1)$ and the line $\Pi_2(P_2)$, then this point $P$ can be viewed as the location of the anomaly $D$.*

In order to provide more insight on the above location search method, we let $u_0$ be the voltage in the absence of the anomaly $D$. With the same injection current $g = \mathbf{a} \cdot \mathbf{n}$, the voltage $u_0$ satisfies $\nabla u_0|_\Omega = \mathbf{a}/\sigma_0$. If we denote by $f_0$ the boundary voltage of $u_0$, it follows from (1.20) that

$$0 \approx H(\mathbf{r}; g, f_0) \quad \text{for } \mathbf{r} \in \mathbb{R}^3 \setminus \bar{\Omega}. \quad (1.26)$$

Subtracting (1.26) from (1.25) gives

$$H(\mathbf{r}; g, f_0) - H(\mathbf{r}; g, f) \approx \frac{1}{\lambda_1} \int_D \nabla_{\mathbf{r}'} \Phi(\mathbf{r} - \mathbf{r}') \cdot \nabla H(\mathbf{r}'; g, f) d\mathbf{r}' \quad \text{for } \mathbf{r} \in \mathbb{R}^3 \setminus \bar{\Omega}.$$

Since $H(\mathbf{r}; g, f_0) = 0$ for $r \in \mathbb{R}^3 \setminus \bar{\Omega}$,

$$H(\mathbf{r}; g, f) \approx \frac{-1}{\lambda_1} \int_D \nabla_{\mathbf{r}'} \Phi(\mathbf{r} - \mathbf{r}') \cdot \nabla H(\mathbf{r}'; g, f) d\mathbf{r}' \quad \text{for } \mathbf{r} \in \mathbb{R}^3 \setminus \bar{\Omega}. \quad (1.27)$$

Due to the special injection current $g = \mathbf{a} \cdot \mathbf{n}$, $\nabla H(\mathbf{r}; g, f_0) = \mathbf{a}/\sigma_0$. Using the assumption that the anomaly $D$ is relatively small and situated away from $\partial\Omega$, we obtain

$$\nabla H(\mathbf{r}; g, f) \quad \approx \quad \nabla H(\mathbf{r}; g, f_0) = \mathbf{a}/\sigma_0 \qquad \text{for } \mathbf{r} \in D.$$

Hence, (1.27) is reduced to

$$H(\mathbf{r}; g, f) \approx \frac{1}{4\pi\sigma_0\lambda_1} \int_D \frac{(\mathbf{r} - \mathbf{r}') \cdot \mathbf{a}}{|\mathbf{r} - \mathbf{r}'|^3} d\mathbf{r}' \qquad \text{for } \mathbf{r} \in \mathbb{R}^3 \setminus \bar{\Omega}. \qquad (1.28)$$

Examining the integrand of (1.28), we can see that the sign of $H(\cdot; g, f)$ is determined by $\mathbf{a}$, $D$ and the sign of $\lambda_1$. Indeed, the identity (1.28) leads to the crucial observation that for $P_1 \in \mathbb{R}^n \setminus \bar{\Omega}$ with $H(P_1; g, f) = 0$, the plane or the line $\Pi_1(P_1) := \{\mathbf{r} \ ; \ \mathbf{a} \cdot (\mathbf{r} - P_1) = 0\}$ divides the domain $D$.

### Total Size Estimation

The total size estimation of anomalies $D = \cup_{j=1}^M D_j$ also uses the projection current $g = \mathbf{a} \cdot \mathbf{n}$, where $\mathbf{a}$ is a unit constant vector. In this section, we assume that the conductivity values of all anomalies are the same constant $\sigma_1$ so that $\sigma(\mathbf{r}) = \sigma_0$ in the background region $\Omega \setminus \overline{\cup_{j=1}^M D_j}$ and $\sigma = \sigma_1$ in the anomalies $\cup_{j=1}^M D_j$. We may assume $\Omega$ contains the origin. Define the scaled domain $\Omega_t = \{t\mathbf{r} : \mathbf{r} \in \Omega\}$ for a scaling factor $t > 0$. Let $v_r$ be the solution of the problem

$$\begin{cases} \nabla \cdot ((\sigma_0 \chi_{\Omega \setminus \Omega_t} + \sigma_1 \chi_{\Omega_t}) \nabla v_t) = 0 \quad \text{in } \Omega \\ \sigma_0 \, \mathbf{n} \cdot \nabla v_t = g \quad \text{on } \partial\Omega, \qquad \int_{\partial\Omega} v_t = 0 \end{cases}$$

where $\chi_{\mathcal{T}}$ is the indicator function of the domain $\mathcal{T}$.

The total size of $\cup_{j=1}^M D_j$ is very close to the size of the domain $\Omega_{t_0}$ where $t_0$, $0 < t_0 < 1$, is determined uniquely from

$$\int_{\partial\Omega} v_{t_0} \, g \, ds = \int_{\partial\Omega} u \, g \, ds. \qquad (1.29)$$

Various numerical experiments indicate that the algorithm gives a nearly exact estimate for arbitrary multiple anomalies [31] even though some restrictions on anomalies are necessary in its rigorous proof.

First, we show why $t_0$ is uniquely determined in the interval $(0, 1)$. Let $\eta(t) := \int_{\partial\Omega} v_t g \, ds$ as a function of $t$ defined in the interval $(0, 1)$. If $t_1 < t_2$, it follows from integration by parts that

$$
\begin{aligned}
\eta(t_1) - \eta(t_2) &= \int_{\partial\Omega} (v_{t_1} - v_{t_2}) g \, ds \\
&= \int_{\Omega} (\sigma_0 \chi_{\Omega\setminus\Omega_{t_1}} + \sigma_1 \chi_{\Omega_{t_1}}) |\nabla (v_{t_1} - v_{t_2})|^2 \, d\mathbf{r} \\
&\quad + (\sigma_1 - \sigma_0) \int_{\Omega_{t_2}\setminus\Omega_{t_1}} |\nabla v_{t_2}|^2 \, d\mathbf{r}, \\
\eta(t_1) - \eta(t_2) &= -\int_{\Omega} (\sigma_0 \chi_{\Omega\setminus\Omega_{t_2}} + \sigma_1 \chi_{\Omega_{t_2}}) |\nabla (v_{t_1} - v_{t_2})|^2 \, d\mathbf{r} \\
&\quad + (\sigma_1 - \sigma_0) \int_{\Omega_{t_2}\setminus\Omega_{t_1}} |\nabla v_{t_1}|^2 \, d\mathbf{r}.
\end{aligned}
$$

These identities give a monotonicity of $\eta(t)$:

$$
\eta(t_1) < \eta(t_2) \quad \text{if } \sigma_1 - \sigma_0 < 0 \qquad \text{and} \qquad \eta(t_1) > \eta(t_2) \quad \text{if } \sigma_1 - \sigma_0 > 0.
$$

Since $D \subset \Omega$, a similar monotonicity argument leads to the following inequalities;

$$
\begin{aligned}
\eta(0) &< \int_{\partial\Omega} ug \, ds < \eta(1) \quad \text{if } \sigma_1 - \sigma_0 < 0, \\
\eta(0) &> \int_{\partial\Omega} ug \, ds > \eta(1) \quad \text{if } \sigma_1 - \sigma_0 > 0.
\end{aligned}
$$

Since $\eta(t)$ is continuous, there exists a unique $t_0$ so that $\eta(t_0) = \int_{\partial\Omega} ug \, ds$.

Next, we try to provide an explanation on the background idea of the following size estimation:

$$
\text{volume of } \Omega_{t_0} \quad \approx \quad \text{total volume of } \cup_{j=1}^{M} D_j \, . \tag{1.30}
$$

We begin with the following identities which can be obtained easily from integrating by parts;

$$
\begin{aligned}
\int_{\partial\Omega} (u - v_t) g \, d\sigma &= \int_{\Omega} \sigma \nabla (u - v_t)|^2 \, d\mathbf{r} + (\sigma_1 - \sigma_0) \int_{\Omega_t} |\nabla v_t|^2 \, d\mathbf{r} \\
&\quad - (\sigma_1 - \sigma_0) \int_{D} |\nabla v_t|^2 \, d\mathbf{r}, \\
\int_{\partial\Omega} (u - v_t) g \, d\sigma &= -\int_{\Omega} \sigma_t |\nabla (u - v_t)|^2 \, d\mathbf{r} + (\sigma_1 - \sigma_0) \int_{\Omega_t} |\nabla u|^2 \, d\mathbf{r} \\
&\quad - (\sigma_1 - \sigma_0) \int_{D} |\nabla u|^2 \, d\mathbf{r}
\end{aligned}
$$

where $\sigma = \sigma_0 \chi_{\Omega \setminus \bar{D}} + \sigma_1 \chi_D$ and $\sigma_t = \sigma_0 \chi_{\Omega \setminus \bar{\Omega}_t} + \sigma_1 \chi_{\Omega_t}$. By adding the above two identities, we obtain

$$\begin{aligned}
2\int_{\partial\Omega}(u - v_\iota) g \, d\sigma - (\sigma_1 - \sigma_0)\left[\int_D |\nabla(u \quad v_t)|^2 \, d\mathbf{r} + \int_{\Omega_t} |\nabla v_t|^2 + |\nabla u|^2 \, d\mathbf{r}\right] \\
- (\sigma_1 - \sigma_0)\left[\int_{\Omega_t} |\nabla(u - v_t)|^2 \, d\mathbf{r} + \int_D |\nabla v_t|^2 + |\nabla u|^2 \, d\mathbf{r}\right] \\
= 2(\sigma_1 - \sigma_0)\left[\int_{\Omega_t} \nabla u \cdot \nabla v_t \, d\mathbf{r} - \int_D \nabla u \cdot \nabla v_t \, d\mathbf{r}\right].
\end{aligned}$$

According to the choice of $t_0$,

$$\int_{\Omega_{t_0}} \nabla u \cdot \nabla v_{t_0} \, d\mathbf{r} \quad = \quad \int_D \nabla u \cdot \nabla v_{t_0} \, d\mathbf{r}.$$

The above identity is possible when the volume of $\Omega_{t_0}$ is close to the total volume of $D = \cup_{j=1}^{M} D_j$.

### Experimental Settings and Results

In order to test the feasibility of the location search and size estimation methods, Kwon et al. carried out phantom experiments [31]. They used a circular phantom with 290 mm diameter as a container and it was filled with NaCl solution of $0.69\,\mathrm{S\,m^{-1}}$ conductivity. Anomalies with different conductivity values, shapes and sizes were placed inside the phantom. Equally spaced 32 electrodes were attached on the surface of the phantom. Using a 32-channel EIT system, they applied the algorithms described in the previous section to the measured boundary current–voltage data to detect anomalies.

In this section, we describe one example of applying the algorithms described in the previous section. The circular phantom can be regarded as a unit disk $\Omega := B_1(0,0)$ by normalizing the length scale. In order to demonstrate how the location search and size estimation algorithm works, we place four insulators $D = \cup_{j=1}^{4} D_j$ into the phantom as shown in Fig. 1.19:

$$\begin{aligned}
D_1 &= B_{0.1138}\big(0.5172, 0.5172\big), & D_2 &= B_{0.1759}\big(-0.5172, 0.5172\big), \\
D_3 &= B_{0.1828}\big(-0.5172, -0.5172\big), & D_4 &= B_{0.2448}\big(0.1724, -0.1724\big).
\end{aligned}$$

We inject a projection current $g = \mathbf{a} \cdot \mathbf{n}$ with $\mathbf{a} = (0,1)$ and measure the boundary voltage $f$.

For the location search described in Sect. 80, we choose two observation lines,

$$\Sigma_1 := \{(-1.5, s) | s \in \mathbb{R}\} \quad \text{and} \quad \Sigma_2 := \{(s, -1.5) | s \in \mathbb{R}\}.$$

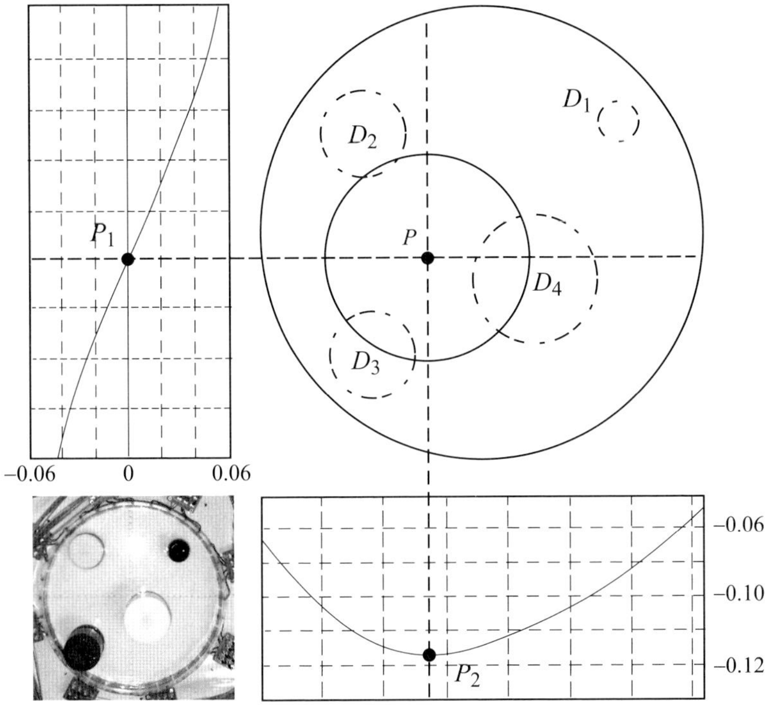


**Fig. 1.19** Illustration of the location and size estimation process. Four anomalies are all insulators and the conductivity of the saline is $0.69\,\mathrm{S\,m^{-1}}$. *Lower-left:* configuration of anomalies. *Upper-left:* $H$-plot on $\Sigma_1$. *Lower-right:* $H$-plot on $\Sigma_2$. *Upper-right:* estimation of the center of four anomalies and the total size. See [29, 31] for the details

We evaluate the two-dimensional version of $H(\mathbf{r}; f, g)$ defined in (1.21) with $\Phi$ replaced by $\Phi(\mathbf{r}) = \frac{1}{2\pi} \log \sqrt{x^2 + y^2}$. In Fig. 1.19, the upper-left plot is the graph of $H(\mathbf{r}; f, g)$ on $\Sigma_1$ and the lower-right plot is the graph of $H(\mathbf{r}; f, g)$ on $\Sigma_2$. We find the zero point of $H(\mathbf{r}; f, g)$ on $\Sigma_1$ and the maximum point of $|H(\mathbf{r}; f, g)|$ on $\Sigma_2$ denoted by dots in Fig. 1.19. The intersecting point was calculated as $P(-0.1620, -0.0980)$ which is close to the center of mass $P_M(-0.1184, -0.0358)$. For the case of a single anomaly or a cluster of multiple anomalies, the intersecting point furnishes a meaningful location information.

For the size estimation, we use (1.29) and (1.30) to compute the total volume of $\cup_{j=1}^{M} D_j$ [31]. The estimated total size was 0.4537 compared with the true total size of 0.4311. In Fig. 1.19, the corresponding disk with the size of 0.4537 centered at $P(-0.1620, -0.0980)$ is drawn with a solid line. The relative error of the estimated size was about 5.24%.

### 1.3.2 Trans-Admittance Scanner

The trans-admittance scanner (TAS) is a device for detecting anomaly whose conductivity is significantly different from the conductivity of surrounding normal tissues. On the surface of a region of interest, we place a scanning probe with a planar array of electrodes kept at the ground potential. We apply a sinusoidal voltage $V_0 \sin \omega t$ between a distant reference electrode and the probe to make electrical current travel through the region of interest. The resulting electric potential at a position $\mathbf{r} = (x, y, z)$ and time $t$ can be expressed as the real part of $u(\mathbf{r})e^{i\omega t}$ where the complex potential $u(\mathbf{r})$ is governed by the equation $\nabla \cdot ((\sigma + i\omega\epsilon)\nabla u(\mathbf{r})) = 0$ in the subject. Using the scanning probe equipped with a planar array of electrodes, we measure a distribution of exit currents $g = (\sigma + i\omega\epsilon)\frac{\partial u}{\partial \mathbf{n}}$ which reflects the electrical properties of tissues under the scan probe. Here, $\frac{\partial u}{\partial \mathbf{n}}$ is the normal derivative of $u$.

The inverse problem of TAS is to detect a suspicious abnormality in the region of interest underneath the probe from the measured Neumann data $g$ which is basically same as the trans-admittance data (Fig. 1.20). In order for the reconstruction to be practical, we must take account of the followings. First, since the data $g$ is available only on a small portion instead of the whole surface of the subject, the reconstruction algorithm should be robust against any change in the geometry of the domain. Second, since the background conductivity is usually unknown in practice, it should be robust against some small perturbation of the complex conductivity distribution inside the region of the interest and any large change outside the region of interest. Third, the inhomogeneous complex conductivity of a specific normal breast is unknown and it is very difficult to calculate the reference Neumann data $g^*$ without any anomaly. Fourth, we must deal with the ill-posedness of this inverse problem so that the reconstruction method is well-posed.

All of previous anomaly detection methods utilize a difference $g - g^*$ which can be viewed as a kind of background subtraction to make the anomaly apparently visible [4, 53]. However, since it is almost impossible to have reliable difference data $g - g^*$, any reconstruction algorithm using the difference is far from practical applicability. In order to deal with this

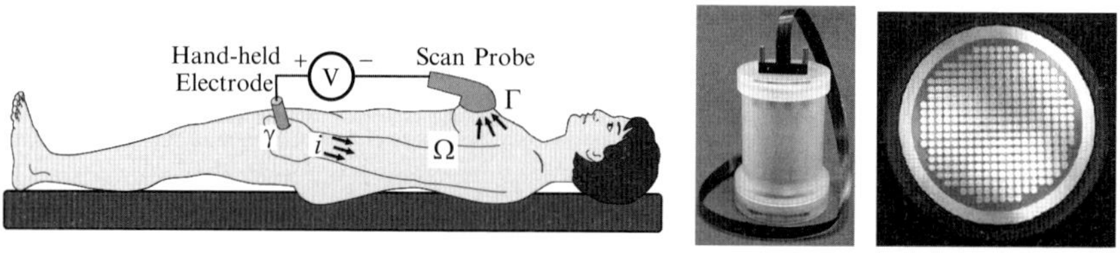


**Fig. 1.20** TAS setup. Voltage is applied between the hand-held electrode and the planar array of electrodes in the scan probe. Exit currents through the scan probe are measured to provide trans-admittance data

problem, we propose a multi-frequency TAS method which uses a frequency difference of trans-admittance data measured at a certain moment [23, 43]. In the multi-frequency TAS, we apply voltage with two different frequencies and measure two sets of corresponding exit currents at the same time. From a mathematical analysis of the model,the imaginary part of a weighted frequency-difference trans-admittance map has been found to be proportional to $|D|\frac{2\xi_3^2-(x-\xi_1)^2-(y-\xi_2)^2}{4\pi|\mathbf{x}-\xi|^5}$ where $|D|$ and $\xi$ are the size and location of the anomaly, respectively, and $\mathbf{x} = (x, y, z)$ is a position vector. Based on this relation, a novel multi-frequency anomaly estimation algorithm was proposed to provide both its size $|D|$ and location $\xi$ estimates.

## Frequency-Difference TAS Model

Let the human body occupy a three-dimensional domain $\Omega$ with a smooth boundary $\partial\Omega$. Let $\Gamma$ and $\gamma$ be portions of $\partial\Omega$, denoting the probe plane placed on the breast and the surface of the metallic reference electrode, respectively. For simplicity, we let $z$ be the axis normal to $\Gamma$ and let the center of $\Gamma$ be the origin. Hence, the probe region $\Gamma$ can be approximated as a two-dimensional region $\Gamma = \{(x, y, 0) : \sqrt{x^2 + y^2} < L\}$ where $L$ is the radius of the scan probe. We set the region of interest inside the breast as a half ball $\Omega_L = \Omega \cap B_L$ shown in Fig. 1.21 where $B_L$ is the ball with the radius $L$ and centered at the origin. We suppose that there is a cancerous lesion $D$ inside $\Omega_L$. Through $\gamma$, we apply a sinusoidal voltage of $V_0 \sin \omega t$ with its frequency $f = \omega/2\pi$ in a range of 50 Hz to 500 kHz. Then the corresponding complex potential $u_\omega$ at $\omega$ satisfies the following mixed boundary value problem:

$$\begin{cases} \nabla \cdot ((\sigma + i\omega\epsilon)\nabla u_\omega(\mathbf{r})) = 0 \quad \text{in } \Omega \\ u_\omega(\mathbf{r}) = 0, \quad x \in \Gamma \\ u_\omega(\mathbf{r}) = V_0, \quad x \in \gamma \\ (\sigma + i\omega\epsilon)\nabla u_\omega(\mathbf{r}) \cdot \mathbf{n}(\mathbf{r}) = 0, \quad x \in \partial\Omega \setminus (\Gamma \cup \gamma) \end{cases}$$

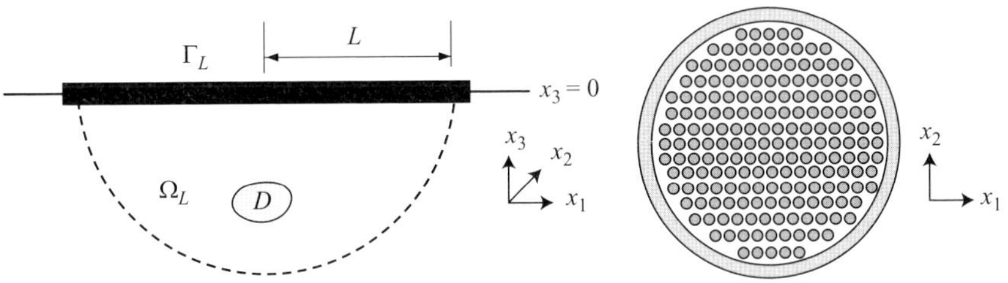


**Fig. 1.21** Simplified model of the breast region with a cancerous lesion $D$ under the scan probe

where $\mathbf{n}$ is the unit outward normal vector to the boundary $\partial\Omega$. Denoting the real and imaginary parts of $u_\omega$ by $v_\omega = \Re u_\omega$ and $h_\omega = \Im u_\omega$, respectively, the mixed boundary value problem can be expressed as the following coupled system:

$$\begin{cases} \nabla\cdot(\sigma\nabla v_\omega) - \nabla\cdot(\omega\epsilon\nabla h_\omega) = 0 & \text{in } \Omega \\ \nabla\cdot(\omega\epsilon\nabla v_\omega) + \nabla\cdot(\sigma\nabla h_\omega) = 0 & \text{in } \Omega \\ v_\omega = 0 \quad \text{and} \quad h_\omega = 0 & \text{on } \Gamma \\ v_\omega = V_0 \quad \text{and} \quad h_\omega = 0 & \text{on } \gamma \\ \mathbf{n}\cdot\nabla v_\omega = 0 \quad \text{and} \quad \mathbf{n}\cdot\nabla h_\omega = 0 & \text{on } \partial\Omega\setminus(\Gamma\cup\gamma). \end{cases}$$

The scan probe $\Gamma$ consists of a planar array of electrodes $\mathcal{E}_1, \cdots, \mathcal{E}_m$ and we measure exit current $g_\omega$ through each electrode $\mathcal{E}_j$. We can measure the real and imaginary parts of the Neumann data $g_\omega(\mathbf{r}) := \Re g_\omega(\mathbf{r}) + i\Im g_\omega(\mathbf{r})$ where

$$\begin{aligned} \Re g_\omega &= \mathbf{n}\cdot(-\sigma\nabla v_\omega(\mathbf{r}) + \omega\epsilon\nabla h_\omega(\mathbf{r})) \\ \Im g_\omega &= \mathbf{n}\cdot(-\sigma\nabla h_\omega(\mathbf{r}) - \omega\epsilon\nabla v_\omega(\mathbf{r})). \end{aligned}$$

In the multi-frequency TAS, we apply voltage at two different frequencies $f_1 = \omega_1/2\pi$ and $f_2 = \omega_2/2\pi$ with $50\,\text{Hz} \le f_1 < f_2 \le 500\,\text{kHz}$ and measure two sets of corresponding Neumann data $g_{\omega_1}$ and $g_{\omega_2}$ through $\Gamma$ at the same time. Assume that there is a breast tumor $D$ beneath the probe $\Gamma$ so that $\sigma + i\omega\epsilon$ changes abruptly across $\partial D$. Since both $\sigma$ and $\epsilon$ depend on $\omega$ and $\mathbf{r}$, $\sigma(x,\omega_1) \neq \sigma(\mathbf{r},\omega_2)$ and $\epsilon(\mathbf{r},\omega_1) \neq \epsilon(\mathbf{r},\omega_2)$. To distinguish them, we denote $\tau(\mathbf{r}) := \frac{\sigma(\mathbf{r},\omega_2)}{\sigma(\mathbf{r},\omega_1)}$, $\kappa(\mathbf{r}) := \frac{\epsilon(\mathbf{r},\omega_2)}{\epsilon(\mathbf{r},\omega_1)}$, $\sigma(\mathbf{r}) := \sigma(\mathbf{r},\omega_1)$, $\epsilon(\mathbf{r}) := \epsilon(\mathbf{r},\omega_1)$. In the multi-frequency TAS model, we use a weighted frequency-difference of Neumann data $g_{\omega_2} - \alpha g_{\omega_1}$ instead of a time-difference $g_\omega - g_\omega^*$ where $\alpha$ is chosen as $\alpha := \frac{\tau\sigma + i\omega_2\kappa\epsilon}{\sigma + i\omega_1\epsilon}|_\Gamma$.

With a careful analysis, the weighted difference $g_{\omega_2} - \alpha g_{\omega_1}$ and the anomaly has the following approximate relation: for $(x,y,0)\in\Gamma$,

$$\begin{aligned} \frac{1}{2}\Im\left(g_2 - \alpha g_1\right)(x,y) \approx \beta\left(\frac{\sigma_c}{\sigma_n} - \frac{\kappa_c\epsilon_c}{\kappa_n\epsilon_n}\right) \\ \times |D|\frac{2\xi_3^2 - (x-\xi_1)^2 - (y-\xi_2)^2}{4\pi\left[(x-\xi_1)^2 + (y-\xi_2)^2 + \xi_3^2\right]^{5/2}} \end{aligned} \tag{1.31}$$

where $\beta = \frac{\omega_2\kappa_n\epsilon_n}{\sigma_n}\frac{(3\sigma_c)^2}{(2\sigma_c+\sigma_n)^2}g_0(\xi_1,\xi_2)$,

$$\sigma = \begin{cases} \sigma_n \text{ in } \Omega_L\setminus\overline{D} \\ \sigma_c \text{ in } D \end{cases} \quad \text{and } \epsilon = \begin{cases} \epsilon_n \text{ in } \Omega_L\setminus\overline{D} \\ \epsilon_c \text{ in } D. \end{cases}$$

Figure 1.22 shows an example of the real and imaginary parts of the difference image $g_{\omega_2} - \alpha g_{\omega_1}$ on the probe region $\Gamma$.

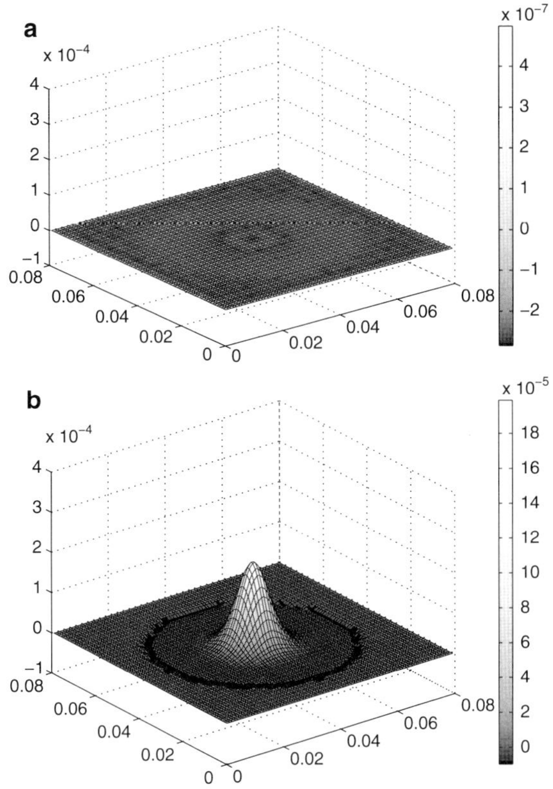


**Fig. 1.22** Real and imaginary parts of the difference of trans-admittance data $g_1 - \alpha g_2$: (**a**) $\Re\{g_1 - \alpha g_2\}$ and (**b**) $\Im\{g_1 - \alpha g_2\}$

Hence, the weighted difference $g_{\omega_2} - \alpha g_{\omega_1}$ provides the location and size information of anomaly. However, in practice, we do not know the constant $\alpha$ since the background complex conductivities at each frequency is not available. Hence, we should determine $\alpha$ using measured data $g_{\omega_1}$ and $g_{\omega_2}$ only. It is crucial to observe that $\alpha_n$ can be expressed by

$$\alpha = \frac{\int_\Gamma g_{\omega_2} ds}{\int_\Gamma g_{\omega_1} ds} + \frac{|D|}{|\Omega_L|} \frac{\frac{1}{|D|}\int_D (\alpha\eta_{\omega_1} - \eta_{\omega_2})\nabla u_{\omega_2} \cdot \nabla u_{\omega_1} d\mathbf{r}}{\frac{1}{|\Omega_L|}\int_\Omega \eta_{\omega_1} |\nabla u_{\omega_1}|^2 d\mathbf{r}} \tag{1.32}$$

where $|D|$ is the volume of the anomaly $D$ and $\eta_\omega(\mathbf{r}) = \sigma(r,\omega) + i\omega\epsilon(\mathbf{r},\omega)$. Since the ratio $\frac{|D|}{|\Omega_L|}$ is small, $\alpha$ is approximated by

$$\alpha \approx \frac{\int_\Gamma g_{\omega_2}\, ds}{\int_\Gamma g_{\omega_1}\, ds}. \tag{1.33}$$

The identity (1.32) is proved in [23].

## TAS Reconstruction Algorithm

Figure 1.21 shows a scan probe with 320 current-sensing electrodes. The diameter of each electrode is 2 mm. Let $\mathbf{x}^j := (x^j, y^j)$ be the center of the $j$-th electrode. The distance between the neighboring centers is around 3 mm. We denote by $F(j)$ the weighted difference of the exit current at the $j$-th electrode:

$$F(j) := Im\,(g_{\omega_2} - \alpha g_{\omega_1})\,(\mathbf{x}^j), \quad j = 1, \cdots, N \quad (N = 320).$$

Using the data $F(j), j = 1, \cdots, N$, we try to determine the location $\xi = (\xi_1, \xi_2, \xi_3)$ and the size $|D|$. For robust detection of an anomaly, we must take account of the well known inherent ill-posedness and the insensitivity of the data for a distant anomaly from the probe. Figure 1.23 shows the experimental data of the imaginary part of the frequency-difference admittance map, which are noisy. We speculate that a reliable detection range could be $0 \leq \xi_3 \leq 20$ mm. We also observe that it would not be appropriate to determine $\xi$ and $|D|$ simultaneously using the standard least square method minimizing the following misfit functional:

$$\mathcal{M}(|D|, \xi) := \sum_{j=1}^{N} |\Phi(j, |D|, \xi) - F(j)|^2$$

where $\Phi(j, |D|, \xi)$ is the right side of (1.31) at $\mathbf{x}^j$. The graph of $\mathcal{M}(j, |D|, \xi)$ in Fig. 1.24 explains why the standard gradient descent method does not work in this misfit functional.

In our detection method, we first determine the transversal position $(\xi_1, \xi_2)$ which is robust and reliable:

1. Find the index $j_0$ of the electrode at which the weighted difference $g_{\omega_2} - \alpha g_{\omega_1}$ has the maximum value, that is, $F(j_0) = \sup_j F(j)$.
2. Take $j_1, \cdots, j_8$ that are indices of the electrodes neighboring to $j_0$.
3. Determine the transversal location $(\xi_1, \xi_2)$ from the formula

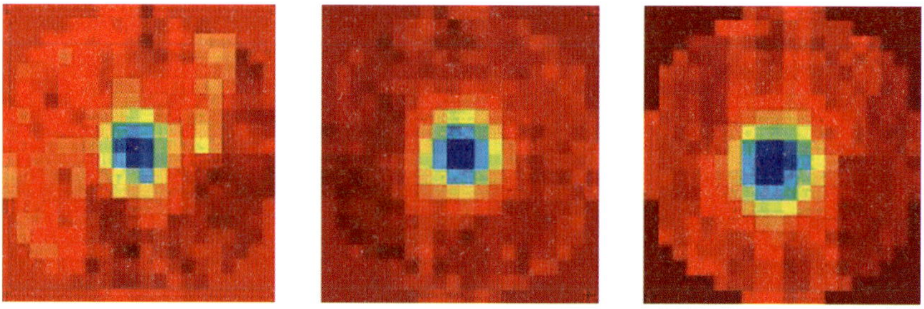

**Fig. 1.23** Imaginary part of the frequency-difference trans-admittance map, $g_1 - g_2$ from a saline phantom with a cubic anomaly at 5 mm depth

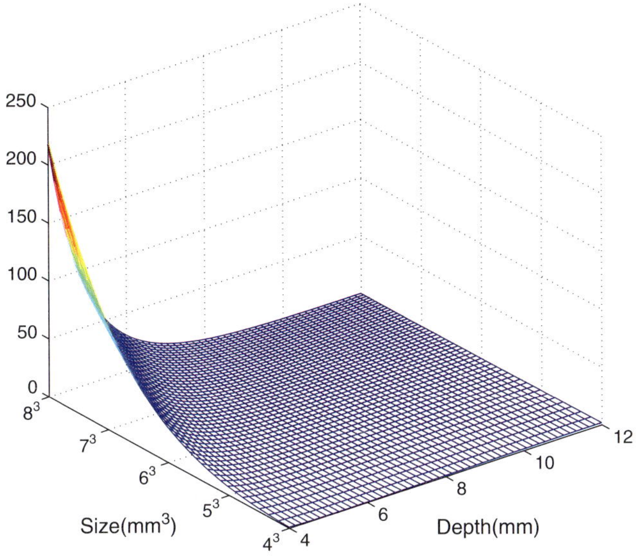


**Fig. 1.24** Graph of the misfit functional $\mathcal{M}(|D|, \xi)$ in terms of the size $|D|$ and the depth $\xi_3$ with fixed $\xi_1$ and $\xi_2$. The minimum occurs at the depth $\xi_3 = 8\,\text{mm}$ and the size $|D| = 6^3\,\text{mm}^3$. It is very difficult to robustly determine the minimum point using the gradient descent method

$$(\xi_1, \xi_2) = (1 - \sum_{k=1}^{8} \lambda_j)\mathbf{x}^{j_O} + \sum_{k=1}^{8} \lambda_k \mathbf{x}^{j_k}$$

where $\lambda_k = \frac{|F(j_k)|}{a + \sum_{k=0}^{8} |F(j_k)|}$ and $0 < a < 1$.

Next, we determine the depth $\xi_3$ of the anomaly:

1. Choose the set $A := \{j \ : \ \theta_1 F(j_O) < F(j) < \theta_2 F(j_O)\}$ where $0 < \theta_1 < \frac{1}{2} < \theta_2 < 1$.
2. For each $j \in A$, compute $z_j$ with $0 < z_j < 20\,\text{mm}$ satisfying

$$\frac{1}{F(j_O)} \frac{2z_j^2 - r_{j_O}^2}{\left(z_j^2 + r_{j_O}^2\right)^{5/2}} = \frac{1}{F(j)} \frac{2z_j^2 - r_j^2}{\left(z_j^2 + r_j^2\right)^{5/2}}$$

   where $r_j := \left|\mathbf{x}^j - (\xi_1, \xi_2)\right|$.
3. Determine the depth $\xi_3$ by the formula

$$\xi_3 = \frac{1}{\sum_{j \in A} F(j)^{-1}} \sum_{j \in A} F(j)^{-1} z_j.$$

Finally, we can determine the size $|D|$ from the knowledge of the location $\xi$ and the approximate formula.

## 1.4 Magnetic Resonance Electrical Impedance Tomography

Magnetic resonance electrical impedance tomography (MREIT) is a new medical imaging technique combining electrical impedance tomography (EIT) and magnetic resonance imaging (MRI). Noting that EIT suffers from the ill-posed nature of the corresponding inverse problem, we introduce MREIT which utilizes internal information on the induced magnetic field in addition to the boundary voltage subject to an injection current to produce three-dimensional images of conductivity and current density distributions. Since 2000, imaging techniques in MREIT have been advanced rapidly and now are at the stage of in vivo animal experiments. In both EIT and MREIT, we inject currents through electrodes placed on the surface of a subject. While EIT is limited by the boundary measurements of current–voltage data, MREIT utilizes the internal magnetic flux density $\mathbf{B}$ obtained using an MRI scanner. This is the main reason why MREIT could eliminate the ill-posedness of EIT. Though MREIT can produce conductivity images with high spatial resolution and accuracy, disadvantages of MREIT over EIT may include the lack of portability, potentially long imaging time and requirement of an expensive MRI scanner.

In the early MREIT system, all three components of $\mathbf{B} = (B_x, B_y, B_z)$ have been used and their measurements required mechanical rotations of the subject within the MRI scanner [20, 30, 32, 40, 64]. Assuming the knowledge of the full components of $\mathbf{B}$, we can directly compute the current density $\mathbf{J} = \nabla \times \mathbf{B}/\mu_0$ where $\mu_0$ is the magnetic permeability of the free space and biological tissues in the human body. Kwon et al. [30] developed the $J$-substitution algorithm that reconstructs high-resolution conductivity images from the magnitude of $\mathbf{J}$. This $J$-substitution algorithm has been applied in experimental studies using saline phantoms to produce conductivity images of $64 \times 64$ pixels with a voxel size of $0.6 \times 0.6 \times 10$ mm$^3$ using a 0.3-T experimental MRI scanner [20, 32]. In the papers [22, 24], we also provided the corresponding mathematical theory including the uniqueness. However, this early MREIT method using $\mathbf{B} = (B_x, B_y, B_z)$ as measured data sets has serious technical difficulties in its clinical applications due to the requirement of subject rotations within the main magnet of the MRI scanner. Therefore, in order for MREIT to be practical, we must reconstruct cross-sectional images of a conductivity distribution $\sigma$ from only $B_z$ that can be obtained without rotating the subject. Looking back on the early day, we thought that the attempt for $B_z$-based MREIT was hopeless since we didn't appreciate the powerful role of $\Delta B_z$.

In 2002, the first constructive $B_z$-based MREIT algorithm called the harmonic $B_z$ algorithm has been proposed by Seo–Yoon–Woo–Kwon [56]. Figure 1.25 shows a schematic diagram of the harmonic $B_z$ algorithm and typical MREIT images of a conductivity phantom including three chunks of

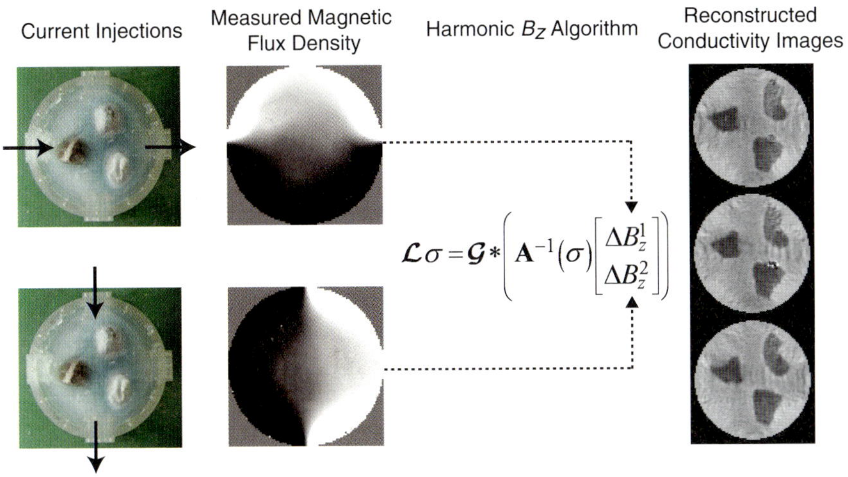


**Fig. 1.25** Overview of the harmonic $B_z$ algorithm

biological tissues having different conductivity values inside a cylindrical container. After the harmonic $B_z$-algorithm, various reconstruction algorithms for the $B_z$-based MREIT model have been developed [41, 42, 44, 45, 54, 55]. Recent published numerical simulations and phantom experiments show that conductivity images with high spatial resolution are achievable when the contrast of the conductivity is not very high.

In this section, we provide the rationale of pursuing MREIT research that requires an expensive MRI scanner. Then, recent progress in MREIT providing conductivity and current density images with high spatial resolution and accuracy will be presented including mathematical theory, algorithms and experimental methods. With numerous potential applications in mind, some future researches in MREIT will be proposed.

### *1.4.1 Fundamentals in MREIT*

In order to image a conductivity distribution inside the subject, we must inject current (or apply voltage) using a pair of electrodes which generates the internal current density $\mathbf{J}$, electrical field $\mathbf{E}$ and magnetic flux density $\mathbf{B}$. If we let $z$ be the axis that is parallel to the direction of the main magnetic field in an MRI scanner, we can measure the $z$-component of $\mathbf{B}$, that is, $B_z$. In MREIT, we try to express the conductivity distribution $\sigma$ in terms of $B_z$ data. We begin with setting up an exact mathematical model of MREIT that agrees with a planed medical imaging system. To simplify the model, let us make several assumptions which should not go astray from the practical model. Let the subject to be imaged occupy a three-dimensional bounded domain $\Omega \subset \mathbb{R}^3$ with a smooth connected boundary $\partial\Omega$ and each $\Omega_{z_0} := \Omega \cap \{z = z_0\} \subset \mathbb{R}^2$, the slice of $\Omega$ cut by the plane $\{z = z_0\}$, has a smooth connected boundary. We assume that the conductivity distribution $\sigma$ in $\Omega$ is isotropic. We attach a pair of copper electrodes $\mathcal{E}^+$ and $\mathcal{E}^-$ on $\partial\Omega$ in order to inject current, and let $\mathcal{E}^+ \cup \mathcal{E}^-$ be the portion of the surface $\partial\Omega$ where electrodes are attached. The injection current $I$ produces a current density $\mathbf{J} = (J_x, J_y, J_z)$ inside $\Omega$ satisfying the following problem:

$$\begin{cases} \nabla \cdot \mathbf{J} = 0 & \text{in } \Omega \\ I = -\int_{\mathcal{E}^+} \mathbf{J} \cdot \mathbf{n} ds = \int_{\mathcal{E}^-} \mathbf{J} \cdot \mathbf{n} ds, & \mathbf{J} \times \mathbf{n} = 0 \text{ on } \mathcal{E}^+ \cup \mathcal{E}^- \\ \mathbf{J} \cdot \mathbf{n} = 0 & \text{on } \partial\Omega \setminus \overline{\mathcal{E}^+ \cup \mathcal{E}^-}, \end{cases} \tag{1.34}$$

where $\mathbf{n}$ is the outward unit normal vector on $\partial\Omega$ and $ds$ the surface area element. The condition of $\mathbf{J} \times \mathbf{n} = 0$ on $\mathcal{E}^+ \cup \mathcal{E}^-$ comes from the fact that copper electrodes are almost perfect conductors. Since $\mathbf{J}$ is expressed as $\mathbf{J} = -\sigma \nabla u$ where $u$ is the corresponding electrical potential, (1.34) can be converted to

$$\begin{cases} \nabla \cdot (\sigma \nabla u) = 0 & \text{in } \Omega \\ I = \int_{\mathcal{E}^+} \sigma \frac{\partial u}{\partial \mathbf{n}} ds = -\int_{\mathcal{E}^-} \sigma \frac{\partial u}{\partial \mathbf{n}} ds, & \nabla u \times \mathbf{n} = 0 \text{ on } \mathcal{E}^+ \cup \mathcal{E}^- \\ \sigma \frac{\partial u}{\partial \mathbf{n}} = 0 & \text{on } \partial\Omega \setminus \overline{\mathcal{E}^+ \cup \mathcal{E}^-}, \end{cases} \tag{1.35}$$

where $\frac{\partial u}{\partial \mathbf{n}} = \nabla u \cdot \mathbf{n}$. The above nonstandard boundary value problem (1.35) is well-posed and has a unique solution up to a constant. Let us briefly discuss the boundary conditions that are essentially related with the size of electrodes. The condition $\nabla u \times \mathbf{n}|_{\mathcal{E}^\pm} = 0$ ensures that each of $u|_{\mathcal{E}^+}$ and $u|_{\mathcal{E}^-}$ is a constant, since $\nabla u$ is normal to its level surface. The term $\pm I = \int_{\mathcal{E}^\pm} \sigma \frac{\partial u}{\partial \mathbf{n}} ds$ means that the total amount of injection current through the electrodes is $I$ mA. Let us denote $g := -\sigma \frac{\partial u}{\partial \mathbf{n}}|_{\partial\Omega}$. In practice, it is difficult to specify the Neumann data $g$ in a point-wise sense because only the total amount of injection current $I$ is known. It should be noticed that the boundary condition in (1.35) leads $|g| = \infty$ on $\partial\mathcal{E}^\pm$, singularity along the boundary of electrodes, and $g \notin L^2(\partial\Omega)$. But, fortunately $g \in H^{-1/2}(\partial\Omega)$, which also can be proven by the standard regularity theory in PDE.

The exact model (1.35) can be converted into the following standard problem of elliptic equation with mixed boundary conditions. Indeed, it is easy to see [35] that

$$u = \frac{I}{\int_{\partial\mathcal{E}^+} \sigma \frac{\partial \tilde{u}}{\partial \mathbf{n}} ds} \tilde{u} \quad \text{in } \Omega \qquad \text{(up to a constant)} \tag{1.36}$$

where $\tilde{u}$ is the solution of the mixed boundary value problem:

$$\begin{cases} \nabla \cdot (\sigma \nabla \tilde{u}) = 0 \quad \text{in} \quad \Omega \\ \tilde{u}|_{\mathcal{E}^+} = 1, \ \tilde{u}|_{\mathcal{E}^-} = 0 \\ -\sigma \frac{\partial \tilde{u}}{\partial \mathbf{n}} = 0 \quad \text{on} \quad \partial\Omega \setminus (\mathcal{E}^+ \bigcup \mathcal{E}^-). \end{cases} \tag{1.37}$$

The presence of an internal current density $\mathbf{J}$ in $\Omega$ gives rise to a magnetic flux density $\mathbf{B}^J$ via the Biot–Savart law:

$$\mathbf{B}^J(\mathbf{r}) = \frac{\mu_0}{4\pi} \int_\Omega \mathbf{J}(\mathbf{r}') \times \frac{\mathbf{r} - \mathbf{r}'}{|\mathbf{r} - \mathbf{r}'|^3} d\mathbf{r}'.$$

Similarly, an external current $I$ along lead wires produces a magnetic flux density $\mathbf{B}^I$. Hence, the total magnetic flux density due to the internal current density $\mathbf{J}$ and external current $I$ is $\mathbf{B} = \mathbf{B}^J + \mathbf{B}^I$. Using an MRI machine in which the subject $\Omega$ is located, we can measure the $z-$component $B_z$ of $\mathbf{B}$.

We have the following relation between $B_z$ and $\sigma$. From the Ampere law $\mathbf{J} = \frac{1}{\mu_0} \nabla \times \mathbf{B}$, we have

$$\mu_0 \nabla \times \mathbf{J} = \nabla \times \nabla \times \mathbf{B} = -\nabla^2 \mathbf{B} + \nabla \underbrace{\nabla \cdot \mathbf{B}}_{=0} = -\nabla^2 \mathbf{B}.$$

On the other hand, we have

$$\nabla \times \mathbf{J} = -\nabla \times [\sigma \nabla u] = -\nabla \sigma \times \nabla u - \underbrace{\sigma \nabla \times \nabla u}_{=0}.$$

Combining the above two identities lead to the following key identity:

$$\frac{1}{\mu_0} \nabla^2 B_z = z\text{-component of } \nabla \sigma \times \nabla u = \underbrace{\nabla u \mathbb{L} \cdot \nabla \sigma}_{\text{directional derivative}} \tag{1.38}$$

where

$$\mathbb{L} = \begin{bmatrix} 0 & -1 & 0 \\ 1 & 0 & 0 \\ 0 & 0 & 0 \end{bmatrix}.$$

**Observation 1.2.** *The identity (1.38) means that $\nabla^2 B_z$ holds the information of the transversal change of $\sigma$ in the direction $\nabla u \mathbb{L}$, while it is blind to its orthogonal direction $\nabla u$.*

This is because the transversal gradient of $\sigma$ can be decomposed as

$$\begin{pmatrix} \frac{\partial \sigma}{\partial x} \\ \frac{\partial \sigma}{\partial x} \end{pmatrix} = \frac{1}{\beta} \begin{bmatrix} -u_y \\ u_x \\ 0 \end{bmatrix} \cdot \nabla \sigma \begin{bmatrix} -u_y \\ u_x \end{bmatrix} + \frac{1}{\beta} \begin{bmatrix} u_x \\ u_y \\ 0 \end{bmatrix} \cdot \nabla \sigma \begin{bmatrix} u_x \\ u_y \end{bmatrix}$$

where $\beta = u_x^2 + u_y^2$.

Observation 1.2 provides a way to reconstruct the conductivity distribution from $B_z$. Define the Neumann-to-$B_z$ map $\Lambda_\sigma : H^{-1/2}(\partial\Omega) \to H^1(\Omega)$ by

$$\Lambda_\sigma(g)(\mathbf{r}) := \frac{\mu_0}{4\pi}\int_\Omega \frac{\sigma(\mathbf{r}')\,[(x-x')\frac{\partial u}{\partial y}(\mathbf{r}') - (y-y')\frac{\partial u}{\partial x}(\mathbf{r}')]}{|\mathbf{r}-\mathbf{r}'|^3}\,d\mathbf{r}'$$

where $g$ is the corresponding Neumann data in (1.35).

**Observation 1.3.** *Since $B_z$ is blind to the transversal direction of $\nabla u$, there are infinitely many $\tilde\sigma$ such that*

$$\Lambda_{\tilde\sigma}[g] = \Lambda_\sigma[g] \quad in\ \Omega. \tag{1.39}$$

*In particular, for any increasing function $\phi : [\min u, \max u] \to \mathbb{R}$ such that $\phi'(\min u) = 1 = \phi'(\max u)$, $\phi(\min u) = \phi(\min u)$, and $\phi(\max u) = \phi(\max u)$, $\tilde\sigma(\mathbf{r}) := \frac{\sigma(\mathbf{r})}{\phi'(u(\mathbf{r}))}$ satisfies (1.39).*

Note that $\tilde u(\mathbf{r}) = \phi(u(\mathbf{r}))$ and $u$ have the same equipotential lines and the same Neumann data. Moreover, $\nabla\cdot(\tilde\sigma\nabla\tilde u) = 0$ with the Neumann data $\mathbf{n}\cdot(\tilde\sigma\nabla\tilde u) = \mathbf{n}\cdot(\tilde\sigma\nabla\tilde u)$ on the boundary $\partial\Omega$. The above non-uniqueness follows form the facts that

$$\Lambda_\sigma(g)(\mathbf{r}) = \frac{\mu_0}{4\pi}\int_\Omega \frac{-1}{|\mathbf{r}-\mathbf{r}'|}\begin{vmatrix}\frac{\partial\sigma}{\partial x} & \frac{\partial\sigma}{\partial y}\\ \frac{\partial u}{\partial x} & \frac{\partial u}{\partial y}\end{vmatrix}d\mathbf{r}' + \frac{\mu_0}{4\pi}\int_{\partial\Omega}\frac{1}{|\mathbf{r}-\mathbf{r}'|}\,\mathbf{n}\cdot(\sigma\mathbb{L}\nabla u)\,ds$$

and

$$\begin{vmatrix}\frac{\partial\sigma}{\partial x} & \frac{\partial\sigma}{\partial y}\\ \frac{\partial u}{\partial x} & \frac{\partial u}{\partial y}\end{vmatrix} = \begin{vmatrix}\frac{\partial\tilde\sigma}{\partial x} & \frac{\partial\tilde\sigma}{\partial y}\\ \frac{\partial\tilde u}{\partial x} & \frac{\partial\tilde u}{\partial y}\end{vmatrix}.$$

Figure 1.26 shows two different conductivities having the same $B_z$ data subject to the same injection current. Observation 1.3 is crucial to understand why we need at least two linearly independent injection currents.

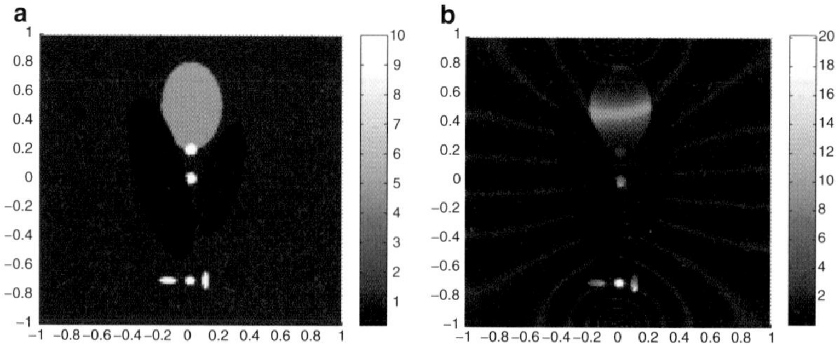


**Fig. 1.26** Distribution of (**a**) $\sigma$ and (**b**) $\tilde\sigma := \frac{\sigma(\mathbf{r})}{\phi'(u(\mathbf{r}))}$ where $\phi$ is chosen so that $\phi(t) = t + \frac{1}{100\pi}\sin(200\pi t)$. Here, $\Omega = (-1,1)\times(-1,1)$, $g(\mathbf{r}) = \delta(\mathbf{r}-(0,1)) - \delta(\mathbf{r}-(0,-1))$, $x \in \partial\Omega$

### Noise Analysis for $B_z$

The measurement of $B_z$ due to an injection current has been studied in [17, 50, 51]. This requires an MRI scanner as a tool to capture internal magnetic flux density images. To extract $B_z$, we my use the spin echo pulse sequence shown in Fig. 1.27. Using the MRI scanner, we obtain the following complex k-space data $\mathcal{S}^{\pm}$ after injection of positive and negative currents, $I^+$ and $I^-$, respectively, as shown in Fig. 1.27.

$$\mathcal{S}^{\pm}(m,n) = \int\int_{-\infty}^{\infty} M(x,y)e^{j\delta(x,y)}e^{\pm j\gamma B_z(x,y)T_c}e^{j(xm\Delta k_x+yn\Delta k_y)}dxdy \tag{1.40}$$

where $\delta$ is any systematic phase artifact, $\gamma = 26.75 \times 10^7\,\mathrm{rad\,T^{-1}s^{-1}}$ the gyromagnetic ratio of hydrogen and $T_c$ the current pulse width in seconds. Taking two-dimensional discrete Fourier transformations, we obtain the following complex signals:

$$\mathcal{M}^{\pm}(x,y) = M(x,y)e^{j\delta(x,y)}e^{\pm j\gamma B_z(x,y)T_c} \tag{1.41}$$

where $\pm B_z$ are the induced magnetic flux densities obtained with injection currents of, $I^+$ or $I^-$, respectively. Note that $|\mathcal{M}^+| = |\mathcal{M}^-| = M$ is proportional to the size (volume) of voxels. The induced magnetic flux density $B_z$ is embedded in the following incremental phase change in MR data:

$$\Psi(x,y) = \arg\left(\frac{\mathcal{M}^+(x,y)}{\mathcal{M}^-(x,y)}\right) = 2\gamma B_z(x,y)T_c \tag{1.42}$$

where we assume that the operator $\arg(\cdot)$ includes any necessary phase unwrapping [11]. In MREIT, it is essential to maximize this phase change to

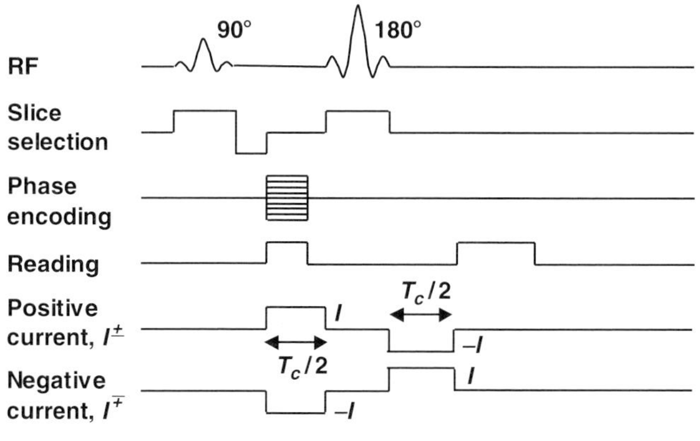


**Fig. 1.27** Spin echo pulse sequence for MREIT

obtain larger $B_z$ signals. Note that by using two current injections of $I^+$ and $I^-$, we reject any systematic phase artifact $\delta$ and double the phase change expressed in (1.42).

Once we obtain $\Psi$, we compute $B_z$ as

$$B_z(x,y) = \frac{\Psi(x,y)}{2\gamma T_c} = \frac{1}{2\gamma T_c} \arg\left(\frac{\mathcal{M}^+(x,y)}{\mathcal{M}^-(x,y)}\right). \tag{1.43}$$

The measured $\mathcal{M}^\pm$ contain independent and identically distributed complex Gaussian random noise $\mathcal{Z}^\pm$, respectively. That is, measured signals may be described by

$$\mathcal{M}^\pm + \mathcal{Z}^\pm \text{ where } \mathcal{Z}^\pm = z_r^\pm + i z_i^\pm$$

and $z_r^+, z_r^-, z_i^+$ and $z_i^-$ are identically distributed Gaussian random variables with zero mean and a variance of $s^2$. Figure 1.28 shows an example of $\mathcal{M}^-$ and $\mathcal{Z}^-$. Without loss of generality, we may set the local coordinate of $\mathcal{Z}^-$ with its real axis parallel to the direction of $\mathcal{M}^-$. Then, the noise $\mathcal{Z}^-$ can be understood as a vector located at the local origin O, having a random magnitude and direction. In [51], Scott et al. defined the signal-to-noise ratio (SNR), $\Upsilon$ in a noisy magnitude image $|\mathcal{M}^- + \mathcal{Z}^-|$ as

$$\Upsilon = \frac{\text{mean}\left(|\mathcal{M}^- + \mathcal{Z}^-|\right)}{\text{sd}\left(\mathcal{M}^- + \mathcal{Z}^-\right)} = \frac{|\mathcal{M}^-|}{\text{sd}\left(\mathcal{Z}^-\right)} = \frac{M}{\sqrt{2}s} \tag{1.44}$$

where mean$(\cdot)$ and sd$(\cdot)$ denote the mean and standard deviation, respectively. They derived an expression for the standard deviation in measured $B_z$ data as

$$\text{sd}\left(B_z\right) = \frac{1}{2\gamma T_c \Upsilon}. \tag{1.45}$$

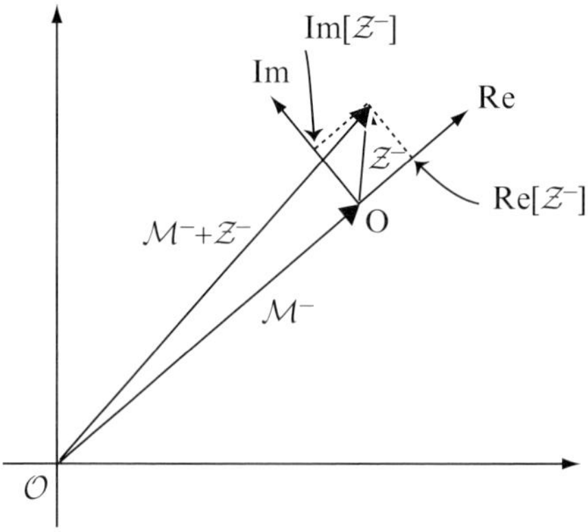


**Fig. 1.28** An example of $\mathcal{M}^-$ and $\mathcal{Z}^-$. Note the definition of the local axis for $\mathcal{Z}^-$

In [47], we found that it is more convenient to define the SNR, $\Upsilon_M$ of the magnitude image as

$$\Upsilon_M = \frac{\text{mean}\,(|\mathcal{M}^- + \mathcal{Z}^-|)}{\text{sd}\,(|\mathcal{M}^- + \mathcal{Z}^-|)} = \frac{M}{\text{sd}\,(|\mathcal{M}^- + \mathcal{Z}^-|)} = \frac{M}{s}. \tag{1.46}$$

We derive an expression for the noise standard deviation in measured $B_z$ using the magnitude image SNR definition in (1.46).

Including noise in (1.43), we now have $B_z$ signals proportional to

$$\begin{aligned} \arg\left(\tfrac{\mathcal{M}^+ + \mathcal{Z}^+}{\mathcal{M}^- + \mathcal{Z}^-}\right) &= \arg\left(\frac{\mathcal{M}^+\left(1+\frac{\mathcal{Z}^+}{\mathcal{M}^+}\right)}{\mathcal{M}^-\left(1+\frac{\mathcal{Z}^-}{\mathcal{M}^-}\right)}\right) \\ &= \arg\left(\tfrac{\mathcal{M}^+}{\mathcal{M}^-}\right) + \arg\left(1 + \tfrac{\mathcal{Z}^+}{\mathcal{M}^+}\right) - \arg\left(1 + \tfrac{\mathcal{Z}^-}{\mathcal{M}^-}\right). \end{aligned} \tag{1.47}$$

Assuming that $|\mathcal{M}^\pm| \gg |\mathcal{Z}^\pm|$, we have $\arg\left(1 + \frac{\mathcal{Z}^\pm}{\mathcal{M}^\mp}\right) \approx \text{Im}\left(\frac{\mathcal{Z}^\pm}{\mathcal{M}^\mp}\right)$. This can also be understood from Fig. 1.28, since the imaginary part of $\mathcal{Z}^-$ perturbs the argument of $\mathcal{M}^-$. Now, the standard deviation of the argument in (1.47) is

$$\begin{aligned} \text{sd}\left[\arg\left(\tfrac{\mathcal{M}^+ + \mathcal{Z}^+}{\mathcal{M}^- + \mathcal{Z}^-}\right)\right] &\approx \text{sd}\left[\text{Im}\left(\tfrac{\mathcal{Z}^+}{\mathcal{M}^+}\right) - \text{Im}\left(\tfrac{\mathcal{Z}^-}{\mathcal{M}^-}\right)\right] \\ &= \sqrt{2}\text{sd}\left[\text{Im}\left(\tfrac{\mathcal{Z}^-}{\mathcal{M}^-}\right)\right] = \tfrac{\sqrt{2}}{|\mathcal{M}^-|}\text{sd}\left[\text{Im}\left(\mathcal{Z}^-\right)\right] \\ &= \tfrac{\sqrt{2}}{|\mathcal{M}^-|\sqrt{2}}\text{sd}\left(\mathcal{Z}^-\right) = \tfrac{1}{M}\text{sd}\left(\mathcal{Z}^-\right). \end{aligned} \tag{1.48}$$

We now estimate the standard deviation of the magnitude image, $|\mathcal{M}^- + \mathcal{Z}^-|$. As shown in Fig. 1.28, the real part of the complex Gaussian random noise $\mathcal{Z}^-$ mainly perturbs the magnitude of $\mathcal{M}^-$. This can be seen by setting, without loss of generality, $\mathcal{M}^- = 1$ and $\mathcal{Z}^- = a + ib$ with $a, b \ll 1$. Then,

$$|\mathcal{M}^- + \mathcal{Z}^-| = \sqrt{(1+a)^2 + b^2} = \sqrt{1 + 2a + a^2 + b^2} = 1 + a + \frac{a^2 + b^2}{2} + .... \approx 1 + a.$$

Therefore, we have

$$\text{sd}\left(\left|\mathcal{M}^- + \mathcal{Z}^-\right|\right) \approx \text{sd}\left(\text{Re}\left[\mathcal{Z}^-\right]\right) = \frac{1}{\sqrt{2}}\text{sd}\left(\mathcal{Z}^-\right). \tag{1.49}$$

Finally, we obtain an expression for the standard deviation in measured magnetic flux density $B_z$ as

$$\text{sd}\left(B_z\right) = \frac{1}{2\gamma T_c\left|\mathcal{M}^-\right|}\text{sd}(\mathcal{Z}^-) = \frac{\sqrt{2}}{2\gamma T_c M}\text{sd}\left(\left|\mathcal{M}^- + \mathcal{Z}^-\right|\right). \tag{1.50}$$

With the definition of SNR, $\Upsilon_M$ from (1.46), we conclude that

$$\text{sd}\left(B_z\right) = \frac{1}{\sqrt{2}\gamma T_c \Upsilon_M}. \tag{1.51}$$

## Two-Dimensional Current Density Imaging in MREIT

In this section, we consider the problem whether we can determine $\mathbf{J}$ from only $B_z$. In general, a single set of $B_z$ data is not sufficient for the reconstruction of $\mathbf{J}$. This means that there are two different current densities $\mathbf{J}$ and $\tilde{\mathbf{J}}$ producing the same $B_z$. However, under the assumption that out-of-plane current density $J_z$ is negligible, $B_z$ data is enough to determine $\mathbf{J}$ approximately [54].

First, we presents a way of recovering a transversal current density $\mathbf{J}$ having $J_z = 0$ within a thin slice to be imaged from measured $B_z$ data. It must be noticed that this transversal current $\mathbf{J} = (J_x, J_y, 0)$ can not be computed directly from $B_z$, that is, $J_x \neq \frac{1}{\mu_0}\partial_y B_z$ and $J_y \neq -\frac{1}{\mu_0}\partial_x B_z$. To explain this more precisely, let $\Omega_s$ be a transversal thin slice of $\Omega$, truncated by two $xy$-planes. We try to reconstruct the current density within $\Omega_s$. Now we assume that the conductivity distribution $\sigma$ does not change much in the $z$-direction. With an appropriate choice of injection current and electrodes, the resulting current density $\mathbf{J}$ in $\Omega_s$ could be approximately a transversal current, that is, $J_z \approx 0$ in $\Omega_s$ although it does not hold in the entire subject $\Omega$. In this special case, we can reconstruct $\mathbf{J}$ in $\Omega_s$ from only $B_z$.

Let $D_t$ denote a cut of the subject $\Omega$ by a $xy$-plane $\{z = t\}$. A thin slice $\Omega_s$ to be imaged could be $\Omega_s = \cup_{-\delta<t<\delta} D_t$ for some small $\delta > 0$. Since a human body is locally cylindrical in its shape, $D_t \approx D_0$ for $-\delta < t < \delta$ and therefore $\Omega_s \approx D_0 \times (-\delta, \delta)$. If the conductivity of the subject $\Omega$ does not change much in the $z$-direction, we could produce approximately a transversal internal current density $\mathbf{J}$, that is, $\mathbf{J} \approx (J_x, J_y, 0)$ in the cylindrical chop $\Omega_s$ using long longitudinal electrodes. Note that $\mathbf{J}$ could have non-zero $z$-components in the exterior $\Omega \setminus \Omega_s$ of the thin chop $\Omega_s$. The transversal current $\mathbf{J} = (J_x, J_y, 0)$ in $\Omega_s$ satisfies the following mixed boundary value problem:

$$\begin{cases} \frac{\partial}{\partial x} J_x + \frac{\partial}{\partial y} J_y = 0 \quad \text{in} \quad \Omega_s \\ \mathbf{J} \cdot \nu = g \quad \text{on} \quad \partial D_0 \times (-\delta, \delta) \\ \frac{\partial}{\partial z} J_x = 0 = \frac{\partial}{\partial z} J_y \quad \text{on} \quad D_{-\delta} \cup D_\delta . \end{cases} \tag{1.52}$$

Here, $D_{-\delta}$ and $D_\delta$ indicate the top and bottom surface of the cylinder $\Omega_s$, respectively. Definitely, the injection current $g$ must be independent of $z$-variable along the lateral boundary $\partial D_0 \times (-\delta, \delta)$ of $\Omega_s$. In order to reconstruct $\mathbf{J}$, we use the Biot–Savart law:

$$B_z(\mathbf{r}) = -\frac{\mu_0}{4\pi} \int_\Omega \frac{(y - y')J_x(\mathbf{r}') - (x - x')J_y(\mathbf{r}')}{|\mathbf{r} - \mathbf{r}'|^3} d\mathbf{r}' + B_z^I(\mathbf{r}), \quad \mathbf{r} \in \Omega_s. \tag{1.53}$$

Here, the magnetic flux density $\mathbf{B}^I$ due to the current in external lead wires can be expressed as

$$\mathbf{B}^I(\mathbf{r}) = \frac{\mu_0}{4\pi}\int_L \nabla_{\mathbf{r}'}\frac{1}{|\mathbf{r}-\mathbf{r}'|} \times \mathbf{I}(\mathbf{r}')dl_{\mathbf{r}'}$$

where $\mathbf{I}$ is the current in the lead wire $L$ in the direction of the lead wire and $dl$ is a line element. It must be noticed that $B_z$ changes along the $z$-direction in $\Omega_s$ even if $\mathbf{J}$ is independent of $z$ in $\Omega_s$.

We divide $\mathbf{B}^J$ into two parts, one is the magnetic flux density due to $\mathbf{J}$ in $\Omega_s$ and the other is due to $\mathbf{J}$ in $\Omega \setminus \Omega_s$. Then, the $B_z$ in (1.53) can be rewritten as

$$B_z(\mathbf{r}) = -\frac{\mu_0}{4\pi}\int_{\Omega_s}\frac{(y-y')J_x(\mathbf{r}')-(x-x')J_y(\mathbf{r}')}{|\mathbf{r}-\mathbf{r}'|^3}d\mathbf{r}' + G(\mathbf{r}) + B_z^I(\mathbf{r}) \quad (1.54)$$

for $\mathbf{r} \in \Omega_s$. Here, $G$ is the $z$-component of the magnetic flux density due to $\mathbf{J}$ in $\Omega \setminus \Omega_s$:

$$G(\mathbf{r}) := -\frac{\mu_0}{4\pi}\int_{\Omega\setminus\Omega_s}\frac{(y-y')J_x(\mathbf{r}')-(x-x')J_y(\mathbf{r}')}{|\mathbf{r}-\mathbf{r}'|^3}d\mathbf{r}'.$$

Since the lead wire is located outside of $\Omega$, it is easy to see that $B_z^I$ is harmonic in $\Omega_s$:

$$\nabla^2 B_z^I = 0 \quad \text{in } \Omega_s.$$

Similarly, $G$ also satisfies the Laplace equation:

$$\nabla^2 G = 0 \quad \text{in } \Omega_s.$$

Since $\nabla \times (-J_y, J_x, 0) = 0$ in $\Omega_s$, there is a function $w$ in $\Omega_s$ such that

$$\nabla w(\mathbf{r}) = (-J_y, J_x, 0), \qquad \text{in } \Omega_s.$$

Substituting $\mathbf{J} = \nabla w$ into (1.54) yields

$$B_z(\mathbf{r}) = -\frac{\mu_0}{4\pi}\int_{\Omega_s}\frac{(\mathbf{r}-\mathbf{r}')\cdot\nabla w(\mathbf{r}')}{|\mathbf{r}-\mathbf{r}'|^3}d\mathbf{r}' + G(\mathbf{r}) + B_z^I(\mathbf{r}). \quad (1.55)$$

Integrating by parts yields

$$B_z(\mathbf{r}) = \mu_0 w(\mathbf{r}) - \frac{\mu_0}{4\pi}\int_{\partial\Omega_s}\frac{(\mathbf{r}-\mathbf{r}')\cdot\nu(\mathbf{r}')}{|\mathbf{r}-\mathbf{r}'|^3}\phi(\mathbf{r}')dS\mathbf{r}' + G(\mathbf{r}) + B_I(\mathbf{r}) \quad (1.56)$$

for $\mathbf{r} \in \Omega_s$.

Since the integral in the right side of the above equality satisfies the Laplace equation, we obtain

$$\nabla^2 B_z = \mu_0 \nabla^2 w \quad \text{in } \Omega_s.$$

Set $H := w - \frac{1}{\mu_0} B_z$. Since $H$ satisfies the Laplace equation in $\Omega_s$, it can be computed explicitly from its boundary condition. This requires to know the boundary condition for $w$. Since $\frac{\partial w}{\partial z} = 0$,

$$\frac{\partial}{\partial z} H(x, y, \pm\delta) = \frac{-1}{\mu_0} \frac{\partial}{\partial z} B_z(x, y, \pm\delta), \quad (x, y) \in D_0. \tag{1.57}$$

Next, it follows from the boundary condition (1.52) for $\mathbf{J}$ that

$$\left( -\frac{\partial w}{\partial y}, \frac{\partial w}{\partial x}, 0 \right) \cdot \nu = \mathbf{J} \cdot \nu = g \quad \text{on } \partial D_0.$$

The above boundary condition can be understood as the tangential derivative of $w$ along the curve $\partial D_0$. Hence, if we write the boundary $\partial D$ as $\gamma(t), 0 < t < 1$, then we compute $w|_{\partial D}$ in the way that

$$w(\gamma(t)) - w(\gamma(0)) = \int_0^t g(\gamma(t)) |\gamma'(t)| dt := \phi(\gamma(t)).$$

Hence, $H$ has the following boundary condition:

$$H = \phi - \frac{1}{\mu_0} B_z \quad \text{in } \partial D_0 \times (-\delta, \delta). \tag{1.58}$$

Hence, $H$ can be computed by solving the Laplace equation with the boundary condition (1.57) and (1.58). Then, we can compute $\mathbf{J}$ as

$$J_x = \frac{\partial}{\partial y} H - \frac{1}{\mu_0} \frac{\partial}{\partial y} B_z \quad \text{and} \quad J_y = -\frac{\partial}{\partial x} H + \frac{1}{\mu_0} \frac{\partial}{\partial x} B_z.$$

Next, we explain why a single set of $B_z$ data is not sufficient for the reconstruction of $\mathbf{J}$. This can be done by showing two different current densities $\mathbf{J}$ and $\tilde{\mathbf{J}}$ that produce the same $B_z$. Let $\varphi$ be a function such that $(0, 0, \varphi) \cdot \mathbf{n} = 0$ on $\partial\Omega$ and $\int\int_{\{(x,y) \mid (x,y,t) \in \Omega\}} \frac{\partial \varphi}{\partial z} dx dy = 0$ for each $t$. Set

$$\tilde{\mathbf{J}} = (\tilde{J}_x, \tilde{J}_y, \tilde{J}_z) := \mathbf{J} + \left(\frac{\partial w}{\partial x}, \frac{\partial w}{\partial y}, -\varphi\right)$$

where $w$ satisfies the following boundary value problem:

$$\begin{cases} \frac{\partial^2 w}{\partial x^2} + \frac{\partial^2 w}{\partial y^2} \quad = \frac{\partial \varphi}{\partial z} & \text{in} \quad \Omega, \\ \left( \frac{\partial w}{\partial x}, \frac{\partial w}{\partial y}, 0 \right) \cdot \mathbf{n} = 0 & \text{on} \quad \partial\Omega. \end{cases} \tag{1.59}$$

The composition of the added vector field $(\frac{\partial w}{\partial x}, \frac{\partial w}{\partial y}, -\varphi)$ is chosen so that $\nabla \cdot (\frac{\partial w}{\partial x}, \frac{\partial w}{\partial y}, -\varphi) = 0$. Hence, $\tilde{\mathbf{J}}$ also satisfies

$$\nabla \cdot \tilde{\mathbf{J}} = 0 \quad \text{in } \Omega, \tag{1.60}$$

$$\tilde{\mathbf{J}} \cdot \mathbf{n} = g \quad \text{on } \partial\Omega. \tag{1.61}$$

Let $\tilde{\mathbf{B}}$ be the magnetic flux density generated by $\tilde{\mathbf{J}}$. Then, its $z$-component $\tilde{B}_z$ can be expressed as

$$\tilde{B}_z(\mathbf{r}) = B_z(\mathbf{r}) - \mu_0 \int_\Omega \nabla\Phi(\mathbf{r} - \mathbf{r}') \cdot (\frac{\partial w}{\partial y}, -\frac{\partial w}{\partial x}, 0)\, d\mathbf{r}'.$$

Since $(\frac{\partial w}{\partial y}, -\frac{\partial w}{\partial x}, 0) = \nabla \times (0, 0, w)$, the divergence theorem leads to

$$\tilde{B}_z(\mathbf{r}) - B_z(\mathbf{r}) = -\mu_0 \int_{\partial\Omega} \Phi(\mathbf{r} - \mathbf{r}')\, \mathbf{n} \cdot (\frac{\partial w}{\partial y}, -\frac{\partial w}{\partial x}, 0)\, ds.$$

The right side of the above identity is harmonic in $\Omega$ and it is zero provided $\mathbf{n} \cdot (\frac{\partial w}{\partial y}, -\frac{\partial w}{\partial x}, 0) = 0$ on $\partial\Omega$. In general, we may choose infinitely many $\varphi$ such that the corresponding $w$ satisfies $\mathbf{n} \cdot (\frac{\partial w}{\partial y}, -\frac{\partial w}{\partial x}, 0) = 0$ on $\partial\Omega$. For the case of a cylindrical domain $\Omega = \{\mathbf{r} \mid \sqrt{x^2 + y^2} < 1,\ |z| < 1\}$, we may choose

$$\varphi(x, y, z) = (z^2 - 1)\left(2 - 3\sqrt{x^2 + y^2}\right)$$

so that the corresponding $w$ depends only on $\sqrt{x^2 + y^2}$ and $z$. With this $\varphi$, $\mathbf{n} \cdot (\frac{\partial w}{\partial y}, -\frac{\partial w}{\partial x}, 0) = 0$ on the cylindrical boundary $\partial\Omega$, and therefore $B_z = \tilde{B}_z$. This shows that two different $\mathbf{J}$ and $\tilde{\mathbf{J}}$ may produce the same $B_z$.

### *1.4.2 Mathematical Framework of MREIT*

Based on the Observations 1.2 and 1.3, we set up a mathematical framework of MREIT. Recall that the relation between the measurable quantity $B_z$ and the unknown $\sigma$ is governed by the Biot–Savart law:

$$B_z(\mathbf{r}) = \frac{\mu_0}{4\pi} \int_\Omega \frac{\langle \mathbf{r} - \mathbf{r}'\,,\ \sigma(\mathbf{r}')\mathbb{L}\nabla u(\mathbf{r}')\rangle}{|\mathbf{r} - \mathbf{r}'|^3}\, d\mathbf{r}' \qquad \text{for } \mathbf{r} \in \Omega, \tag{1.62}$$

where $\mathbb{L} = \begin{pmatrix} 0 & 1 & 0 \\ -1 & 0 & 0 \\ 0 & 0 & 0 \end{pmatrix}$. Here, we must read $u$ as a nonlinear function of $\sigma$. According to the Observations 1.2 and 1.3, the unique determination of $\sigma$ requires us to inject at least two linearly independent input currents $I_1$ and $I_2$.

Now, we are ready to explain the exact MREIT model. We sequentially inject electrical currents $I_1$ and $I_2$ through two pairs of surface electrodes $\mathcal{E}_1^{\pm}$ and $\mathcal{E}_2^{\pm}$, respectively. Let $u_j$ and $B_z^j$ be the potential and magnetic flux density corresponding to $I_j$ with $j = 1, 2$.

For the measured data $B_z^1, B_z^2$ corresponding to two input currents $I_1, I_2$ and a given constant $\alpha > 0$, we try to reconstruct $\sigma$ satisfying the following conditions for $j = 1, 2$:

$$\begin{cases} \nabla \cdot (\sigma \nabla u_j) = 0 \quad \text{in } \Omega \\ I = \int_{\mathcal{E}^+} \sigma \frac{\partial u_j}{\partial \mathbf{n}}\, ds = -\int_{\mathcal{E}^+} \sigma \frac{\partial u_j}{\partial \mathbf{n}}, \quad \nabla u_j \times \mathbf{n}|_{\mathcal{E}^+ \cup \mathcal{E}^-} = 0 \\ \sigma \frac{\partial u_j}{\partial \mathbf{n}} = 0 \quad \text{on } \partial\Omega \setminus \overline{\mathcal{E}^+ \cup \mathcal{E}^-} \\ B_z^j(\mathbf{r}) = \frac{\mu_0}{4\pi} \int_\Omega \frac{\langle \mathbf{r} - \mathbf{r}', \sigma \mathbb{L} \nabla u_j \rangle}{|\mathbf{r} - \mathbf{r}'|^3}\, d\mathbf{r}', \quad \mathbf{r} \in \Omega \\ \left| u_1|_{\mathcal{E}_2^+} - u_1|_{\mathcal{E}_2^-} \right| = \alpha. \end{cases} \tag{1.63}$$

The last condition regarding $\alpha$ is necessary for fixing the scaling uncertainty of $\sigma$. Without this condition, whenever $\sigma$ and $u_j$ satisfy the other four relations in (1.63), so do $c\sigma$ and $\frac{u_j}{c}$ for any positive constant $c$. Here, we should avoid measuring the voltage difference between the pair of electrodes used for current injection since any electrode contact impedance may cause measurement errors. Therefore, in practice, we usually use the other pair of electrodes for the voltage measurement.

## Uniqueness of MREIT in Two-Dimension

Although $B_z$-based MREIT has made a significant progress last five years, its uniqueness is still an open problem in three-dimension. In the two-dimensional case, we have uniqueness using the index theory of critical points as in [1, 52]. Let us state the two-dimensional uniqueness result in [46].

**Theorem 1.4.** *Let $\Omega$ be a smooth domain in $\mathbb{R}^2$ with a connected boundary and $\sigma, \tilde\sigma \in C^1(\bar\Omega)$. (Here, the smoothness assumption for the conductivity is only for simplicity.) Assume we place the electrodes in the order of $\mathcal{E}_1^+, \mathcal{E}_2^+, \mathcal{E}_1^-$, and $\mathcal{E}_2^-$ counterclockwise. Let $g_j$ be the corresponding Neumann data using the pair $\mathcal{E}_\pm^j$. If $\Lambda_\sigma[g_j] = \Lambda_\sigma[g_j]$ for $j = 1, 2$, then*

$$\sigma = \tilde\sigma \quad \text{in } \Omega.$$

*Proof.* Since $\Lambda_\sigma[g_j] = \Lambda_\sigma[g_j]$ for $j = 1, 2$, there exist $\phi^j, j = 1, 2$ which satisfy the following

$$\Delta \phi^j = 0, \; \nabla \phi^j = \sigma \nabla u_j - \tilde\sigma \nabla \tilde u_j \qquad \text{in } \Omega.$$

Moreover, $\phi^j$ satisfies the zero Neumann boundary condition:

$$\nabla \phi^j \cdot \nu = \sigma \nabla u_j \cdot \nu - \tilde{\sigma} \nabla \tilde{u}_j \cdot \nu = g_j - g_j = 0 \quad \text{on } \partial \Omega.$$

Since $\phi^j$ is harmonic with zero Neumann boundary condition, it is just a constant function in $\Omega$. This implies that $\mathbf{J}^j - \tilde{\mathbf{J}}^j = -\sigma \nabla u_j + \tilde{\sigma} \nabla \tilde{u}_j = -\nabla \phi^j = 0$, that is, $\mathbf{J}^j = \tilde{\mathbf{J}}^j$. Since $\nabla \times \nabla u_j = 0$,

$$\begin{aligned} 0 = \nabla \times \left( \frac{\sigma \nabla u_j}{\sigma} \right) &= \nabla \frac{1}{\sigma} \times (\sigma \nabla u_j) + \frac{1}{\sigma} \nabla \times (\sigma \nabla u_j) \\ &= \frac{1}{\sigma} (-\nabla \log \sigma \times (\sigma \nabla u_j) + \nabla \times (\sigma \nabla u_j)). \end{aligned}$$

Hence, we obtain $\nabla \times (\sigma \nabla u_j) = \nabla \log \sigma \times (\sigma \nabla u_j)$. Similarly, we have $\nabla \times (\tilde{\sigma} \nabla \tilde{u}_j) = \nabla \log \tilde{\sigma} \times (\tilde{\sigma} \nabla \tilde{u}_j)$. Since $\sigma \nabla u_j = \tilde{\sigma} \nabla \tilde{u}_j$ in $\Omega$,

$$\begin{aligned} 0 &= \nabla \times (\sigma \nabla u_j) - \nabla \times (\tilde{\sigma} \nabla \tilde{u}_j) \\ &= \nabla \log \sigma \times (\sigma \nabla u_j) - \nabla \log \tilde{\sigma} \times (\tilde{\sigma} \nabla \tilde{u}_j) \\ &= \nabla \log \sigma \times (\sigma \nabla u_j) - \nabla \log \tilde{\sigma} \times (\sigma \nabla u_j) \\ &= \nabla \log (\frac{\sigma}{\tilde{\sigma}}) \times (\sigma \nabla u_j). \end{aligned}$$

This can be rewritten as the following matrix form:

$$\begin{pmatrix} \sigma \frac{\partial u_1}{\partial x} & \sigma \frac{\partial u_1}{\partial y} \\ \sigma \frac{\partial u_2}{\partial x} & \sigma \frac{\partial u_2}{\partial y} \end{pmatrix} \begin{pmatrix} -\frac{\partial}{\partial y} \log \frac{\sigma}{\tilde{\sigma}} \\ \frac{\partial}{\partial x} \log \frac{\sigma}{\tilde{\sigma}} \end{pmatrix} = \begin{pmatrix} 0 \\ 0 \end{pmatrix} \qquad \text{in } \Omega.$$

Since $0 < \sigma < \infty$, the above identity can be rewritten as

$$\begin{pmatrix} \frac{\partial u_1}{\partial x} & \frac{\partial u_1}{\partial y} \\ \frac{\partial u_2}{\partial x} & \frac{\partial u_2}{\partial y} \end{pmatrix} \begin{pmatrix} -\frac{\partial}{\partial y} \log \frac{\sigma}{\tilde{\sigma}} \\ \frac{\partial}{\partial x} \log \frac{\sigma}{\tilde{\sigma}} \end{pmatrix} = \begin{pmatrix} 0 \\ 0 \end{pmatrix} \qquad \text{in } \Omega. \tag{1.64}$$

Hence, $\sigma = \tilde{\sigma}$ provided

$$|\det \mathbb{A}[\sigma]| := \begin{vmatrix} \frac{\partial u_1}{\partial x}(\mathbf{r}) & \frac{\partial u_1}{\partial y}(\mathbf{r}) \\ \frac{\partial u_2}{\partial x}(\mathbf{r}) & \frac{\partial u_2}{\partial y}(\mathbf{r}) \end{vmatrix} \neq 0, \qquad \mathbf{r} \in \Omega.$$

We will prove this by deriving a contradiction. Assume that there exists a point $\mathbf{r}_0 \in \Omega$ such that

$$\nabla u_1(\mathbf{r}_0) \times \nabla u_2(\mathbf{r}_0) = 0.$$

Then, there exit a scalar $\alpha$ such that

$$\nabla u_1(\mathbf{r}_0) = \alpha \nabla u_2(\mathbf{r}_0).$$

Let $\eta := u_1 - \alpha u_2$. Then $\nabla \eta(\mathbf{r}_0) = 0$ and $\eta$ satisfies the following boundary value problem:

$$\begin{cases} \nabla \cdot (\sigma \nabla \eta) = 0 \quad \text{in} \quad \Omega \\ \eta|_{\mathcal{E}_1^+} = a_1, \quad \eta|_{\mathcal{E}_1^+} = -\alpha a_2 \text{ and } \quad \eta|_{\mathcal{E}_1^- \cup \mathcal{E}_2^-} = 0 \\ \sigma \nabla \eta \cdot \mathbf{n} = 0 \quad \text{on} \quad \partial\Omega \setminus \mathcal{E}_j^+ \cup \mathcal{E}_j^- \\ \int_{\mathcal{E}_1^\pm} \sigma \nabla \eta \cdot \mathbf{n}\, ds = \pm I_1, \quad \int_{\mathcal{E}_2^\pm} \sigma \nabla \eta \cdot \mathbf{n}\, ds = \mp \alpha I_j \end{cases} \tag{1.65}$$

where $a_j = u_j|_{\mathcal{E}_j^+}$ is a positive constant. Since $\nabla \eta(\mathbf{r}_0) = 0$, it follows from maximum principle that $\mathbf{r}_0$ is a saddle point. Now, we adopt the method used in [1, 52]. At the point $\mathbf{r}_0$, the level curve $\Upsilon := \{\mathbf{r} \in \Omega \ : \ \eta(\mathbf{r}) - \eta(\mathbf{r}_0)\}$ separates $\Omega$ into at least four regions $\Omega_1, \Omega_2, \Omega_3, \Omega_4, \cdots$ in the order of counter clockwise such that

$$\Omega_1 \cup \Omega_3 \subset \{\mathbf{r} \in \Omega \ : \ \eta(\mathbf{r}) > \eta(\mathbf{r}_0)\} \text{ and } \Omega_2 \cup \Omega_4 \subset \{\mathbf{r} \in \Omega \ : \ \eta(\mathbf{r}) < \eta(\mathbf{r}_0)\}.$$

From, the Hopf's boundary point lemma, there exist $P_j \in \partial\Omega \cap \partial\Omega_j$ such that

$$\sigma \frac{\partial \eta}{\partial \mathbf{n}}(P_1) > 0, \quad \sigma \frac{\partial \eta}{\partial \mathbf{n}}(P_2) < 0, \quad \sigma \frac{\partial \eta}{\partial \mathbf{n}}(P_3) > 0, \quad \sigma \frac{\partial \eta}{\partial \mathbf{n}}(P_4) < 0.$$

Since $P_1, P_2, P_3, P_4$ are lying in the order along $\partial\Omega$, $\sigma \frac{\partial \eta}{\partial \mathbf{n}}$ changes its sign at least twice along $\partial\Omega$. This is a contradiction. □

### Harmonic $B_z$ Algorithm

The harmonic $B_z$ algorithm [54] is the first realistic image reconstruction algorithm in $B_z$-based MREIT. It successfully reconstructs conductivity images with high spatial resolution and accuracy in numerical simulations and experimental studies. In spite of great performance of the iteration algorithm, there is little progress on its convergence and stability till now.

To explain the algorithm, we define

$$\mathcal{L}_{z_0} \sigma(x, y) := \sigma(x, y, z_0) + \frac{1}{2\pi} \int_{\partial\Omega_{z_0}} \frac{(x - x',\ y - y') \cdot \nu(x', y')}{|x - x'|^2 + |y - y'|^2} \sigma(x', y', z_0)\, dl, \tag{1.66}$$

where $\nu$ is the unit outward normal vector to $\partial\Omega_{z_0}$ and $\Omega_{z_0} := \Omega \cap \{z = z_0\} \subset \mathbb{R}^2$. For a vector-valued function $F = (F_1, F_2)$ defined on $\Omega$, we define

$$\mathcal{G}_{z_0} * F(x,y) := \frac{1}{2\pi\mu_0}\int_{\Omega_{z_0}} \frac{(x-x'\ ,\ y-y')}{|x-x'|^2+|y-y'|^2}\cdot F(x',y',z_0)\,dx'dy'. \quad (1.67)$$

Let $u_j[\sigma]$ be a solution to the direct problem (1.35) corresponding to $I_j$ satisfying

$$\begin{vmatrix} \frac{\partial u_1}{\partial x} & \frac{\partial u_1}{\partial y} \\ \frac{\partial u_2}{\partial x} & \frac{\partial u_2}{\partial y} \end{vmatrix} \neq 0 \quad \text{in } \Omega \quad (1.68)$$

and set

$$\mathbb{A}[\sigma] := \begin{bmatrix} \frac{\partial u_1[\sigma]}{\partial y} & -\frac{\partial u_1[\sigma]}{\partial x} \\ \frac{\partial u_2[\sigma]}{\partial y} & -\frac{\partial u_2[\sigma]}{\partial x} \end{bmatrix}. \quad (1.69)$$

Now let us state the explicit relation between $\nabla\sigma$ and $B_z^j$, on which the harmonic $B_z$ algorithm is based.

$$\sigma^0 = 1, \qquad \begin{bmatrix} \frac{\partial\sigma}{\partial x} \\ \frac{\partial\sigma}{\partial y} \end{bmatrix} = \frac{1}{\mu_0}\mathbb{A}^{-1}[\sigma]\begin{bmatrix} \Delta B_z^1 \\ \Delta B_z^2 \end{bmatrix} \quad \text{for } n = 0,1,2,\ldots. \quad (1.70)$$

From the above relation, we have the following implicit representation formula.

**Theorem 1.5.** *Suppose that $|\nabla\sigma|$ is compactly supported in $\Omega$ and $u_j[\sigma]$ for $j = 1,2$ satisfy (1.68). Then the following identity*

$$\mathcal{L}_z\sigma(x,y) = \mathcal{G}_z * \left(\mathbb{A}[\sigma]^{-1}\begin{bmatrix} \nabla^2 B_z^1 \\ \nabla^2 B_z^2 \end{bmatrix}\right)(x,y), \qquad (x,y)\in\Omega_z \quad (1.71)$$

*holds for each $z$. Moreover, $\mathcal{L}_z : H_*^{1/2}(\Omega_z) \to H_*^{1/2}(\Omega_z)$ is invertible where $H_*^{1/2}(\Omega_z) := \{\eta \in H^{1/2}(\Omega_z) : \int_{\partial\Omega_z}\eta = 0\}$.*

For the proof, see [35].

### Convergence Properties of Harmonic $B_z$ Algorithm in Special Cases

In this section, we carry out a convergence analysis for two simplest cases and obtain interesting observations. For details, we refer to [56]. Throughout this section, we only consider a two-dimensional model which is the case where $\frac{\partial\sigma}{\partial z} = 0$ and $\frac{\partial u_j[\sigma]}{\partial z} = 0$. We assume the followings:

- Target domain is $\Omega = \{(x,y) : -1 < x,y < 1\}$, a square.
- Inject $I_1 = 2\,\text{mA}$ using two electrodes $\mathcal{E}_1^{\pm} = \{(\pm 1, y) : |y| < 1\}$.
- Inject $I_2 = 2\,\text{mA}$ using two electrodes $\mathcal{E}_2^{\pm} = \{(x, \pm 1) : |x| < 1\}$.
- Target conductivity $\sigma^*$ satisfies $0 < \sigma^* < \infty$.

Recall that, in the iteration algorithm, we construct the sequence $\{\sigma^n\}$ using

$$\sigma^0 = 1, \qquad \begin{bmatrix} \frac{\partial \sigma^{n+1}}{\partial x} \\ \frac{\partial \sigma^{n+1}}{\partial y} \end{bmatrix} = \frac{1}{\mu_0} \mathbb{A}^{-1}[\sigma^n] \begin{bmatrix} \Delta B_z^1 \\ \Delta B_z^2 \end{bmatrix} \quad \text{for } n = 0, 1, 2, \cdots, \tag{1.72}$$

where $\mathbb{A}^{-1}[\sigma]$ is the $2 \times 2$ matrix defined in (1.69). We denote by $u_j^n$ and $u_j^*$ the solution of (1.63) with $\sigma$ replaced by $\sigma^n$ and $\sigma^*$, respectively.

**Observation 1.6.** *If the target conductivity distribution $\sigma^*$ is depending only on $x$-variable (or $y$-variable), then $\nabla \sigma^1 = \nabla \sigma^*$ in $\Omega$. This means that the contrast of $\sigma^*$ is recovered after only one iteration.*

*Proof.* We only consider the case $\sigma^*(x, y) = \sigma^*(x)$.

It is not so hard to see that the true solutions $u_1^*$ and $u_2^*$ of (1.63) are

$$u_1^*(x, y) = \int_{-1}^{x} \frac{1}{\sigma^*(t)}\, dt \qquad u_2^*(x, y) = y + 1.$$

According to (1.70), the corresponding $\Delta B_z^1$ and $\Delta B_z^2$ satisfy

$$\begin{bmatrix} \frac{\partial \sigma^*}{\partial x} \\ 0 \end{bmatrix} = \frac{1}{\mu_0} \begin{bmatrix} 0, & -\frac{1}{\sigma^*} \\ 1, & 0 \end{bmatrix}^{-1} \begin{bmatrix} \Delta B_z^1 \\ \Delta B_z^2 \end{bmatrix} \tag{1.73}$$

and therefore

$$\Delta B_z^1(x, y) = 0, \qquad \Delta B_z^2(x, y) = \mu_0 \frac{\partial \sigma^*}{\partial x}.$$

Since $\sigma^0 = 1$, we have $u_1^0 = x + 1$ and $u_2^0 = y + 1$ and

$$\mathbb{A}[\sigma^0] = \begin{bmatrix} 0, & -1 \\ 1, & 0 \end{bmatrix}.$$

Therefore, $\nabla \sigma^1$ can be expressed as

$$\begin{bmatrix} \frac{\partial \sigma^1}{\partial x} \\ \frac{\partial \sigma^1}{\partial y} \end{bmatrix} = \frac{1}{\mu_0} \mathbb{A}^{-1}[\sigma^0] \begin{bmatrix} \Delta B_z^1 \\ \Delta B_z^2 \end{bmatrix} = \mathbb{A}^{-1}[\sigma^0] \begin{bmatrix} 0 \\ \frac{\partial \sigma^*}{\partial x} \end{bmatrix} = \begin{bmatrix} \frac{\partial \sigma^*}{\partial x} \\ 0 \end{bmatrix}.$$

This proves $\nabla \sigma^* = \nabla \sigma^1$.

**Observation 1.7.** *Suppose that $\sigma^*$ is a small perturbation of a constant such that*

$$\|\sigma^* - 1\|_{C^1(\Omega)} \le \epsilon \quad \textit{and} \quad \sigma^*(\mathbf{r}) = 1 \quad \textit{in } \Omega \setminus \Omega_1,$$

*where $\Omega_1 = \{\mathbf{r} \in \Omega : dist(\mathbf{r}, \partial\Omega) > 0.1\}$. For a sufficiently small $\epsilon$, the sequence $\{\sigma^n\}$ constructed by (1.72) with $\sigma_n|_{\partial\Omega} = 1$ converges to the true conductivity $\sigma^*$. Moreover,*

$$\|\nabla(\sigma^n - \sigma^*)\|_{L^\infty(\Omega)} = O(\epsilon^{n+1}), \quad n = 1, 2, \cdots.$$

*Proof.* Note that $\|\nabla\sigma^*\|_{L^\infty(\Omega)} \le \epsilon$ from the assumption, so the true solutions $u_1^*$ and $u_2^*$ can be expressed as

$$u_1^* = x + 1 + \epsilon w_1, \qquad u_2^* = y + 1 + \epsilon w_2,$$

where $w_1$ and $w_2$ are solutions of

$$\begin{cases} \nabla\cdot(\sigma^*\nabla(\epsilon w_1)) = -\frac{\partial\sigma^*}{\partial x} \\ w_1|_{\mathcal{E}_1^-} = 0, \ \frac{\partial w_1}{\partial y}|_{\mathcal{E}_2^+\cup\mathcal{E}_2^-} = 0 \\ w_1|_{\mathcal{E}_1^+} = 0 \end{cases} \tag{1.74}$$

and

$$\begin{cases} \nabla\cdot(\sigma^*\nabla(\epsilon w_2)) = -\frac{\partial\sigma^*}{\partial y} \\ w_2|_{\mathcal{E}_2^-} = 0, \ \frac{\partial w_2}{\partial x}|_{\mathcal{E}_1^+\cup\mathcal{E}_1^-} = 0 \\ w_2|_{\mathcal{E}_2^+} = 0. \end{cases} \tag{1.75}$$

From the boundary conditions, we have

$$\int_\Omega \epsilon\sigma^*|\nabla w_1|^2 d\mathbf{r} = \int_\Omega \frac{\partial\sigma^*}{\partial x} w_1 \, d\mathbf{r}\,, \qquad \int_\Omega \epsilon\sigma^*|\nabla w_2|^2 \, d\mathbf{r} = \int_\Omega \frac{\partial\sigma^*}{\partial y} w_2 \, d\mathbf{r}. \tag{1.76}$$

Due to $w_1|_{\mathcal{E}_1^-} = 0$, we have Poincaré's inequality

$$\|w_1\|_{L^p(\Omega)} = \left(\int_\Omega \left|\int_{-1}^x \frac{\partial w_1}{\partial x}(t,y)dt\right|^p d\mathbf{r}\right)^{1/p} \le 2^{1/p}\|\nabla w_1\|_{L^p(\Omega)} \tag{1.77}$$

for $1 \le p < \infty$. It follows from (1.77), (1.76), and the assumption $\|\nabla\sigma^*\|_{L^\infty(\Omega)} \le \epsilon$ that

$$\begin{aligned} \int_\Omega \epsilon\sigma^*|\nabla w_1|^2 &\le \epsilon\int_\Omega |w_1| = \epsilon\int_\Omega \left|\int_{-1}^x \frac{\partial w_1}{\partial x}(t,y)dt\right| d\mathbf{r} \\ &\le 2\epsilon\int_\Omega |\nabla w_1| \le 4\epsilon\|\nabla w_1\|_{L^2(\Omega)}. \end{aligned}$$

Hence, for $0 < \epsilon < \frac{1}{2}$, we have

$$\|\nabla w_1\|_{L^2(\Omega)} \le \frac{4}{1-\epsilon} \le 8. \tag{1.78}$$

Using (1.78), we can show that

$$\|\nabla w_1\|_{L^\infty(\Omega_1)} \le C_0 \tag{1.79}$$

where $C_0$ is a constant independent of $\sigma^*$ whenever $\|\sigma^* - 1\|_{C^1(\Omega)} < \frac{1}{2}$. For the sake of clarity, we include the proof of (1.79). For $\eta \in C_0^2(\Omega)$ with $\eta = 1$ in $\Omega_1$, we have

$$|\nabla w_1(\mathbf{r})| \le \int_\Omega \frac{1}{2\pi|\mathbf{r}-\mathbf{r}'|}|\Delta(w_1\eta)|d\mathbf{r}', \qquad \mathbf{r} \in \Omega_1. \tag{1.80}$$

Since $-\Delta w_1 = \frac{\nabla\sigma^*\cdot\nabla w_1}{\sigma^*} + \frac{\sigma_x^*}{\epsilon\sigma^*}$, it follows from the assumption $\|\sigma^* - 1\|_{C^1(\Omega)} \le \epsilon < \frac{1}{2}$ and Poincaré's inequality (1.77) that

$$\begin{aligned} \|\Delta(w_1\eta)\|_{L^p(\Omega)} &\le \|\eta\|_{C^2(\Omega)} \left((4+\epsilon)\|\nabla w_1\|_{L^p(\Omega)} + \tfrac{8}{1-\epsilon}\right) \\ &\le \|\eta\|_{C^2(\Omega)} \left(5\|\nabla w_1\|_{L^p(\Omega)} + 16\right) \end{aligned} \tag{1.81}$$

for $1 \le p < \infty$.

Application of Young's inequality to (1.80) and using (1.81) yields

$$\|\nabla w_1\|_{L^{10}(\Omega_1)} \le 56\|\eta\|_{C^2(\Omega)}. \tag{1.82}$$

From (1.80), (1.81) and (1.82), we have

$$\|\nabla w_1\|_{L^\infty(\Omega_1)} \le \left(\int_\Omega |\mathbf{r}|^{-10/9} d\mathbf{r}'\right)^{9/10} \|\Delta(w_1\eta)\|_{L^{10}(\Omega)} \le C_0.$$

This completes the proof of (1.79). Similarly, we have

$$\|\nabla w_2\|_{L^\infty(\Omega_1)} \le C_0. \tag{1.83}$$

Now, we try to find the relation between $\Delta B_z^j$ and $\sigma^*$. Assume that $\epsilon$ is so small that $\mathbb{A}^{-1}[\sigma^*]$ exists. According to (1.70), the corresponding $\Delta B_z^1$ and $\Delta B_z^2$ satisfy

$$\begin{bmatrix} \frac{\partial\sigma^*}{\partial x} \\ \frac{\partial\sigma^*}{\partial y} \end{bmatrix} = \frac{1}{\mu_0} \begin{bmatrix} \epsilon\frac{\partial w_1}{\partial y}, & -1 - \epsilon\frac{\partial w_1}{\partial x} \\ 1 + \epsilon\frac{\partial w_2}{\partial y}, & -\epsilon\frac{\partial w_2}{\partial x} \end{bmatrix}^{-1} \begin{bmatrix} \Delta B_z^1 \\ \Delta B_z^2 \end{bmatrix}. \tag{1.84}$$

Since $\nabla\sigma^* = 0$ in $\Omega \setminus \Omega_1$, we have $\Delta B_z^j = 0 \quad$ in $\Omega \setminus \Omega_1$. Moreover, it follows that

$$\|\frac{1}{\mu_0}\Delta B_z^2 - \frac{\partial\sigma^*}{\partial x}\|_{L^\infty(\Omega_1)} \le 2C_0\epsilon^2, \qquad \|\frac{1}{\mu_0}\Delta B_z^1 + \frac{\partial\sigma^*}{\partial y}\|_{L^\infty(\Omega_1)} \le 2C_0\epsilon^2$$

from direct computation and (1.79), (1.83). Now we proceed the iteration process as that in the previous section. If $\sigma^0 = 1$, then $\sigma^1$ satisfies

$$\begin{bmatrix} \frac{\partial\sigma^1}{\partial x} \\ \frac{\partial\sigma^1}{\partial y} \end{bmatrix} = \frac{1}{\mu_0} \begin{bmatrix} 0, & -1 \\ 1, & 0 \end{bmatrix}^{-1} \begin{bmatrix} \Delta B_z^1 \\ \Delta B_z^2 \end{bmatrix} = \begin{bmatrix} \frac{\partial\sigma^*}{\partial x} \\ \frac{\partial\sigma^*}{\partial y} \end{bmatrix} + O(\epsilon^2),$$

which generates from the above estimate and $\sigma^1 = \sigma^*$ on $\partial\Omega$ that

$$\|\sigma^1 - \sigma^*\|_{C^1(\Omega)} \le 4C_0\epsilon^2. \tag{1.85}$$

For $\sigma^1$, the solution $u_j^1$ with $j = 1, 2$ can be expressed as

$$u_1^1 = u_1^* + \epsilon^2 w_1^1, \quad u_2^1 = u_2^* + \epsilon^2 w_2^1 \tag{1.86}$$

due to (1.85), where each $w_j^1$ satisfies

$$\nabla \cdot (\sigma^1 \nabla(\epsilon^2 w_j^1)) = -\nabla \cdot ((\sigma^1 - \sigma^*)\nabla u_j^*)$$

with the same boundary condition of $w_j^1$ as (1.74) and (1.75).

Following the same procedure for deriving (1.79) and (1.83), we obtain

$$\|\nabla w_1^1\|_{L^\infty(\Omega_1)} + \|\nabla w_2^1\|_{L^\infty(\Omega_1)} \le C.$$

Now we are ready to estimate $\|\nabla(\sigma^2 - \sigma^*)\|_{L^\infty(\Omega)}$. It follows from (1.72) and (1.84) that

$$\begin{bmatrix} \frac{\partial}{\partial x}(\sigma^2 - \sigma^*) \\ \frac{\partial}{\partial y}(\sigma^2 - \sigma^*) \end{bmatrix} = \frac{1}{\mu_0}\left(\mathbb{A}^{-1}[\sigma^1] - \mathbb{A}^{-1}[\sigma^*]\right)\begin{bmatrix} \Delta B_z^1 \\ \Delta B_z^2 \end{bmatrix}.$$

If $\epsilon$ is so small that $C_0\epsilon(1 + C_0\epsilon) < \frac{1}{4}$, it follows from (1.84), (1.79), (1.83) and the estimates on $\| \nabla w_j^1 \|_{L^\infty}$ that

$$\det \mathbb{A}[\sigma^1] \ge \frac{1}{2}, \qquad \det \mathbb{A}[\sigma^*] \ge \frac{1}{4}$$

in $\Omega_1$ from the direct computation, noticing the expression

$$\mathbb{A}[\sigma^1] = \mathbb{A}[\sigma^*] + \epsilon^2 \begin{bmatrix} \frac{\partial w_1^1}{\partial y} & -\frac{\partial w_1^1}{\partial x} \\ \frac{\partial w_2^1}{\partial y} & -\frac{\partial w_2^1}{\partial x} \end{bmatrix}.$$

Hence, by using triangle inequality and the lower bounds of $\det \mathbb{A}[\sigma^1]$, det $\mathbb{A}[\sigma^*]$ we have

$$\|\nabla\sigma^2 - \nabla\sigma^*\|_{L^\infty(\Omega_1)} \le C\epsilon \sum_{j=1}^{2} \|\nabla u_j^1 - \nabla u_j^*\|_{L^\infty(\Omega_1)} \le C\epsilon^3, \tag{1.87}$$

since $\|\nabla u_j^1 - \nabla u_j^*\|_{L^\infty(\Omega_1)} = \epsilon^2 \|\nabla w_j^1\|_{L^\infty(\Omega_1)} \le C\epsilon^2$.

If we continue this process with sufficiently small $\epsilon$, we obtain

$$\|\nabla\sigma^n - \nabla\sigma^*\|_{L^\infty(\Omega_1)} \le C\epsilon^{n+1}.$$

This completes the proof.

### 1.4.3 Other Algorithms

After the introduction of the harmonic $B_z$ algorithm, there has been an effort to improve its performance especially in terms of the way we numerically differentiate the measured noisy $B_z$ data. Based on a novel analysis utilizing the Helmholtz decomposition, Park et al. [45] suggested the gradient $B_z$ decomposition algorithm. This method seems good, but we recently realized that the use of the harmonic $B_z$ algorithm locally is better in practical situations. This issue will be explained at the end of this section.

In order to explain the algorithm, let us assume that $\Omega = D \times [-\delta, \delta] = \{\mathbf{r} = (x, y, z) | (x, y) \in D,\ -\delta < z < \delta\}$ is an electrically conducting subject where $D$ is a two-dimensional smooth simply connected domain. Let $u$ be the solution of the Neumann boundary value problem with the Neumann data $g$. We parameterize $\partial D$ as $\partial D := \{(x(t), y(t)) : 0 \le t \le 1\}$ and define $\tilde g(x(t), y(t), z) := \int_0^t g((x(t), y(t), z))\sqrt{|x'(t)|^2 + |y'(t)|^2}dt$ for $(x, y, z) \in \partial D \times (-\delta, \delta)$.

The gradient $B_z$ decomposition algorithm is based on the following key identity:

$$\sigma = \frac{\left| -\left(\frac{\partial H}{\partial y} + \Lambda_x[u]\right)\frac{\partial u}{\partial x} + \left(\frac{\partial H}{\partial x} + \Lambda_y[u]\right)\frac{\partial u}{\partial y}\right|}{(\frac{\partial u}{\partial x})^2 + (\frac{\partial u}{\partial y})^2} \quad \text{in } \Omega \tag{1.88}$$

where

$$\Lambda_x[u] := \frac{\partial \psi}{\partial y} - \frac{\partial W_z}{\partial x} + \frac{\partial W_x}{\partial z} \quad \text{and} \quad \Lambda_y[u] := \frac{\partial \psi}{\partial x} + \frac{\partial W_z}{\partial y} - \frac{\partial W_y}{\partial z} \quad \text{in } \Omega$$

and

$$H = \phi + \frac{1}{\mu_0}B_z, \qquad W(\mathbf{r}) := \int_{\Omega_\delta} \frac{1}{4\pi|\mathbf{r} - \mathbf{r}'|}\frac{\partial(\sigma\nabla u(\mathbf{r}'))}{\partial z}d\mathbf{r}'.$$

Here, $\phi$ and $\psi$ are solutions of the following equations:

$$\begin{cases} \nabla^2\phi = 0 \quad \text{in } \Omega \\ \phi = \tilde g - \frac{1}{\mu_0}B_z \quad \text{on } \partial\Omega_{side} \\ \frac{\partial\phi}{\partial z} = -\frac{1}{\mu_0}\frac{\partial B_z}{\partial z} \quad \text{on } \partial\Omega_{tb} \end{cases}$$

and

$$\begin{cases} \nabla^2 \psi = 0 & \text{in } \Omega \\ \nabla \psi \cdot \tau = \nabla \times W \cdot \tau & \text{on } \partial\Omega_{side} \\ \frac{\partial \psi}{\partial z} = -\nabla \times W \cdot \mathbf{e}_z & \text{on } \partial\Omega_{tb} \end{cases}$$

where $\mathbf{e}_z = (0,0,1)$, $\partial\Omega_{side} = \partial D \times (-\delta, \delta)$, $\Omega_{tb}$ is the top and bottom surfaces of $\Omega$, and $\tau := (-n_y, n_x, 0)$ is the tangent vector on the lateral boundary $\partial D \times (-\delta, \delta)$.

Since the term $u$ in (1.88) is a highly nonlinear function of $\sigma$, the identity (1.88) can be viewed as an implicit reconstruction formula for $\sigma$. It should be noticed that we can not identify $\sigma$ with a single $g$ using (1.88). Hence, we may use an iterative reconstruction scheme with multiple Neumann data $g_j, j = 1, \cdots, N$ to find $\sigma$. Let $u_j^m$ be the solution of the Neumann BVP with $\sigma = \sigma_m$ and $g_j$. Then, the reconstructed $\sigma$ is the limit of a sequence $\sigma_m$ that is obtained by the following formula:

$$\sigma_{m+1} = \frac{\sum_{i=1}^N \left| -\left(\frac{\partial H_i}{\partial y} + \Lambda_x[u_i^m]\right) \frac{\partial u_i^m}{\partial x} + \left(\frac{\partial H_i}{\partial x} + \Lambda_y[u_i^m]\right) \frac{\partial u_i^m}{\partial y} \right|}{\sum_{i=1}^N \left[ \left(\frac{\partial u_i^m}{\partial x}\right)^2 + \left(\frac{\partial u_i^m}{\partial y}\right)^2 \right]}.$$

This algorithm using single differentiation of $B_z$ looks good in mathematical point of view and works well in phantom experiments having a homogeneous background. Noting the noise amplification property of the algorithm, various other methods were also developed to improve the image quality [26, 44, 45]. All these methods show good performance in the image reconstruction of a conductivity distribution with a relatively low contrast. However, if their conductivity distributions are very inhomogeneous with a large contrast, then the image quality using these methods is significantly lower than what we expected from a low-contrast case. We recently realized that, for a robust algorithm, it is desirable to maximize the influence of $B_z$, while minimizing the influence of the structure of $u_j$. Our experience show that we should use the fine structure of $B_z$ locally, while we try to use rough structure of $u_j$ instead of using the detailed structure in the algorithm. We will not go into too much details on this complicated issue in this chapter.

### Anisotropic Conductivity Reconstruction in MREIT (Unstable Algorithm)

Now, we turn our attention to the anisotropic conductivity image reconstruction problem. Some biological tissues are known to have anisotropic conductivity values. The ratio of the anisotropy depends on the type of tissue and the human skeletal muscle, for example, shows an anisotropy of up to ten between the longitudinal and transversal direction. Hence, clinical applications of MREIT require us to develop an anisotropic conductivity image reconstruction method. Lately, a new algorithm that can handle anisotropic conductivity distributions has been suggested in [53]. This algorithm requires at least seven different injection currents, while the previous two algorithms for the isotropic case require at least two different injection currents. Recently, we realized that this anisotropic algorithm could be unstable without having additional information. Although this anisotropic algorithm may not be practically useful, it is worthwhile to explain the algorithm for future studies.

Let us state the main ingredients for the anisotropic case:

- Biot–Savart law: $B_z(\mathbf{r}) = \frac{\mu_0}{4\pi}\int_\Omega \frac{(y-y')J_x(\mathbf{r}')-(x-x')J_y(\mathbf{r}')}{|\mathbf{r}-\mathbf{r}'|^3}d\mathbf{r}'$.
- $\nabla\cdot\mathbf{J}=0, \quad \nabla\cdot\mathbf{B}=0, \quad \mathbf{J}=\frac{1}{\mu_0}\nabla\times\mathbf{B}$.
- $\mathbb{J}[\underline{\underline{\sigma}},g] = -\underline{\underline{\sigma}}\nabla u[\underline{\underline{\sigma}},g] : \quad \begin{cases} \nabla\cdot(\underline{\underline{\sigma}}\nabla u) = 0 & \text{in } \Omega \\ -\underline{\underline{\sigma}}\nabla u\cdot\mathbf{n} = g & \text{on } \partial\Omega. \end{cases}$
- $\underline{\underline{\sigma}} = \begin{pmatrix} \sigma_{11} & \sigma_{12} & \sigma_{13} \\ \sigma_{12} & \sigma_{22} & \sigma_{23} \\ \sigma_{13} & \sigma_{23} & \sigma_{33} \end{pmatrix}$ is a positive-definite symmetric matrix.

The reconstruction algorithm begins with $\frac{1}{\mu_0}\nabla^2 B_z = \partial_y J_x - \partial_x J_y$ to get

$$\mathbf{U}\mathbf{s} = \mathbf{b} \tag{1.89}$$

where

$$\mathbf{b} = \frac{1}{\mu_0}\begin{bmatrix} \nabla^2 B_z^1 \\ \vdots \\ \nabla^2 B_z^N \end{bmatrix}, \qquad \mathbf{s} = \begin{bmatrix} -\partial_y\sigma_{11}+\partial_x\sigma_{12} \\ -\partial_y\sigma_{12}+\partial_x\sigma_{22} \\ -\partial_y\sigma_{13}+\partial_x\sigma_{23} \\ \sigma_{12} \\ -\sigma_{11}+\sigma_{22} \\ \sigma_{23} \\ \sigma_{13} \end{bmatrix}$$

and

$$\mathbf{U} = \begin{bmatrix} u_x^1 & u_y^1 & u_z^1 & u_{xx}^1 - u_{yy}^1 & u_{xy}^1 & u_{xz}^1 & -u_{yz}^1 \\ \vdots & \vdots & \vdots & \vdots & \vdots & \vdots & \vdots \\ u_x^N & u_y^N & u_z^N & u_{xx}^N - u_{yy}^N & u_{xy}^N & u_{xz}^N & -u_{yz}^N \end{bmatrix}.$$

We do not know the true $\underline{\underline{\sigma}}$ and therefore the matrix $\mathbf{U}$ is unknown. This requires us to use an iterative procedure to compute $\mathbf{s}$ in (1.89). For now, let us assume that we have computed all seven terms of $\mathbf{s}$. From $\mathbf{s}$, we can immediately determine $\sigma_{12}(\mathbf{r}) = s_4(\mathbf{r})$, $\sigma_{13}(\mathbf{r}) = s_7(\mathbf{r})$ and $\sigma_{23}(\mathbf{r}) = s_6(\mathbf{r})$. To determine $\sigma_{11}$ and $\sigma_{22}$ from $\mathbf{s}$, we use the relation between $\mathbf{s}$ and $\underline{\underline{\sigma}}$:

$$\frac{\partial\sigma_{11}}{\partial x} = s_2 - \frac{\partial s_5}{\partial x} + \frac{\partial s_4}{\partial y} \quad \text{and} \quad \frac{\partial\sigma_{11}}{\partial y} = -s_1 + \frac{\partial s_4}{\partial x}. \tag{1.90}$$

Using the above relation, we can derive the following expression of $\sigma_{11}$:

$$\begin{aligned}\sigma_{11}(a,b,z) = &\int_{\Omega_z}\left[\left(-s_2+\frac{\partial s_5}{\partial x}-\frac{\partial s_4}{\partial y}\right)\frac{\partial}{\partial x}\Psi(x-a,y-b)\right]dxdy\\ &+\int_{\Omega_z}\left[\left(s_1-\frac{\partial s_4}{\partial x}\right)\frac{\partial}{\partial y}\Psi(x-a,y-b)\right]dxdy\\ &+\int_{\partial\Omega_z}\tilde{\sigma}_{11}(x,y,z)\left[\tilde{\mathbf{n}}\cdot\nabla_{x,y}\Psi(x-a,y-b)\right]dl_{x,y},\end{aligned} \tag{1.91}$$

where $\Omega_z = \{(x,y) : (x,y,z) \in \Omega\}$, $\Psi(x,y) = \frac{1}{2\pi}\log\sqrt{x^2+y^2}$ and $\tilde{\sigma}_{11}$ is $\sigma_{11}$ restricted at the boundary $\partial\Omega_z$. We can compute $\tilde{\sigma}_{11}$ on $\partial\Omega_z$ by solving the corresponding singular integral and therefore we obtain the representation of the internal $\sigma_{11}$. Similarly, we can have the representation for the component $\sigma_{22}$. The last component $\sigma_{33}$ whose information is missing in (1.89) can be obtained by using the physical law $\nabla \cdot \mathbf{J} = 0$.

## *1.4.4 Challenges in MREIT and Open Problems for Its Achievable Spatial Resolution*

In order to apply MREIT to the human subject, one of the important remaining problems to be solved is to reduce the amount of injection current to a level that the human subject can tolerate. The amount of the injection current corresponds to the $L^1$-norm of the Neumann data, that is, $\|g\|_{L^1(\partial\Omega)}$. By limiting it to a low level, the measured MR data such as the $z$-component of magnetic flux density $B_z$ tend to have a low SNR and get usually degraded in their accuracy due to the non-ideal data acquisition system of an MR scanner.

The major difficulty in dealing with the convergence and the numerical stability of the MREIT algorithm comes from the inverse matrix $\mathbb{A}^{-1}[\sigma]$ in the iteration procedure. Indeed, if the condition number of $\mathbb{A}[\sigma]$ is large, then a small error in the measured data $B_z^j$ results in a large error in the reconstructed conductivity. The inverse matrix $\mathbb{A}^{-1}[\sigma]$ is affected by the lower bound of $|\nabla u_1^\sigma|$ and $|\nabla u_2^\sigma|$ and the angle between them where $u_j^\sigma$ is the solution of

$$\begin{cases}\nabla\cdot(\sigma\nabla u_j) = 0 \text{ in } \quad \Omega\\ -\sigma\nabla u_j\cdot\ \mathbf{n}|_{\partial\Omega} = g_j,\ \int_{\partial\Omega} u_j = 0.\end{cases}$$

The area of the parallelogram determined by the vectors $\nabla u_1^\sigma$ and $\nabla u_2^\sigma$ depends not only on the electrode configuration (determining Neumann data) but also on the conductivity distribution that is unknown. Since quality of the reconstructed conductivity $\sigma$ depends heavily on the area of parallelogram $|\mathbf{e}_3 \cdot \nabla u_1 \times \nabla u_2|$, we need to define a function estimating its achievable spatial resolution.

We begin with defining admissible class of conductivities having "the maximal multi-scale anisotropy number" $\lambda$. Let

$$\Upsilon_\lambda := \{\sigma \in C(\bar{\Omega}) \ : \ \sup_{B_r \subset \Omega} \frac{\|\log \sigma\|_{TV(B_r)}}{r^2} < \lambda\}.$$

Note that $\Upsilon_0$ is the set of homogeneous material and the anisotropic conductivity may be included in $\Upsilon_\infty$. The precise meaning of $\lambda$ will be discussed in future work. In order to estimate an achievable spatial resolution for MREIT image, we define the MREIT-resolution function having the variables $\Omega_0 \subset \Omega, \lambda, \delta > 0$ and $g_1, g_2 \in H^{1/2}(\partial\Omega)$ by

$$\eta(\lambda, \Omega_0, \delta, g_1, g_2) = \sup\{\rho^{-1} \ ; \ \phi(\rho; \lambda, \Omega_0, g_1, g_2) > \delta\}$$

where

$$\phi(\rho; \lambda, \Omega_0, g_1, g_2) := \inf_{\sigma \in \Upsilon_\lambda} \ \inf_{B_r \subset \Omega_0, r > \rho} \left[ \frac{\int_{B_r} |\nabla u_1^\sigma \times \nabla u_2^\sigma|}{\sqrt{\int_{B_r} |\nabla u_1^\sigma|^2 \int_{B_r} |\nabla u_2^\sigma|^2}} \right].$$

The quantity $\eta(\lambda, \Omega_0, \delta, g_1, g_2)$ is important in MREIT since it is related with the achievable spatial resolution of a reconstructed MREIT image using the harmonic $B_z$ algorithm. However, the problem of estimating the $\eta(\lambda, \Omega_0, \delta, g_1, g_2)$ is wide open.

## References


1. G. Alessandrini and R. Magnanini, *The index of isolated critical points and solutions of elliptic equations in the plane*, Ann. Scoula. Norm. Sup. Pisa Cl Sci. **19** (1992), 567–589.
2. G. Alessandrini, E. Rosset, and J.K. Seo, *Optimal size estimates for the inverse conductivity poblem with one measurements*, Proc. Amer. Math. Soc **128** (2000), 53–64.
3. H. Ammari and H. Kang, *Reconstruction of small inhomogeneities from boundary measurement*, Lecture Notes in Mathematics, vol. 1846, Springer, Berlin, 2004.
4. H. Ammari, O. Kwon, J.K. Seo, and E.J. Woo, *T-scan electrical impedance imaging system for anomaly detection*, IEEE Trans. Biomed. Eng. **65** (2004), 252–266.
5. H. Ammari and J.K. Seo, *An accurate formula for the reconstruction of conductivity inhomogeneity*, Advances in Applied Math. **30** (2003), 679–705.
6. K. Astala and L. Paivarinta, *Calderon's inverse conductivity problem in the plane*, Annals of Mathematics **163** (2006), 265–299.
7. D.C. Barber and B.H. Brown, *Applied potential tomography*, . Phys. E: Sci. Instruments **17** (1984), 723.
8. K. Boone, D. Barber, and B. Brown, *Imaging with electricity: report of the european concerted action on impedance tomography*, J. Med. Eng. Tech **21** (1997), 201–232.
9. A.P. Calderon, *On an inverse boundary value problem, seminar on numerical analysis and its applications to continuum physics(1980), rio de janeiro*, Mat. apl. comput (reprinted) **25** (2006(1980)), 133–138.

10. M. Cheney, D. Isaacson, and J.C. Newell, *Electrical impedance tomography*, SIAM Review **41** (1999), 85–101.
11. D.C. Ghiglia and M.D. Pritt, *Two-dimensional phase unwrapping: Theory, algorithms and software*, Wiley Interscience, New York, 1998.
12. H. Griffiths, H.T. Leung, and R.J. Williams, *Imaging the complex impedance of the thorax*, Clin Phys Physiol Meas. **13 Suppl A** (1992), 77–81.
13. R.P. Henderson and J.G. Webster, *An impedance camera for spatially specific measurements of the thorax*, IEEE Trans. Biomed. Eng. **25** (1978), 250–254.
14. D.S. Holder (ed.), *Electrical impedance tomography: Methods, history and applications*, Series in medical physics and biomedical engineering, Institute of Physics, 2004.
15. Y.Z. Ider and O. Birgul, *Use of the magnetic field generated by the internal distribution of injected currents for electrical impedance tomography (mr-eit)*, Elektrik **6** (1998), 215–225.
16. V. Isakov, *On uniqueness of recovery of a discontinuous conductivity coefficient*, Comm.Pure Appl. Math. **41** (1988), 856–877.
17. M. Joy, G. Scott, and R. Henkelman, *In vivo detection of applied electric currents by magnetic resonance imaging*, Magn. Reson. Imag. **7** (1989), 89.
18. H. Kang and J.K. Seo, *Identification of domains with near-extreme conductivity: global stability and error estimates*, Inverse Problems **15** (1999), 851–867.
19. C.E. Kenig, J. Sjöstrand, and G. Uhlmann, *The calderón problem with partial data*, Annals of Mathematics **165** (2007), 567–591.
20. H.S. Khang, B.I. Lee, S.H. Oh, E.J. Woo, S.Y. Lee, M.H. Cho, O. Kwon, J.R. Yoon, and J.K. Seo, *J-substitution algorithm in magnetic resonance electrical impedance tomography (MREIT): phantom experiments for static resistivity images*, IEEE Trans. Med. Imaging **21** (2002), 695–702.
21. S. Kim, J.K. Seo, E.J. Woo, and H. Zribi, *Multi-frequency trans-admittance scanner: mathematical framework and feasibility*, SIAM J. Appl. Math. **69**(1) (2008), 22–36.
22. S.W. Kim, O. Kwon, J.K. Seo, and J.R. Yoon, *On a nonlinear partial differential equation arising in magnetic resonance electrical impedance tomography*, SIAM J. Math Anal. **34** (2002), 511–526.
23. S.W. Kim, J.K. Seo, E.J. Woo, and H. Zribi, *Frequency difference trans-admittance scanner for breast cancer detection*, preprint.
24. Y.J. Kim, O. Kwon, J.K. Seo, and E.J. Woo, *Uniqueness and convergence of conductivity image reconstruction in magnetic resonance electrical impedance tomography*, Inv. Prob. **19** (2003), 1213–1225.
25. R. Kohn and M. Vogelius, *Determining conductivity by boundary measurements*, Comm.Pure Appl. Math. **37** (1984), 113–123.
26. O. Kwon, C.J. Park, E.J. Park, J.K. Seo, and E.J. Woo, *Electrical conductivity imaging using a variational method in $b_z$-based mreit*, Inverse Problems **21** (2005), 969–980.
27. O. Kwon and J.K. Seo, *Total size estimation and identification of multiple anomalies in the inverse conductivity problem*, Inverse Problems **17** (2001), 59–75.
28. O. Kwon, J.K. Seo, and S.W. Kim, *Location search techniques for a grounded conductor*, SIAM J. Applied Math. **21**, no. 6.
29. O. Kwon, J.K. Seo, and J.R. Yoon, *A real-time algorithm for the location search of discontinuous concuctivites with one measurement*, Comm. Pure Appl. Math. **55** (2002), 1–29.
30. O. Kwon, E.J. Woo, J.R. Yoon, and J.K. Seo, *Magnetic resonance electrical impedance tomography (mreit): simulation study of j-substitution algorithm*, IEEE Trans. Biomed. Eng. **49** (2002), 160–167.
31. O. Kwon, J. R. Yoon, J.K. Seo, E.J. Woo, and Y.G. Cho, *Estimation of anomaly location and size using electrical impedance tomography*, IEEE Trans. Biomed. Eng. **50** (2003), 89–96.
32. B.I. Lee, S.H. Oh, E.J. Woo, S.Y. Lee, M.H. Cho, O. Kwon, J.K. Seo, and W.S. Baek, *Static resistivity image of a cubic saline phantom in magnetic resonance electrical impedance tomography (mreit)*, Physiol. Meas. **24** (2003), 579–589.

33. J. Lee, E.J. Woo, H. Zribi, S.W. Kim, and J.K. Seo, *Multi-frequency electrical impedance tomography (mfeit): frequency-difference imaging and its feasibility*, preprint.
34. J.J. Liu, H.C. Pyo, J.K. Seo, and E.J. Woo, *Convergence properties and stability issues in mreit algorithm*, Contemporary Mathematics **408** (2006), 201–216.
35. J.J. Liu, J.K. Seo, M. Sini, and E.J. Woo, *On the convergence of the harmonic $b_z$ algorithm in magnetic resonance electrical impedance tomography*, SIAM J. Appl. Math. **67**(5) (2007), 1259–1282.
36. P. Metherall, *Three dimensional electrical impedance tomography of the human thorax*, University of Sheffield, Dept. of Med. Phys and Clin. Eng., 1998.
37. P. Metherall, D.C. Barber, R.H. Smallwood, and B.H. Brown, *Three-dimensional electrical imepdance tomography*, Nature **380** (1996), 509–512.
38. A. Nachman, *Reconstructions from boundary measurements*, Ann. Math. **128** (1988), 531–577.
39. ______, *Global uniqueness for a two-dimensional inverse boundary value problem*, Ann. Math. **142** (1996), 71–96.
40. B.M. Eyuboglu O. Birgul and Y.Z. Ider, *Current constrained voltage scaled reconstruction (ccvsr) algorithm for mr-eit and its performance with different probing current patterns*, Phys. Med. Biol. **48** (2003), 653–671.
41. S.H. Oh, B.I. Lee, T.S. Park, S.Y. Lee, E.J. Woo, M.H. Cho, J.K. Seo, and O. Kwon, *Magnetic resonance electrical impedance tomography at 3 tesla field strength*, Magn. Reson. Med. **51** (2004), 1292–1296.
42. S.H. Oh, B.I. Lee, E.J. Woo, S.Y. Lee, M.H. Cho, O. Kwon, and J.K. Seo, *Conductivity and current density image reconstruction using harmonic $b_z$ algorithm in magnetic resonance electrical impedance tomography*, Phys. Med. Biol. **48** (2003), 3101–3116.
43. T.I. Oh, J. Lee, J.K. Seo, S.W. Kim, and E.J. Woo, *Feasibility of breast cancer lesion detection using multi-frequency trans-admittance scanner (tas) with 10hz to 500khz bandwidth*, Physiological Measurement **28** (2007), S71–S84.
44. C. Park, O. Kwon, E.J. Woo, and J.K. Seo, *Electrical conductivity imaging using gradient $b_z$ decomposition algorithm in magnetic resonance electrical impedance tomography (mreit)*, IEEE Trans. Med. Imag. **23** (2004), 388–394.
45. C. Park, E.J. Park, E.J. Woo, O. Kwon, and J.K. Seo, *Static conductivity imaging using variational gradient $b_z$ algorithm in magnetic resonance electrical impedance tomography*, Physiol.Meas. **25** (2004), 257–269.
46. H.C. Pyo, O. Kwon, J.K. Seo, and E.J. Woo, *Mathematical framework for bz-based mreit model in electrical impedance imaging*, Compter and Mathematics with Applications **51** (2006), 817–828.
47. R. Sadleir, S. Grant, S.U. Zhang, B.I. Lee, H.C. Pyo, S.H. Oh, C. Park, E.J. Woo, S.Y. Lee, O. Kwon, and J.K. Seo, *Noise analysis in magnetic resonance electrical impedance tomography at 3 and 11 t field strengths*, Physiol. Meas. **26** (2005), 875–884.
48. F. Santosa and M. Vogelius, *A backprojection algorithm for electrical impedance imaging*, SIAM J. Appl. Math. **50** (1990), 216–243.
49. G.J. Saulnier, R.S. Blue, J.C. Newell, D. Isaacson, and P.M. Edic, *Electrical impedance tomography*, IEEE Sig. Proc. Mag. **18** (2001), 31–43.
50. G. Scott, R. Armstrong M. Joy, and R. Henkelman, *Measurement of nonuniform current density by magnetic resonance*, IEEE Trans. Med. Imag. **10** (1991), 362–374.
51. ______, *Sensitivity of magnetic-resonance current density imaging*, J. Mag. Res. **97** (1992), 235–254.
52. J.K. Seo, *On the uniqueness in the inverse conductivity problem*, J. Fourier Anal. Appl. **2** (1996), 227–235.
53. J.K. Seo, O. Kwon, H. Ammari, and E.J. Woo, *Mathematical framework and anomaly estimation algorithm for breast cancer detection: electrical impedance technique using ts2000 configuration*, IEEE Trans. Biomed. Eng. **51** (2004), 1898–1906.

54. J.K. Seo, O. Kwon, B.I. Lee, and E.J. Woo, *Reconstruction of current density distributions in axially symmetric cylindrical sections using one component of magnetic flux density: computer simulation study*, Physiol. Meas. **24** (2003), 565–577.
55. J.K. Seo, H.C. Pyo, C.J. Park, O. Kwon, and E.J. Woo, *Image reconstruction of anisotropic conductivity tensor distribution in mreit: computer simulation study*, Phys. Med. Biol. **49**(18) (2004), 4371–4382.
56. J.K. Seo, J.R. Yoon, E.J. Woo, and O. Kwon, *Reconstruction of conductivity and current density images using only one component of magnetic field measurements*, 2003, pp. 1121–1124.
57. E. Somersalo, D. Isaacson, and M. Cheney, *Existence and uniqueness for electrode models for electric current computed tomography*, SIAM J. Appl. Math. **52** (1992), 1023–40.
58. J. Sylvester and G. Uhlmann, *A uniqueness theorem for an inverse boundary value problem in electrical prospection*, Comm. Pure Appl. Math. **39** (1986), 91–112.
59. ———, *A global uniqueness theorem for an inverse boundary value problem*, Ann. Math. **125** (1987), 153–169.
60. A.T. Tidswell, A. Gibson, R.H. Bayford, and D.S. Holder, *Three-dimensional electrical impedance tomography of human brain activity*, Physiol. Meas. **22** (2001), 177–186.
61. G. Uhlmann, *Developments in inverse problems since calderdon's foundational paper*, Harmonic analysis and partial differential equations (Chicago, IL, 1996), Univ. Chicago Press (1999), 295–345.
62. J.G. Webster, *Electrical impedance tomography*, Adam Hilger, Bristol, UK, 1990.
63. E.J. Woo, S.Y. Lee, and C.W. Mun, *Impedance tomography using internal current density distribution measured by nuclear magnetic resonance*, SPIE **2299** (1994), 377–385.
64. S. Onart Y.Z. Ider and W. Lionheart, *Uniqueness and reconstruction in magnetic resonance-electrical impedance tomography (mr-eit) physiol. meas.*, Physiol. Meas. **24** (2003), 591–604.
65. N. Zhang, *Electrical impedance tomography based on current density imaging*, MS Thesis Dept. of Elec. Eng. Univ. of Toronto Toronto Canada (1992).

# Chapter 2
# Time Reversing Waves for Biomedical Applications

**Mickael Tanter and Mathias Fink**

## 2.1 Introduction

Time reversal is a concept that always fascinated the majority of scientists. In fact, this fundamental symmetry of physics, the time reversal invariance, can be exploited in the domain of wave physics, in acoustics and more recently in electromagnetism, leading to a huge variety of experiments and instruments both for fundamental physics and applications. Today, these applications go from medical imaging and therapy to telecommunications, underwater acoustics, seismology or non-destructive testing.

The evolution of electronic components enables today the building of time reversal mirrors that make a wave live back all the steps of its past life. These systems exploit the fact that in a majority of cases the propagation of acoustics waves (sonic or ultrasonic) and electromagnetic waves is a reversible process. Whatever the distortions (diffraction, multiple-scattering, reverberation) suffered in a complex environment by a wave emerging from a point source, there always exists, at least theoretically, a dual wave able to travel in the opposite direction all the complex travel paths and finally converges back to the initial source location, exactly as if the movie of the wave propagation had been played backwards in time. The main interest of a Time Reversal Mirror (TRM) is to experimentally create this dual wave, thanks to an array of reversible transducers (able to work both in transmit and receive modes) driven using A/D and D/A converters and electronic memories. The TRM is thus able to focus the wave energy through very complex media. In Ultrasonics, a TRM consists in a 2D surface covered with piezoelectric transducers that successively play the role of hydrophones and loudspeakers. The ultrasonic wave is emerging from a given source in the medium and recorded by each of the microphones in electronic memories. Then, in a second

M. Tanter (✉) and M. Fink
Laboratoire Ondes et Acoustique, 10 rue Vauquelin, CNRS, E.S.P.C.I, Université Paris 7, 75005, Paris, France
e-mail: mickael.tanter@espci.fr;mathias.fink@espci.fr

H. Ammari, *Mathematical Modeling in Biomedical Imaging I*,
Lecture Notes in Mathematics 1983, DOI 10.1007/978-3-642-03444-2_2,

step (the time reversal step), all memories are read in the reverse direction. More precisely, the chronology of the signals received by each hydrophone is reversed. The signals recorded at later times, are read first. All hydrophones switch synchronously in a transmit mode (loudspeaker) and re-emit the "time reversed" signals coming from the electronic memories. Thus, new initial conditions for the wave propagation are created, and thanks to reversibility, the diffracted wave has no other solution than living back step by step its past life in a reversed way. Of course, this kind of mirror is totally different than a classical mirror. A time reversal mirror builds the real image of the source at its location whereas as a classical mirror builds a virtual image of the source. The great robustness of the time reversal focusing ability has been verified in many scenarios ranging from ultrasonic propagation (millimetric wavelength) in the human body over several centimeters, to ultrasonic propagation (metric wavelength) in the sea over several tens of kilometers and finally since recently the propagation of centimetric electromagnetic waves over several hundreds of meters [3, 9].

## 2.2 Time Reversal of Acoustic Waves: Basic Principles

Let us consider the propagation of an ultrasonic wave in a heterogeneous and non-dissipative medium, whose compressibility $\kappa(r)$ and density $\rho(r)$ vary in space. By introducing the sound speed $c(r) = (\rho(r)\kappa(r))^{-1/2}$, one can obtain the wave propagation equation for a given pressure field $p(r,t)$:

$$\nabla \cdot \frac{1}{\rho}\nabla p - \frac{1}{\rho c^2}\frac{\partial^2 p}{\partial t^2} = 0. \tag{2.1}$$

One can notice the particular behavior of this wave equation regarding the time variable $t$. Indeed, it only contains a second order time derivative operator. This property is the starting point of the time reversal principle. A straightforward consequence of this property is that if $p(r,t)$ is a solution of the wave equation, then $p(r,-t)$ is also solution of the problem. This property illustrates the invariance of the wave equation during a time reversal operation, the so called time reversal invariance. However, this property is only valid in a non-dissipative medium. If wave propagation is affected by dissipation effects, odd order time derivatives appear in the wave equation and the time reversal invariance is lost. Nevertheless, one should here note that if the ultrasonic absorption coefficient is sufficiently small in the frequency bandwidth of the ultrasonic waves used for the experiments, the time reversal invariance remains valid. This beautiful symmetry has been illustrated by Stokes in the classical case of plane wave reflection and transmission through a plane interface between two layers of different wave speeds. Consider an incident plane wave of normalized amplitude 1 propagating from medium 1 to medium 2. It is possible to observe a reflected plane wave with amplitude

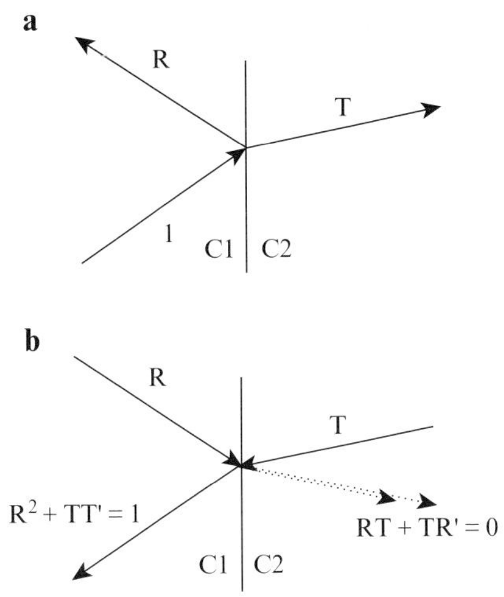


**Fig. 2.1** (**a**) Reflection and transmission of a plane wave at the interface between two media with different sound speed. (**b**) Time reversal of case (**a**)

$R$ and a transmitted plane wave with amplitude $R$ (Fig. 2.1). Starting from this simple configuration in which the total pressure field $p(r,t)$ consists in three plane waves, Stokes tried to verify if this experiment could be "time reversed" or not. He used the fact that a time reversal operation consists in the inversion of the wave vector direction for the simple case of a plane wave. Thus, the time reversed field $p(r,-t)$ can be described by a set of three plane waves: two incident waves of respective amplitude R and T going respectively from medium 1 to 2 and from medium 2 to 1, followed by a transmitted wave of amplitude 1 propagating in medium 1 (see Fig. 2.1). One can easily verify that this new theoretical wavefield is also solution of the wave equation. Indeed, if we define the reflection and transmission coefficients of an incident wave coming from medium 2, the superposition principle shows that the two incident waves generate four plane waves, two propagating in medium 1 with a resulting amplitude $R^2 + TT'$ and two other plane waves propagating in medium 2 with a resulting amplitude $RT + TR'$. An elementary calculation of the reflection and transmission coefficients $R, T, R'$, and $T'$ permits to verify the following equations:

$$R^2 + TT' = 1, \tag{2.2}$$

$$R + R' = 0. \tag{2.3}$$

This example shows that the wave equation can straightforwardly be interpreted as the time reversed version of the previous scene.

In fact, these simple arguments can be generalized to any kind of incident acoustic wave field and any kind of spatial heterogeneities in the medium. It is important to note here that the two previous relations are only valid if the reflected and transmitted plane waves have a real k vector (that is to say a purely propagative wave). In a more general case, the incident field can also contain evanescent components. For example, these evanescent waves can be created by incidences at specific angles or when an ultrasonic beam is diffracted by a medium whose compressibility $\kappa(r)$ contains very high spatial frequencies corresponding to distances smaller than the wavelength. The evanescent waves can not easily be time reversed as their direction is undefined. The superposition of propagative and evanescent waves results in a limitation of the time reversal process. Due to the limited bandwidth of the incident field, a part of the information is lost during the time reversal process.

## 2.3 Time Reversal Cavities and Time Reversal Mirrors

Let us come back to an ideal time reversal experiment for waves. The simplest "thought" experiment consists in visualizing a source of waves totally surrounded by a closed surface covered with piezo-electric transducers. In a first step, the wave field diffracted by the source is received by each transducer and stored in memories. In a second step, one imposes on the same surface the same wave field in its time reversed chronology. The surface, the so called Time Reversal Cavity (TRC) generates a time reversed wavefield, dual solution of the previous wave propagation, that converges exactly towards the initial source location whatever the heterogeneities of the medium (Fig. 2.2).

Thus, one builds experimentally what physicists and mathematicians call the advanced solution of the radiative diffraction problem. This is a dual solution of the "delayed" solution which is the only one observed in usual life. Indeed, it is known that each time ones tries to predict the diffraction pattern of a source, the time reversal symmetry of the wave equation gives serious headaches to physicists. Mathematically, there exist always two possible solutions to this problem: the first one corresponds to a wavefield diffracted by the source that spreads in space form the initial source location like a diverging wave that reaches the observation points only after the moment of initial emission by the source. This solution perfectly suits for physicists as it verifies the causality principle and it is called the "delayed" solution. In a homogeneous medium with uniform sound speed c, this solution can be written as $s(t - r/c)$ where $s(t)$ is a function that depends only of the temporal modulation of the source field and where $r$ is the distance between the source and the observation point. There exits a second solution that does not respects causality: the "advanced" solution where the wavefield diffracted by the source reaches the other regions of the space before the source has been emitting its wavefield and that can be written as $s(t + r/c)$. Although this solution seems to be totally nonsense, it is interesting to notice that most

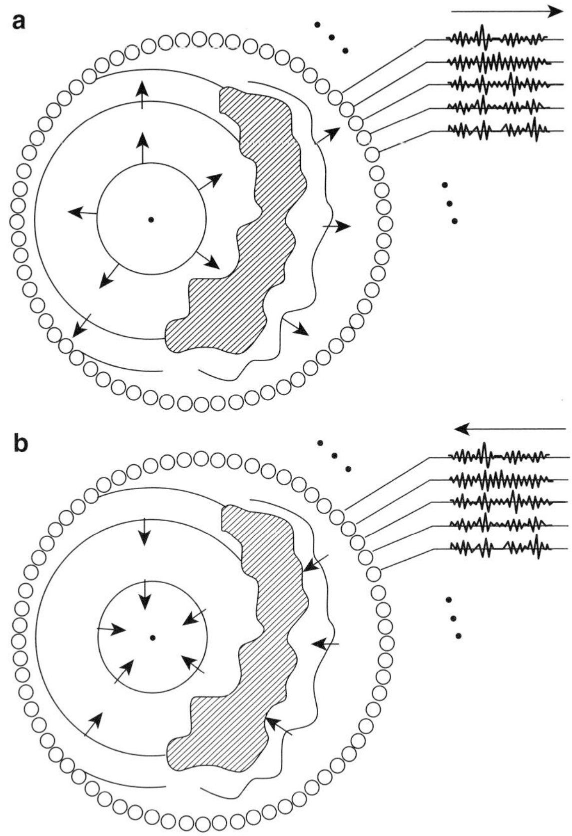


**Fig. 2.2** The concept of a time reversal cavity. (**a**) Forward step: the surface of a closed cavity is covered by piezoelectric transducers working both in transmit and receive modes. A point source emits a wavefield that propagates through the heterogeneities of the medium. The resulting distorted wavefield is received on each element of the cavity. (**b**) Backward or time reversal step: the recorded signals are time reversed and re-emitted by the elements of the cavity. The time reversed wavefield propagates again through the heterogeneities of the medium and refocuses precisely on the location of the initial source

renowned scientists like Feynmann and Wheeler have tried to interpret with some success the theory of electromagnetic radiation by taking into account both "advanced" and "delayed" solutions of the wave equation.

Of course, in our "thought experiment", everything is causal and the arrow of time can not be inverted but we build a solution that resembles the "advanced" solution: it is a wavefield converging towards the source instead of diverging from the source like the "delayed" solution does. However, a short comment is here necessary regarding the building of the "advanced" solution. The wave re-emitted by the time reversal cavity looks like a convergent wavefield during a given time, but a wavefield don't know how to stop (it always has a wave propagation speed) and when the converging wavefield reaches the location of the initial source location, its "collapses" and then continues its propagation as a diverging wavefield. One can then observe around the source location the interference between the converging and diverging waves. This is not anymore exactly the "advanced" wave which is created but this is the "advanced" wave minus the "delayed" wave (indeed the converging wave changes its sign after the collapse and then diverges with an opposite sign). If the wave used in the experiment is monochromatic (narrow band) and contains an important number of wavelengths, one can demonstrate very easily that a constructive interference between both waves is created around the focal point over a distance whose radius is equal to half the wavelength. This is the diffraction limit. More precisely, for an initial point source emitting a short impulse with a large frequency bandwidth, the time reversed wavefield focuses back at the source location, but it generates a focal spot whose dimensions are of the order of half the smallest transmit wavelength.

The process of time reversal focusing acts as an inverse filter of the diffraction transfer function that describes the wave propagation between the source location and the points of the surrounding time reversal cavity. We have already just noticed in the previous paragraph that this inverse filter is not perfect, as it is not possible to refocus the wave on an infinitely small focal spot corresponding to the initial point-like source. One can revisit this diffraction limit using the concept of evanescent waves. Indeed evanescent waves emitted during the forward step by the source are non-propagative waves and suffer a very sharp spatial exponential decay. Consequently, these evanescent waves are not recorded and then time reversed by the elements of the time reversal cavity. Even if we would be able to record them, these evanescent waves are not reversible as they would suffer a second time during backward propagation a spatial exponential decay. Thus, diffraction acts as a low pass filter for the spatial frequencies conveyed by the wavefield. As the spatial frequencies corresponding to evanescent waves ($>1/\lambda$) are lost during propagation: One reaches the classical wave diffraction limit that forbidden to the focal spot to become smaller than half the wavelength. One should notice that in order to be able to generate infinitely small focal spots, much beyond the diffraction limit, it would be necessary to avoid the divergence of the wave after the collapse. This can be done if a new source is introduced at the initial source location, the so-called acoustic sink [11]. This second source emits the time reversed signal $s(-t)$ of the signal emitted initially during the forward step. Instead of absorbing the incident wavefield, this acoustic sing emits, in an active way, at the exact collapse time a new diverging wave which is the exact opposite of the natural divergent wave. This is this noise cancellation wavefield that creates a perfectly destructive interference between the two diverging waves and permits to reach focal spots much smaller than the classical diffraction limit.

In practical situations, a time reversal cavity would be very difficult to build and the time reversal operation is usually performed in practice over a

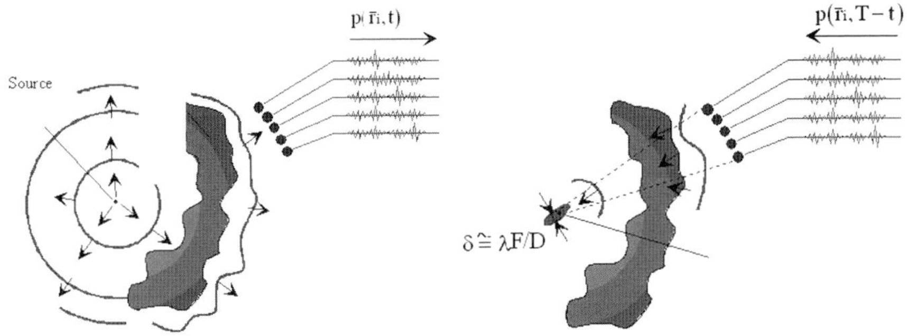


**Fig. 2.3** Principle of the time reversal focusing process over a limited aperture

limited aperture of the cavity, known as time reversal mirror (TRM). Its angular aperture gives rise to another low-pass filter for the angular frequencies that can be transmitted by the array. Its cut-off frequency is more restrictive that the one due to evanescent waves:

In homogeneous medium, it depends of the angular aperture of the array and it results in a focal spot whose lateral width is proportional to the product between the wavelength $\lambda$ and the angular aperture $F/D$ of the array (Fig. 2.3).

In heterogeneous medium, the plane wave decomposition of the wavefield is no more convenient and it becomes very difficult to predict the behaviour of the TRM. It strongly depends of the heterogeneities of the medium. In some given configurations, A. Derode et al. [8] have shown that the focal spot obtained by time reversal focusing can be much thinner than in homogeneous medium when a strongly multiple scattering medium is placed between the MRT and the focal area. In such cases, the time reversal focusing process takes benefit of the reverberations in the medium in order to increase the apparent angular aperture of the array.

## 2.4 Time Reversal Is a Spatial and Temporal Matched Filter of Wave Propagation

In a heterogeneous medium, the time reversal focusing is extremely robust and efficient compared to the other classical focusing techniques based on the application of different time delays on the array aperture. It can be mathematically shown that for any kind of transducers configurations, the time reversal solution is the optimal solution in a non-dissipative when ones tries to maximizes the energy deposit at focus for a given transmit energy on the array. Indeed time reversal focusing performs the spatial and temporal matched filet of the transfer function between the array elements and the targeted focal point. The concept of time reversal is strongly linked to the concept of matched filtering in signal processing. This well known principle describes the fact that the output of a linear system of impulse response $h(t)$ is maximized when we choose as input the time reversed version of the impulse response: $h(-t)$. In that case, the output signal of the linear system is equal to $h(t) \otimes h(-t)$. This is an odd function whose maximal value is reached at $t = 0$ and is given by the energy of the input signal. Let us demonstrate that, for a given transmit energy on the transducers array, the solution obtained using the time reversal process corresponds to the one that maximizes the energy received at the location of the focal point. What is the optimal set of impulse signals transmitted by a set of transducers $E_i$ located at $r_i$ for a focusing at targeted point-like location $r_0$.

As one can notice in Fig. 2.4, the time reversal process is divided into three steps. During the first step, after emission of a wavefield widely spread in space by the array, a strongly reflecting scatterer located in $r_0$ acts as an acoustic source by reflecting the incident field. The electric signal $s_i(t)$ received on each element $E_i$ of the array is given by

$$s_i(t) = h_i^{ae}(t) \otimes h_i(r_0, t), \tag{2.4}$$

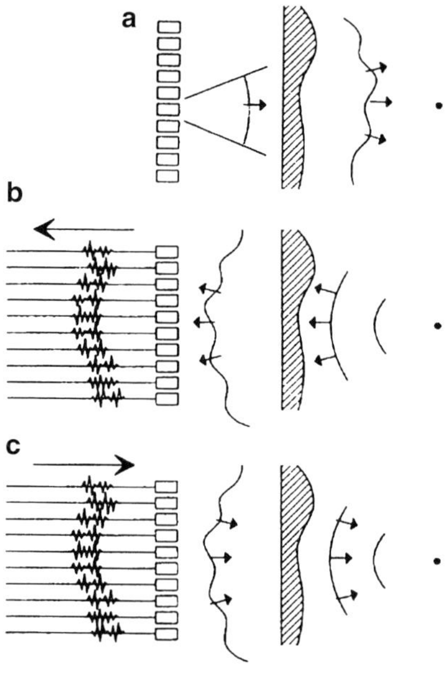


**Fig. 2.4** The three steps of the time reversal focusing process in heterogeneous medium when a strongly scattering point is available at the targeted location. (**a**) A widely spread wavefield is transmitted by the array, (**b**) reflection of the wavefield on the targeted strong scatterer and recording of the backscattered echoes received on the TRM, (**c**) time reversal and re-emission of the signals by the elements of the array

where $h_i^{ae}(t)$ is the acousto-electric impulse response of the transducer that links during transmission the electric input signal to the normal velocity obtained at the output of the transducer and that links, during reception, the acoustic pressure detected by the transducer to the electric signal generated (it is here postulated that this acousto-electric response is identical both for transmit and receive). $h_i(r_0, t)$ is the diffraction impulse response linking the element $E_i$ of the array to the point $r_0$ in the medium. This diffraction impulse response characterizes the propagation of the medium. One should note here that it has been assumed that the spatial reciprocity theorem can be applied to the propagation medium, in other words that the impulse response between point $r_0$ and element $E_i$ is identical to the one between $E_i$ and point $r_0$. During the second step of the time reversal focusing process, these signals are stored in memories by all elements of the array. On should here notice that in a heterogeneous medium, the signals received on each element of the array can have totally different temporal shapes. During the third step of the time reversal process, these signals are time-reversed and a new emission signal is obtained for each transducer:

$$h_i^{ae}(T-t) \otimes h_i(r_0, T-t), \tag{2.5}$$

where $T$ is a time constant introduced in order to ensure the causality of the experiment. These signals are then re-transmitted by the elements of the array. In order to estimate the total acoustic field received at the focal point location $r_0$ of the TRM, one has first to consider the individual contribution of each element $E_i$ at focus. This wavefield results from the convolution product between the electric signal $h_i^{ae}(T-t) \otimes h_i(r_0, T-t)$ applied on the element with the transmit acousto-electric impulse response and the diffraction impulse response characterizing the propagation towards $r_0$ during the third step:

$$h_i^{ae}(T-t) \otimes h_i(r_0, T-t) \otimes h_i(r_0, t) \otimes h_i^{ae}(t). \tag{2.6}$$

Equation (2.6) shows that the electric input signal (the time reversed signal) is the optimal input signal of the linear system defined by the two consecutive impulse responses $h_i^{ae}(t)$ and $h_i(r_0, t)$. The maximal pressure field is received at point $r_0$ at time $T$ independently of the position of element $E_i$. The total pressure field created at point $r_0$ by the TRM is obtained by the simultaneous transmission by all elements of the array of the time reversed signals. Then, one obtains at point $r_0$ the summation of all individual contributions:

$$\sum_i h_i^{ae}(T-t) \otimes h_i(r_0, T-t) \otimes h_i(r_0, t) \otimes h_i^{ae}(t). \tag{2.7}$$

All individual signals reach their maximal value at the same time $t = T$ and interfere constructively. This collective and adaptive matched filtering creates the maximum signal at point $r_0$ for a fixed energy transmitted by the array.

On should note here that this result is valid whatever the geometry or configuration of the array of transducers. For example, an array can have suffered geometric distortions during its design. The time reversal technique is an auto-adaptive technique that permits to correct these artifacts. Beyond that kind of corrections, it should also be noticed that the previous analysis is only based on the spatial reciprocity theorem. So, for this reason, it is valid for any kind of heterogeneous medium whatever complex it is. Consequently, even if the individual contributions can have totally different time profiles, the time reversal process ensures that their maximum will be reached always at the same time $T$ at point $r_0$. We have here demonstrated that the time reversal process performs a temporal matched filter of the impulse transfer function between $r_0$ and the array of transducers. The same kind of demonstration was also given the spatial domain by Tanter et al. [12] and then generalized to a spatio-temporal approach [13, 14].

## 2.5 Iterating the Time Reversal Process

One of the advantages of the time reversal technique lies in the fact that it is very easy to choose the time origin and the duration of the signals to be re-emitted. This is achieved by defining a temporal window that selects in the electronic memories the part of the signals that should be time-reversed. When several scatterers are present in the medium of interest, the time reversal process does not permit to focus only on one selected scatterer. Indeed, if the medium contains for example two targets with different reflectivity coefficients (Fig. 2.5), the time reversal process will produce two wavefronts refocusing simultaneously on both targets with different amplitudes. The mirror produces the real acoustic image of the two targets at their exact locations. The wavefront with the highest amplitude focuses on the target with the highest reflectivity coefficient, whereas at the same time the wavefront of lower amplitude focuses on the second target with lower reflectivity coefficient. We assume here that there is no multiple scattering phenomenons between both scatterers. After the first illumination using time reversal, the wavefront reflected by the weak target will be even weaker compared to the one reflected by the brightest target. Thus, by iterating the time reversal process several times, it is possible to select progressively the brightest target [15]. This iterative process converges and finally produces a single wavefront focusing only on the main target. Once the beam is definitely oriented towards the brightest target, the iteration does not modify anymore the result. One has created a first "invariant" of the time reversal focusing process. Note that this process will converge only if both targets are resolved, that is to say if the first target is located outside of the diffraction focal spot obtained on the second target.

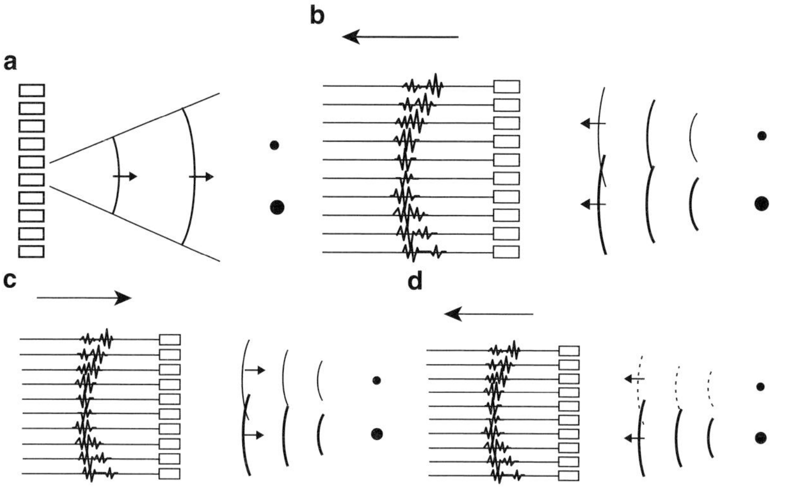


**Fig. 2.5** Basic principle of iterative time reversal processing. (**a**) A first transmit wave insonicates the two scatterers, (**b**) the backscattered echoes are recorded on the array, (**c**) then re-emitted after time reversal, (**d**) the new backscattered waves are received and recorded before a new time reversal iteration. After some iterations, the process converges towards the emission of a single transmit wavefront focusing only on the brightest target

In a second procedure, one can subtract the wavefront corresponding to the first invariant from the global set of backscattered echoes. It avoids any acoustic focusing on the brightest target and a new set of iterations of the time reversal process permits to select the second brightest target. A second invariant of the time reversal process is then found.... All this procedure can be very easily implemented in hardware as it consists in very simple signal processing operations (time reversal and subtractions). It permits to detects very quickly the whole set of targets in a medium and to deduce all the signals to be emitted on the array in order to focus separately on each target. The great interest of this approach is that it performs in real time the optimal corrections of all medium aberrations for different targeted locations [16]. Another approach, the so-called DORT method (Time Reversal Operator Decomposition) developed by C. Prada was proposed in 1994 to solve the same problem. Instead of experimentally performing these iterations in the medium, one can exploit the linearity of the previously described operations in order to achieve the complete processing in software. In a first step, one can measure all the impulse responses between each element of the array. If the array is made of $N$ transmit/receive transducers, one has to measure $N^2$ impulse response signals. In a second step, the DORT method consists in simulating the time reversal operation and the estimation of the time reversal problem invariants by calculating the eigenvectors of a matrix that can be deduced from the set of the $N^2$ impulse responses [17]. Based on all these different concepts, it is possible to develop new approaches for target detection and even in some cases for target destruction. For medical applications, the iterative time reversal method is much more rapid and has been evaluated for breast microcalcifications detection or for kidney stones destruction (lithotripsy). Devices based on that concept permits a real time auto-tracking and auto-focusing on the kidney stone for precise destruction with very high amplitude ultrasonic focused beams (100 MPa) even in the presence of strong motion artifacts due to the breathing of the patient. For target characterization and imaging, the DORT method is more appropriate and it is currently subject of intensive research in underwater acoustics, non-destructive testing of materials and electromagnetism. For example, the study of the frequency spectrum of a complex target gives important information on the target in acoustics and electromagnetism.

## 2.6 Applications of Time Reversal

Thanks to its ability to learn in a very short time how to focus adaptively through very complex media, time reversal was successfully applied in the domain of medical ultrasound both for imaging and therapy [3, 10]. It permits for example to correct the distortions induced during ultrasound therapy of brain tumors on the focused beam by the skull bone heterogeneities. It also gives an elegant solution to the problem of real time tracking and destruction of kidney stones using very high amplitude pressure fields (lithotripsy). It is also able to detect breast microcalcifications with a much better precision than conventional echographic imaging. Finally, the application of time reversal processing to reverberating cavities enables extreme increasing of pressure amplitude generated at focus for remote tissue destruction (histotripsy) thanks to a great temporal compression effect of the time reversed propagation through the reverberating medium.

### *2.6.1 Real Time Tracking and Destruction of Kidney Stones Using Time Reversal*

Extracorporal ultrasonic lithotripsy appeared in 1980 and permits today to treat the majority of kidney stones. However, although the kidney stones can be localized precisely by CT-scans, the current systems are not able to follow in real time the tissues displacements induced by the patient breathing. As an immediate consequence, it is estimated that more than two-third of the ultrasonic shots fail to focus on the target and create damages in surrounding tissues (such as local haemorrhages).

## Correction of Respiratory Motion Using Time Reversal

The real time adaptive focusing ability of time reversal makes it a particularly well suited approach for this problem. Indeed, the destruction of kidney stones is usually done using an ultrasonic focused beam generated by a single spherical shaped transducer and focused always at the same geometrical focus location. The Laboratoire Ondes et Acoustique of ESPCI (Paris, France) has developed a time reversal mirror that is able to follow in real time the organ motion and move electronically the focal spot in order to enforce the beam to remain locked on the target [7]. The kidney stone is used as a passive acoustic source, on which the time reversal mirror learns how to auto-focus (cf. Sects. 2.2 and 2.5).

The basic principle is presented in Fig. 2.6. The high power array of transducers transmits a first wave that is spreading in the whole region of interest (a). Due to its high reflectivity coefficient compared to surrounding tissues, the backscattered echoes coming back from the medium are mainly due to the kidney stone (b). The backscattered signals are time reversed and re-emitted by the array of transducers. The ultrasound beam then refocuses on the kidney stones even if aberrating layers (like fat layers) are present in the medium (c). Due to the very high sound speed in human tissues ($c = 1,500\,\mathrm{m\,s^{-1}}$), this operation can be repeated more than 100 times per second. Tissues displacements due to respiratory motion have much larger time constants. So, the medium has not time enough to move between two time reversal iterations. This ensures that the therapeutic beam remains locked on the target whatever the unwanted motion artifacts.

As one can notice in Fig. 2.7, after some iterations of the time reversal process, the echoes coming back from tissues have been cancelled and the

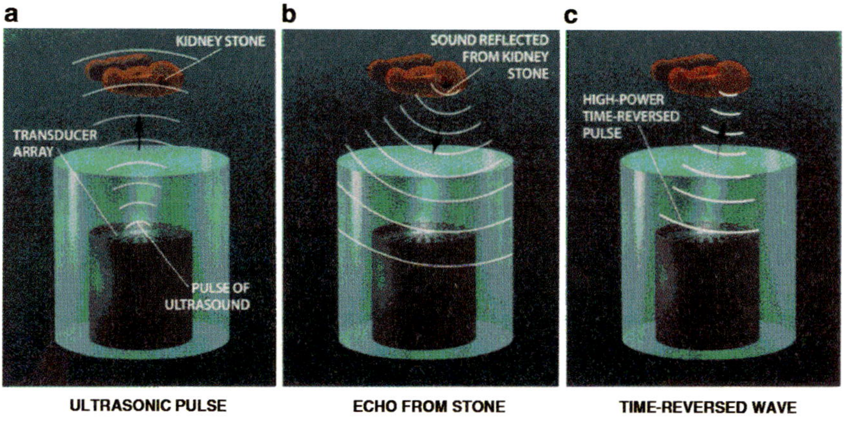


**Fig. 2.6** Application to lithotripsy: Real time tracking and auto-focusing on a kidney stone using time reversal

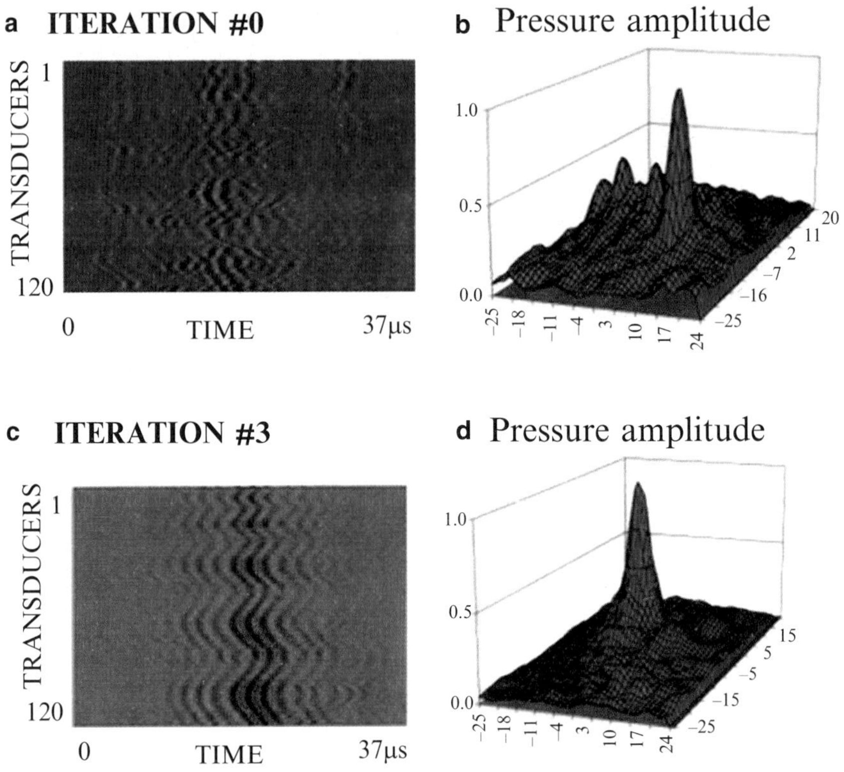


**Fig. 2.7** Ultrasonic echoes received on the transducer array after (**a**) one and (**c**) three iterations of the time reversal process. Spatial beam profile in the focal plane during the first (**b**) and third (**d**) iteration

signature of the kidney stone is highlighted. The focusing beam quality has been greatly improved and the process ensures that the beam remains locked on the target. A first prototype made of 64 electronic channels was developed by the Laboratoire Ondes et Acoustique in Paris during the last decade. Many in vitro experiments were performed on kidney stones and the feasibility of this technique as experimentally demonstrated by this research group. The time reversal mirror is able to locate the stone in less than 40 ms thus ensuring a real time tracking.

Such a system in its final version could soon resemble to the image presented in Fig. 2.8 [2]. Due to the high number of transducers required in order to be able to steer electronically the beam several centimeters away from the geometrical focus, the cost of a time reversal mirror for lithotripsy remained up to recently prohibitive. Next section shows how a very elegant property of time reversal can be used in order to decrease this number of array elements.

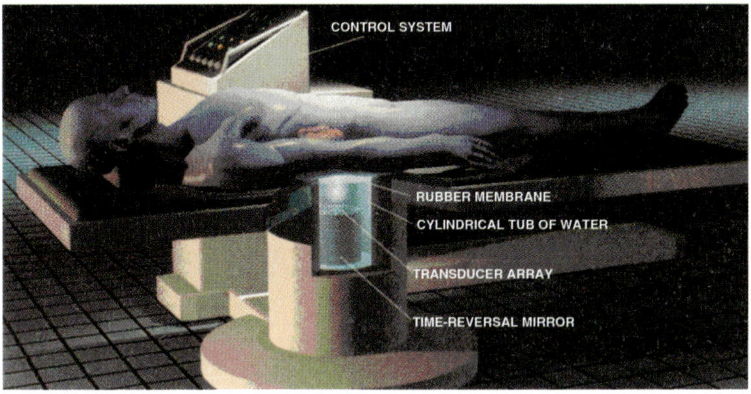


**Fig. 2.8** A schematic view of the concept of time reversal lithotripsy (Scientific American, 1999)

### *2.6.2 Time Reversal and Temporal Pulse Compression*

The ability of time reversal to focusing through a strongly reverberating (dispersive) medium leads to a very interesting property of time reversal: the pulse compression effect. In fact, time reversal processing can take benefit of the multiple reverberations of the ultrasonic wave in a solid waveguide [4]. As we will see in this section, it permits to greatly decrease the number of transducers necessary to reach very high pressure fields in the organs. The system is made of a small number of piezo-electric transducers stuck on one face of a solid metal cavity acting as a waveguide. Another face of the solid waveguide is placed on the skin surface of the patient. We can make a "thought" experiment: if an ultrasonic source is located at the desired focal point (i.e. the kidney stone) and transmits a short impulse, the resulting wavefield will suffer a very strong temporal dispersion during its propagation in the solid waveguide. Experimentally, the signal received by the transducers can last more than 1,000 times more than the initial impulse signal. After time reversal and re-emission of the signals by each transducer, the resulting wave refocuses on the initial source location, exactly as of the movie of the propagation was played backwards in time. Thus, all the energy that was conveyed by a signal lasting several milliseconds is going to recompress at focus in a very short time of the order of the microsecond (Fig. 2.9). It is again the beauty of the matched filter property of time reversal that permits this incredible gain in signal amplitude at focus. This giant amplification of the signal at focus ensures the destruction of the target by mechanical effect. This amplification due to temporal compression of time reversal processing is the acoustic analogous a light amplification in femtosecond lasers. The

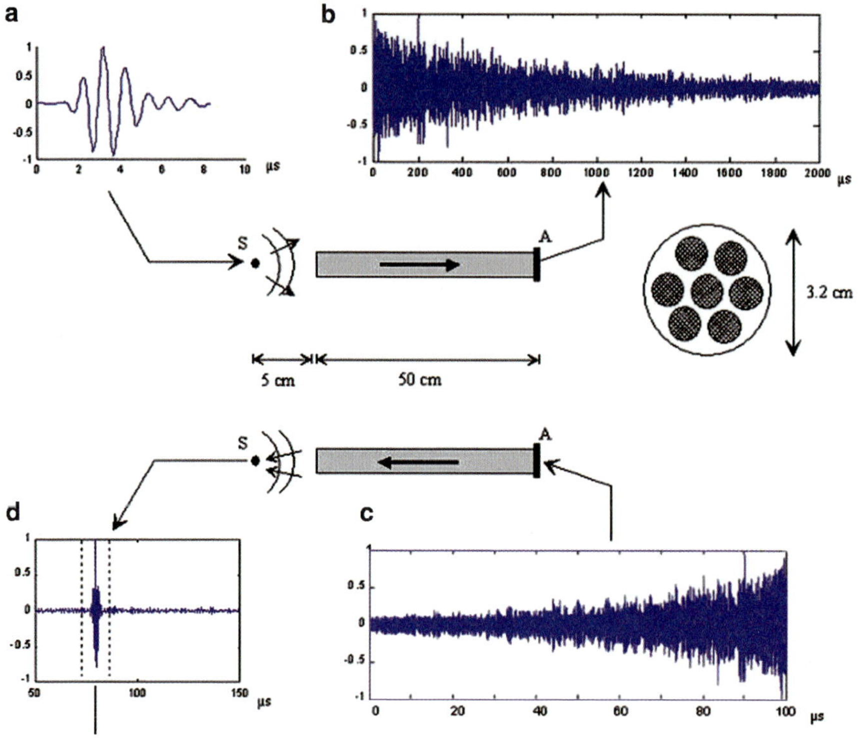


**Fig. 2.9** Basic principle of the giant ultrasonic signal amplification using time reversal. (**a**) and (**b**) During the calibration step, the seven transducers record the signal coming from an acoustic source embedded into a water tank (same medium than human soft tissues). (**b**) and (**c**) The signals are time reversed and re-emitted by the elements of the array. Theses long lasting signals refocus at the focal point location in a very short impulse, thus generating a giant amplification of the maximal pressure field at focus

presence of an acoustic source in the organ is of course not necessary. A first calibration experiment can be done in a water tank (same sound speed as soft human tissues) and echoes recorded on each transducer for each position of the acoustic source in water can be stored in memories. These whole data sets of signals can then be used in vivo for real time focusing at different locations in 3D in the patient organ.

Two prototypes based on this idea were developed at Laboratoire Ondes et Acoustique in Paris, France. Figure 2.10 shows the damage generated locally in chalk pieces using this ultrasonic signal amplification system. Thanks to the used of time reversal reverberations in the waveguide, it should soon be possible to decrease the number of necessary transducers for autotracking lithotripsy by a factor higher than 20! (Fig. 2.11).

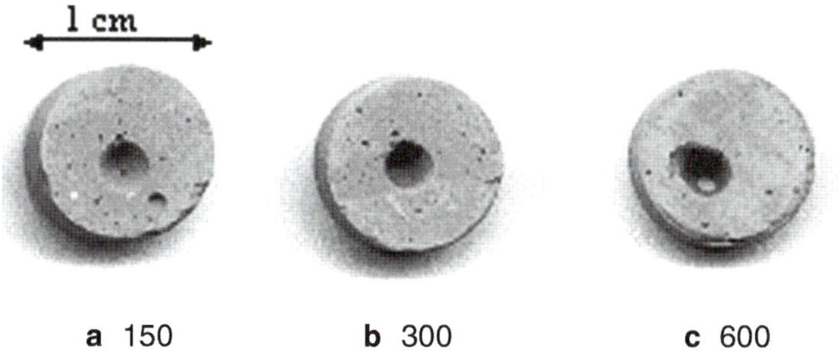


Fig. 2.10 Mechanical damage induced by the seven small transducers system in big chalk pieces using 150, 300 and 600 successive shots

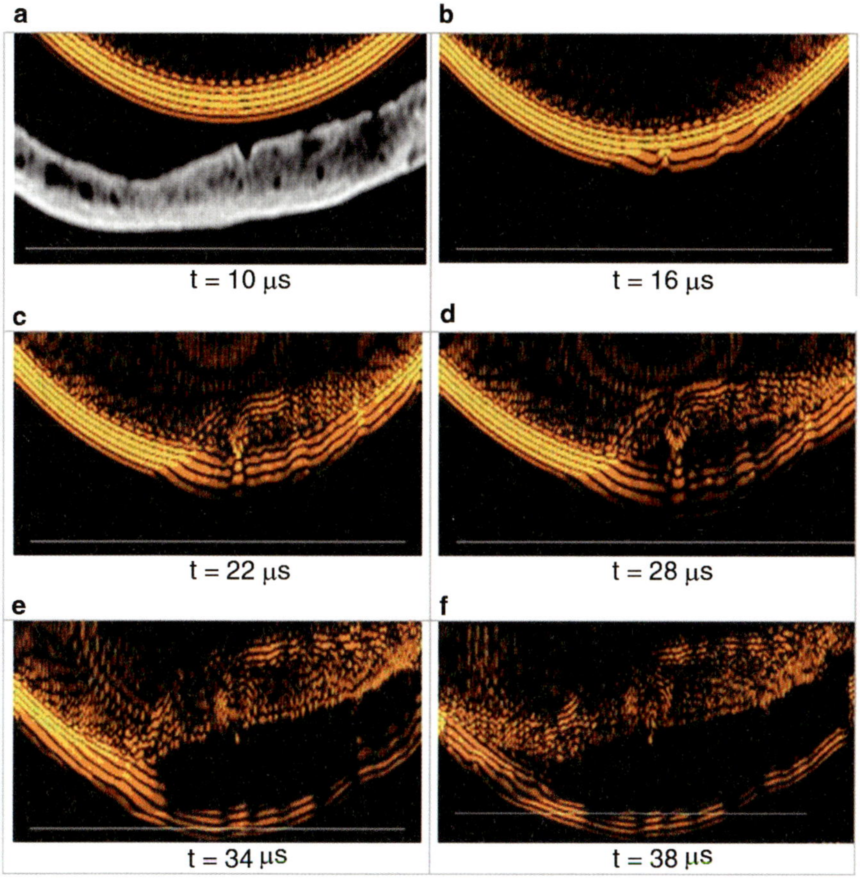


Fig. 2.11 Numerical simulation of the ultrasonic wave propagation through a human skull bone: 2D images of the ultrasonic wavefield at different time steps. The acoustic parameters of the skull bone are deduced from the 3D CT-scan image

### 2.6.3 Transcranial Focusing Using Time Reversal for Brain Tumours Therapy

Extracorporal ultrasound based therapy of brain tumours is surely the medical application that best emphasizes the potential of time reversal for medical science. Non-invasive ultrasound-based systems able to treat tumors difficult to access using conventional surgery are currently under development in many research groups and companies all around the world. Their basic principle is based on the use of an ultrasonic focused beam inducing a temperature elevation in the focal region that locally necroses pathological tissues. This ultrasonic wave is generated by an array made of a important number of piezoelectric transducers covering a spherical surface that surrounds the skull [1,6]. However the building of such an ultrasonic focused beam is made a very difficult task due to the strong heterogeneities induced by the skull bone. The strong discrepancy between sound speed in the skull bone and sound speed in brain tissues combined with the complex internal architecture of the bone leads to a very strong degradation of the ultrasonic beam. The solution to this complex problem requires the use of an adaptive focusing technique. In order to correct the distortions induced by the skull bone, it is necessary to use arrays made of a non-negligible number of elements. On the one hand, it permits to introduce time delays on each element that corrects the beam profile. On the other hand, it permits in a second step to steer electronically the beam if ones wants to treat a complete 3D volume in the brain. Time reversal was shown during the last decade to offer an elegant solution to this complex problem [5].

However, it is important to notice that the time reversal principle is based on time reversal invariance of the wave equation in non-dissipative media. Unfortunately, in addition to sound speed and density heterogeneities, the skull bone introduces also ultrasonic absorption heterogeneities. This ultrasonic absorption degrades the focusing quality of time reversal processing. By taking into account these absorption effects, the concept of time reversal focusing can be generalized to an adaptive focusing technique in absorbing media, the so-called spatial-temporal inverse filter [5].

#### Time Reversal and Biopsy: Adaptive Focusing Using Implanted Hydrophone

The time reversal process requires an acoustic (passive or active) source at the desired focus location. However, contrary to the lithotripsy application where the acoustic reflection on the kidney stone was used as a passive source, one can not rely in the brain of a very bright acoustic scatterer in order to learn how to autofocus on the target. In fact, this reference point can be created during the biopsy that is performed before any important brain surgical act. The biopsy is a very minimally invasive surgical act. During the biopsy it is possible to implant a small hydrophone on the surgeon instrument and acquire the impulse response between each element of the array and the hydrophone. After removal of a small tissue sample, the surgeon instrument containing the hydrophone is removed of the brain. Then, in a second step, a PC computer is calculating the set of signals that will optimally focus on the initial hydrophone location (by combining time reversal and absorption effects corrections). The multi-elements array emits these signals. The ultrasonic beam then refocuses optimally on the initial hydrophone location. The beam can then be steered electronically in order to focus at other locations surrounding the initial autofocus location for the treatment of a large volume. The transcranial focusing quality are almost identical to the ones obtained in homogeneous medium (water) up to $-20\,\mathrm{dB}$ sidelobes level. This contrast is largely sufficient for ultrasonic therapy as the heat deposit is proportional to the square of the pressure field. Thus the heat deposit at focus is more than 100 times higher than in surrounding brain tissues.

### Virtual Time Reversal: Ultrasonic Adaptive Focusing Guided by CT-Scans

It is possible to improve the previous procedure and to get rid of the biopsy step. A way to propose a fully non-invasive focusing technique consists in using the information about the skull bone architecture provided by CT-scans images. Indeed, it is possible to perform the first step of the time reversal experiment (i.e. the forward propagation step) using only numerical computation. One can use the information provided by CT-scans in order to recover the important acoustic parameters that affect wave propagation (sound speed, density and ultrasonic absorption). From these 3D images of acoustic parameters, it is then possible to compute using 3D finite differences simulations the propagation of the ultrasonic wavefield coming from a virtual acoustic source in the numerical brain (Fig. 2.11). The forward propagation from the target to the array is thus performed numerically thanks to the medium knowledge acquired from CT data.

As one can notice in Fig. 2.12, the agreement between numerical simulations and experiments is extremely good. So, it is definitely possible to perform a totally non-invasive and adaptive ultrasonic focusing through the human skull bone and thus, avoid the use of a pre-treatment biopsy.

### High Power Time Reversal for In Vivo Ultrasonic Brain Therapy

A high power time reversal system based on the concepts presented in the previous subsection was developed in the last years at Laboratoire Ondes et Acoustique in Paris, France. The array is made of 300 piezoelectric

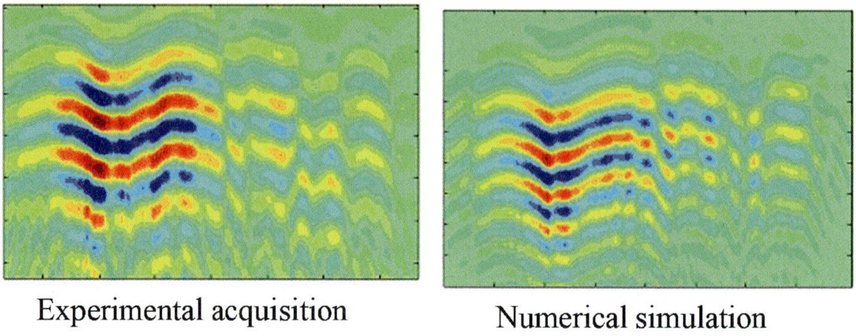


**Fig. 2.12** Temporal wavefronts received on an ultrasonic array after propagation through a human skull bone: Numerical simulation based on CT-scans data compared to the experiment in the same conditions

transducers driven by a high power electronic system relying on 300 independent transmit/receive electronic boards (see Fig. 2.13). This ability of this system to perform non-invasive transcranial brain necrosis has been demonstrated recently in in vivo animal experiments. These initial results are extremely promising and pave the way to human brain tumor treatments.

The skull aberration correction technique was validated experimentally on 20 sheep using this system. These first in vivo validations were using the minimally invasive technique based on the use of the pre-treatment biopsy. Figure 2.14 shows the quality of the focal spot obtained through the skull bone using the time reversal based approach. The focal spot is much sharper with aberrations corrections. Even more important is the acoustic intensity obtained at focus using time reversal compared to conventional focusing. By correcting the skull bone aberration, the time reversal process exhibits a 20 times more intense acoustic field at focus compared to conventional focusing (focal depth 14 cm). The resolution of the focal spot is of the order of 1 mm (Fig. 2.14) resulting in a very precise necrosis of brain tissues even at high depth (Fig. 2.15).

Following these pioneer experiments (based on a minimally invasive technique), a new animal investigation testing a fully non-invasive approach was proposed and validated in 2005 on monkeys (*Macaca fascicularis*). This non-invasive approach consisted in simulating the first step of the time reversal experiment using a priori information about the 3D architecture of the skull bone. First a 3D CT-scan of the monkey was performed as seen in Fig. 2.16a. The Grayscale images obtained in Hounsfield units were converted into local acoustical parameters (speed of sound, density and ultrasonic absorption). Then, the 3D propagation of an ultrasonic wave initiated from the targeted location bone was simulated with a 3D finite differences code, see Fig. 2.16b. After numerical propagation through the 3D skull bone model signals were recorded at the location of virtual HIFU array elements. One week after the CT-scan, the monkey is again fixed into the stereotactic frame, Fig. 2.16c.

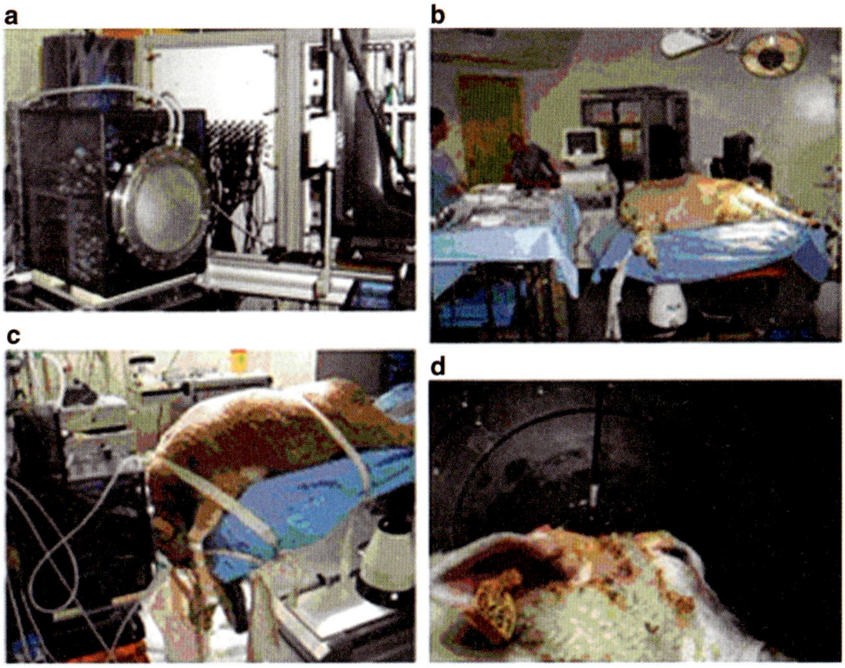


**Fig. 2.13** (**a**) High power time reversal mirror for transcranial ultrasonic brain therapy. The array is made of 300 individual transducers quasi-randomly distributed on a 14 cm radius spherical surface. (**b**) Detailed view of the front face of the array. An ultrasonic imaging probe is embedded in the vicinity of the therapeutic array for skull bone imaging. (**c**) Final system: a latex membrane containing cooled and degazed water permits the coupling of the array to the skin and bone surfaces. (**d**) The system is coupled to a sheep head. A biopsy is performed and a small hydrophone is recording the green function between the array and the targeted location. Finally after the biopsy, several ultrasonic beams are fired during several seconds in order to treat the targeted location and surrounding points by steering electronically the beam at other locations

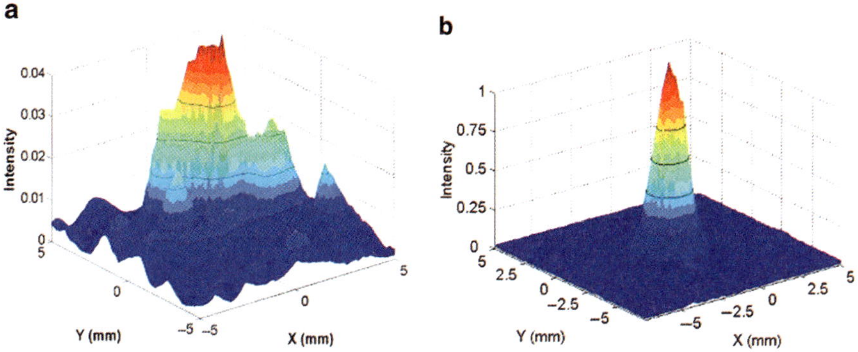


**Fig. 2.14** Spatial beam profile of the transcranial beam in the focal plane (focal distance 12 cm): (**a**) without aberration corrections, (**b**) with aberration corrections. The acoustic energy deposit at focus is 25 times higher than using time reversal adaptive focusing

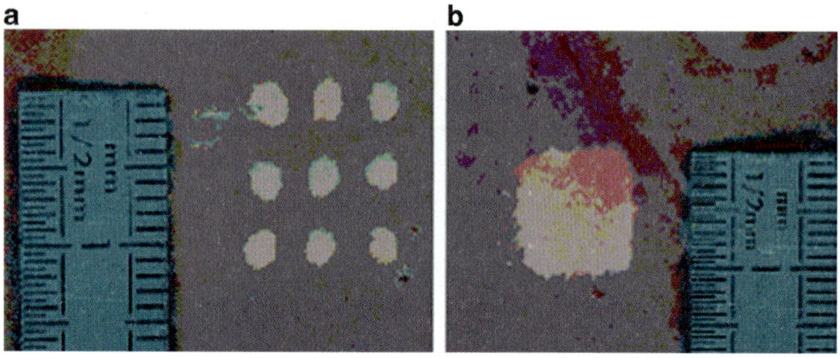


**Fig. 2.15** Ultrasound thermal necrosis induced in liver tissue samples through a human skull bone. (**a**) One can notice that the necrosed area is very sharp (nearly 1 mm diameter). (**b**) The ultrasonic beam can be electronically steered at other locations in order to treat point by point a larger volume. The precision of the necrosis is very high as the focal spot relies on a millimetric resolution

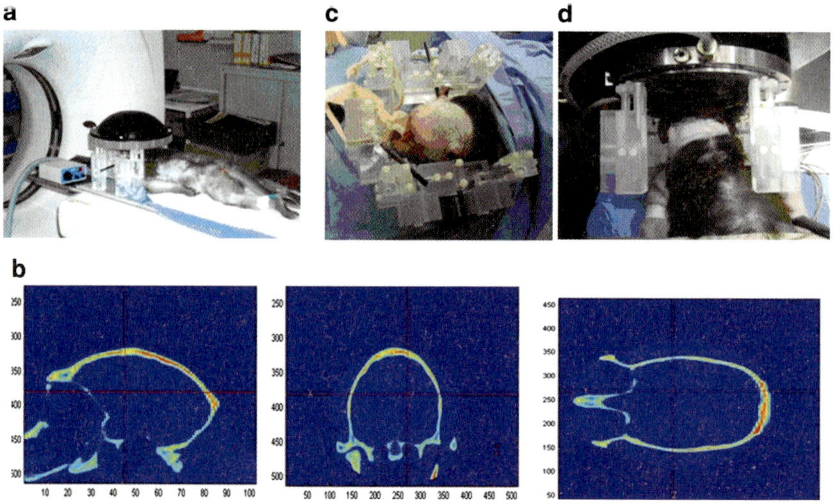


**Fig. 2.16** The different steps of the brain HIFU therapy guided by CT-scans: (**a**) the monkey is imaged into a CT-scanner. Head is fixed on a house-made stereotactic frame. (**b**) A treatment software allows selecting the targeted spot. From these data, numerical simulation of the 3D transcranial ultrasonic propagation enables the estimation and correction of skull aberrations for the HIFU treatment. (**c**) One week after the CT-scan exam, the monkey's head is again fixed into the house-made stereotactic frame (**d**) the HIFU array is coupled to the monkey's head via the stereotactic frame, thus ensuring the concordance of CT-scan reference frames and the HIFU experimental reference frame. The HIFU treatment is then performed using the calculated emission signals

The HIFU array is then coupled to the monkey's head via a water-cooling and coupling system, Fig. 2.16d. The treatment is then performed using phase aberrations deduced from the 3D simulations. Monochromatic signals are emitted by each element of the array using as an emission delay the time-reversed version of the estimated phase aberrations. Thirty sonifications of the therapy beam are achieved using all array elements. Each shot was lasting 10 s and the cooling time between each sonification was fixed to 15 s. Two treatments were performed with the same parameters: first, a treatment sequence was performed in the right hemisphere of the brain using a corrected therapy beam. A second treatment sequence was performed in the contralateral part of the brain with an uncorrected HIFU beam. The treatment duration was identical for both treatments. After the HIFU experiment, the monkey was woken up and a basic neurological score was given showing that no side effects occurred. The monkey's consciousness was normal, its appetite was good and the motor functions were not affected.

One week after the treatment, the monkey was sacrificed. No burns were observed on the skin, on the outer and inner tables of the skull but some petechia on the outer surface of the brain have been induced. Figure 2.17c shows a 7-T MR image of the fixed brain. It clearly shows the thermal necrosis at

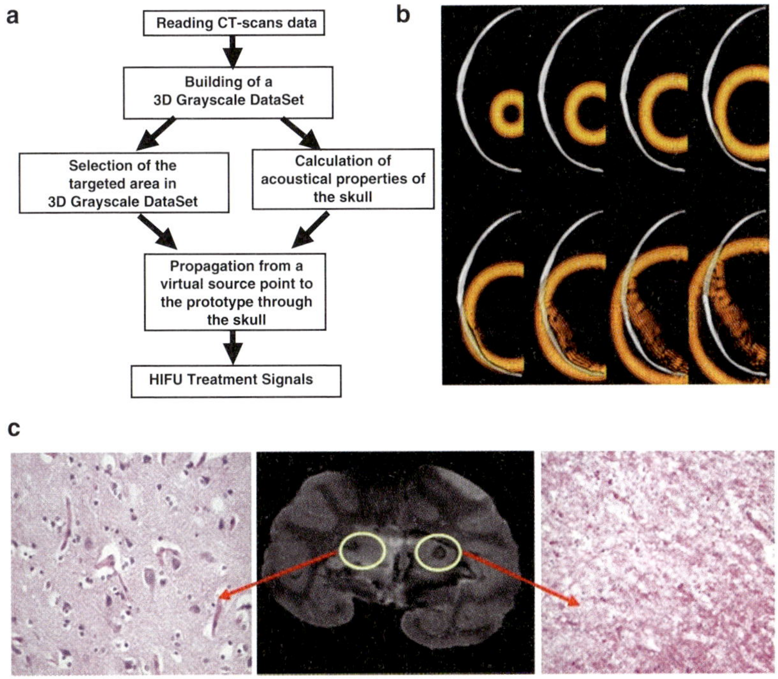


**Fig. 2.17** (**a**) Principle of the transcranial time reversal focusing technique guided by CT-scans. (**b**) 3D simulation of the wave propagation (finite differences): wave field distribution at different time steps (at $t = 0\,\mu s$, $t = 8\,\mu s$, $t = 16\,\mu s$, $t = 24\,\mu s$, $t = 32\,\mu s$, $t = 40\,\mu s$, $t = 48\,\mu s$, $t = 56\,\mu s$, $t = 64\,\mu s$) in a selected plane. (**c**) The monkey was sacrificed one week after treatment: the 7-T MR images of the fixed brain clearly show the thermal necrosis at the targeted location using a Steady State Free Precession (SSFP) sequence. The histological examination clearly show thermal acidophilic necrosis in the area treated with time reversal correction and normal pattern (neurons, oligodendrocytes, capillaries) in the area treated with a non-corrected HIFU beam

the targeted location using a Steady State Free Precession (SSFP) sequence. The histological examination also confirmed thermal acidophilic necrosis in the area treated with the time reversal correction. A normal pattern (neurons, oligodendrocytes, capillaries, i.e. no damage) was observed in the area treated with a non-corrected HIFU beam.

These promising results represent an important step towards human brain ultrasonic therapy. Beyond the thermal ablation application, the validation of this adaptive focusing technique guided by CT-scans will be applicable at lower levels of acoustic energy deposition for the ultrasonic targeting of cerebral gene therapy and sonothrombolysis.

## References


1. M. PERNOT, J.F. AUBRY, M. TANTER, J.L. THOMAS AND M. FINK, High power transcranial beam steering for ultrasonic brain therapy, Phys. Med. Biol., 48(16), 2577–2589, 2003.
2. M. FINK, Time-reversed acoustics, Scientific American, 281 (November), 1–97, 1999.
3. M. FINK, G. MONTALDO AND M. TANTER, Time-reversal acoustics in biomedical engineering, Annu. Rev. Biomed. Eng., 5, 465–497, 2003.
4. G. MONTALDO, P. ROUX, A. DERODE, C. NEGREIRA AND M. FINK, Ultrasound shock wave generator with one-bit time reversal in a dispersive medium, application to lithotripsy, Appl. Phys. Lett., 80 (5), 897–899, 2002.
5. M. TANTER, J.L. THOMAS AND M. FINK, Focusing and steering through absorbing and aberrating layers: Application to ultrasonic propagation through the skull, Journal of Acoustical Society of America, 103 (5), 2403–2410, 1998.
6. G. TER HAAR, Acoustic Surgery, Physics Today, 54 (12), 2001.
7. J.-L. THOMAS, F. WU AND M. FINK, Time reversal focusing applied to lithotripsy, Ultrasonic Imaging, 18, 106–121, 1996.
8. A. DERODE, P. ROUX AND M. FINK, Robust Acoustic time reversal with high order multiple scattering, Physical Review Letters, 75 (23), 4206–4209, 1995.
9. M. FINK, Time reversed Acoustics, Physics Today, 20, 34–40, 1997.
10. M. FINK, G. MONTALDO AND M. TANTER, Time reversal acoustics in biomedical engineering, Annual Review of Biomedical Engineering, 5, 2003.
11. J. DE ROSNY AND M. FINK, Overcoming the diffraction limit in wave physics using a time-reversal mirror and a novel acoustic sink, Physical Review Letters 89 (12), 124301, 2002.
12. M. TANTER, J.L. THOMAS AND M. FINK, Time reversal and the inverse filter, Journal of the Acoustical Society of America, 108 (1), 223–234, 2000.
13. M. TANTER, J.-F. AUBRY, J. GERBER, J.-L. THOMAS AND M. FINK, Optimal focusing by spatio-temporal inverse filter: Part I. Basic principles, Journal of the Acoustical Society of America, 101, 37–47, 2001.
14. J.-F. AUBRY, M. TANTER, J. GERBER, J.-L. THOMAS AND M. FINK, Optimal focusing by spatio-temporal inverse filter: Part II. Experiments, Journal of the Acoustical Society of America, 101, 48–58, 2001.
15. C. PRADA, F. WU AND M. FINK, The iterative time reversal mirror: A solution to self-focusing in the pulse echo mode, Journal of Acoustical Society of America, 90 (2), 1119–1129, 1991

16. G. MONTALDO, M. TANTER AND M. FINK, Revisiting iterative time reversal: real time detection of multiple targets, Journal of the Acoustical Society of America 115 (2), 776–784, 2004
17. C. PRADA AND M. FINK, Eigenmodes of the time reversal operator: a solution to selective focusing in multiple target media, Wave Motion, 20, 151–163, 1994.

# Chapter 3
# The Method of Small-Volume Expansions for Medical Imaging

**Habib Ammari and Hyeonbae Kang**

## 3.1 Introduction

Inverse problems in medical imaging are in their most general form ill-posed [47]. They literally have no solution. If, however, in advance we have additional structural information or supply missing information, then we may be able to determine specific features about what we wish to image with a satisfactory resolution and accuracy. One such type of information can be that the imaging problem is to find unknown small anomalies with significantly different parameters from those of the surrounding medium. These anomalies may represent potential tumors at early stage.

Over the last few years, the method of small-volume expansions has been developed for the imaging of such anomalies. The aim of this chapter is to provide a synthetic exposition of the method, a technique that has proven useful in dealing with many medical imaging problems. The method relies on deriving asymptotics. Such asymptotics have been investigated in the case of the conduction equation, the elasticity equation, the Helmholtz equation, the Maxwell system, the wave equation, the heat equation, and the Stokes system. A remarkable feature of this method is that it allows a stable and accurate reconstruction of the location and of some geometric features of the anomalies, even with moderately noisy data.

In this chapter we first provide asymptotic expansions for internal and boundary perturbations due to the presence of small anomalies. We then apply the asymptotic formulas for the purpose of identifying the location and certain properties of the shape of the anomalies. We shall restrict ourselves to conductivity and elasticity imaging and single out simple fundamental

H. Ammari (✉)
Laboratoire Ondes et Acoustique, 10 rue Vauquelin, CNRS, E.S.P.C.I, Université Paris 7, 75005, Paris, France,
e-mail: habib.ammari@espci.fr

H. Kang
Department of Mathematics, Inha University, Incheon 402-751, Korea
e-mail: hbkang@inha.ac.kr

H. Ammari, *Mathematical Modeling in Biomedical Imaging I*,
Lecture Notes in Mathematics 1983, DOI 10.1007/978-3-642-03444-2_3,

algorithms. We should emphasize that, since biological tissues are nearly incompressible, the model problem in elasticity imaging we shall deal with is the Stokes system rather than the Lamé system. The method of small-volume expansions also applies to the optical tomography and microwave imaging. However, these techniques are not discussed here. We refer the interested reader to, for instance, [2].

Applications of the method of small-volume expansions in medical imaging are described in this chapter. In particular, the use of the method of small-volume expansions to improve a multitude of emerging imaging techniques is highlighted. These imaging modalities include electrical impedance tomography (EIT), magnetic resonance elastography (MRE), impediography, magneto-acoustic imaging, infrared thermography, and acoustic radiation force imaging.

EIT uses low-frequency electrical current to probe a body; the method is sensitive to changes in electrical conductivity. By injecting known amounts of current and measuring the resulting electrical potential field at points on the boundary of the body, it is possible to "invert" such data to determine the conductivity or resistivity of the region of the body probed by the currents. This method can also be used in principle to image changes in dielectric constant at higher frequencies, which is why the method is often called "impedance" tomography rather than "conductivity" or "resistivity" tomography. However, the aspect of the method that is most fully developed to date is the imaging of conductivity/resistivity. Potential applications of electrical impedance tomography include determination of cardiac output, monitoring for pulmonary edema, and in particular screening for breast cancer. See, for instance, [35–37, 41, 44–46].

Recently, a commercial system called TransScan TS2000 (TransScan Medical, Ltd, Migdal Ha'Emek, Israel) has been released for adjunctive clinical uses with X-ray mammography in the diagnostic of breast cancer. The mathematical model of the TransScan can be viewed as a realistic or practical version of the general electrical impedance system. In the TransScan, a patient holds a metallic cylindrical reference electrode, through which a constant voltage of 1–2.5 V, with frequencies spanning 100 Hz–100 KHz, is applied. A scanning probe with a planar array of electrodes, kept at ground potential, is placed on the breast. The voltage difference between the hand and the probe induces a current flow through the breast, from which information about the impedance distribution in the breast can be extracted. See [25]. The method of small-volume expansions provides a rigorous mathematical framework for the TransScan. See Chap. 1 for a detailed study of this EIT system.

Since all the present EIT technologies are only practically applicable in feature extraction of anomalies, improving EIT calls for innovative measurement techniques that incorporate structural information. A very promising direction of research is the recent magnetic resonance imaging technique, called current density imaging, which measures the internal current density distribution. See the breakthrough work by Seo and his group described in Chap. 1. See also [52, 53]. However, this technique has a number of disadvantages, among which the lack of portability and a potentially long imaging time. Moreover, it uses an expensive magnetic resonance imaging scanner.

Impediography is another mathematical direction for future EIT research in view of biomedical applications. It keeps the most important merits of EIT (real time imaging, low cost, portability). It is based on the simultaneous measurement of an electric current and of acoustic vibrations induced by ultrasound waves. Its intrinsic resolution depends on the size of the focal spot of the acoustic perturbation, and thus it may provide high resolution images.

In magneto-acoustic imaging, an acoustic wave is applied to a biological tissue placed in a magnetic field. The probe signal produces by the Lorentz force an electric current that is a function of the local electrical conductivity of the biological tissue [59]. We provide the mathematical basis for this magneto-acoustic imaging approach and propose a new algorithm for solving the inverse problem which is quite similar to the one we design for impediography.

Extensive work has been carried out in the past decade to image, by inducing motion, the elastic properties of human soft tissues. This wide application field, called elasticity imaging or elastography, is based on the initial idea that shear elasticity can be correlated with the pathological state of tissues. Several imaging modalities can be used to estimate the resulting tissue displacements.

Magnetic resonance elastography is a recently developed technique that can directly visualize and quantitatively measure the displacement field in tissues subject to harmonic mechanical excitation at low-frequencies. A phase-contrast magnetic resonance imaging technique is used to spatially map and measure the complete three-dimensional displacement patterns. From this data, local quantitative values of shear modulus can be calculated and images that depict tissue elasticity or stiffness can be generated. The inverse problem for magnetic resonance elastography is to determine the shape and the elastic parameters of an elastic anomaly from internal measurements of the displacement field. In most cases the most significant elastic parameter is the stiffness coefficient. See, for instance, [42, 58, 60, 64, 65].

Another interesting approach to assessing elasticity is to use the acoustic radiation force of an ultrasonic focused beam to remotely generate mechanical vibrations in organs. The acoustic force is due to the momentum transfer from the acoustic wave to the medium. This technique is particularly suited for in vivo applications as it allows in depth vibrations of tissues exactly at the desired location. The radiation force acts as a dipolar source at the pushing ultrasonic beam focus. A spatio-temporal sequence of the propagation of the induced transient wave can be acquired, leading to a quantitative estimation of the viscoelastic parameters of the studied medium in a source-free region.

Infrared thermal imaging is becoming a common screening modality in the area of breast cancer. By carefully examining aspects of temperature and blood vessels of the breasts in thermal images, signs of possible cancer or pre-cancerous cell growth may be detected up to 10 years prior to being discovered using any other procedure. This provides the earliest detection of cancer possible. Because of thermal imaging's extreme sensitivity, these temperature variations and vascular changes may be among the earliest signs of breast cancer and/or a pre-cancerous state of the breast. An abnormal infrared image of the breast is an important marker of high risk for developing breast cancer.

## 3.2 Conductivity Problem

In this section we provide an asymptotic expansion of the voltage potentials in the presence of a diametrically small anomaly with conductivity different from the background conductivity.

Let $\Omega$ be a smooth bounded domain in $\mathbb{R}^d, d \geq 2$ and let $\nu_x$ denote the outward unit normal to $\partial\Omega$ at $x$. Define $N(x,z)$ to be the Neumann function for $-\Delta$ in $\Omega$ corresponding to a Dirac mass at $z$. That is, $N$ is the solution to

$$\begin{cases} -\Delta_x N(x,z) = \delta_z \quad \text{in } \Omega, \\ \displaystyle\frac{\partial N}{\partial \nu_x}\Big|_{\partial\Omega} = -\frac{1}{|\partial\Omega|}, \int_{\partial\Omega} N(x,z)\, d\sigma(x) = 0 \quad \text{for } z \in \Omega. \end{cases} \tag{3.1}$$

Note that the Neumann function $N(x,z)$ is defined as a function of $x \in \overline{\Omega}$ for each fixed $z \in \Omega$.

Let $B$ be a smooth bounded domain in $\mathbb{R}^d$, $0 < k \neq 1 < +\infty$, and let $\hat{v} = \hat{v}(B,k)$ be the solution to

$$\begin{cases} \Delta\hat{v} = 0 \quad \text{in } \mathbb{R}^d \setminus \overline{B}, \\ \Delta\hat{v} = 0 \quad \text{in } B, \\ \hat{v}|_- - \hat{v}|_+ = 0 \quad \text{on } \partial B, \\ k\displaystyle\frac{\partial\hat{v}}{\partial\nu}\Big|_- - \frac{\partial\hat{v}}{\partial\nu}\Big|_+ = 0 \quad \text{on } \partial B, \\ \hat{v}(\xi) - \xi \to 0 \quad \text{as } |\xi| \to +\infty. \end{cases} \tag{3.2}$$

Here we denote

$$v|_\pm(\xi) := \lim_{t\to 0^+} v(\xi \pm t\nu_\xi), \quad \xi \in \partial B,$$

and

$$\frac{\partial v}{\partial \nu_\xi}\Big|_\pm(\xi) := \lim_{t\to 0^+} \langle \nabla v(\xi \pm t\nu_\xi), \nu_\xi\rangle \ , \quad \xi \in \partial B,$$

if the limits exist, where $\nu_\xi$ is the outward unit normal to $\partial B$ at $\xi$, and $\langle , \rangle$ is the scalar product in $\mathbb{R}^d$. For ease of notation we will sometimes use the dot for the scalar product in $\mathbb{R}^d$.

Let $D$ denote a smooth anomaly inside $\Omega$ with conductivity $0 < k \neq 1 < +\infty$. The voltage potential in the presence of the set $D$ of conductivity anomalies is denoted by $u$. It is the solution to the conductivity problem

$$\begin{cases} \nabla \cdot \Big( \chi(\Omega \setminus \overline{D}) + k\chi(D) \Big) \nabla u = 0 \quad \text{in } \Omega, \\ \displaystyle \frac{\partial u}{\partial \nu}\bigg|_{\partial\Omega} = g \quad \Big( g \in L^2(\partial\Omega), \int_{\partial\Omega} g \, d\sigma = 0 \Big), \\ \displaystyle \int_{\partial\Omega} u \, d\sigma = 0 \, , \end{cases} \tag{3.3}$$

where $\chi(D)$ is the characteristic function of $D$.

The background voltage potential $U$ satisfies

$$\begin{cases} \Delta U = 0 \quad \text{in } \Omega \, , \\ \displaystyle \frac{\partial U}{\partial \nu}\bigg|_{\partial\Omega} = g \, , \\ \displaystyle \int_{\partial\Omega} U \, d\sigma = 0 \, . \end{cases} \tag{3.4}$$

The following theorem gives asymptotic formulas for both boundary and internal perturbations of the voltage potential that are due to the presence of a conductivity anomaly.

**Theorem 3.1 (Voltage perturbations).** *Suppose that $D = \delta B + z$, $\delta$ being the characteristic size of $D$, and let $u$ be the solution of (3.3), where $0 < k \neq 1 < +\infty$.*

(i) *The following asymptotic expansion of the voltage potential on $\partial\Omega$ holds for $d = 2, 3$:*

$$u(x) \approx U(x) - \delta^d \nabla U(z) M(k, B) \partial_z N(x, z). \tag{3.5}$$

*Here $U$ is the background solution, that is, the solution to (3.4), $N(x, z)$ is the Neumann function, that is, the solution to (3.1), and $M(k, B) = (m_{pq})_{p,q=1}^d$ is the polarization tensor (PT) given by*

$$M(k, B) := (k - 1) \int_B \nabla \hat{v}(\xi) \, d\xi, \tag{3.6}$$

*where $\hat{v}$ is the solution to (3.3).*

(ii) *Let $w$ be a smooth harmonic function in $\Omega$. The weighted boundary measurements $I_w$ satisfies*

$$I_w := \int_{\partial\Omega} (u - U)(x) \frac{\partial w}{\partial \nu}(x) \, d\sigma(x) \approx -\delta^d \nabla U(z) \cdot M(k, B) \nabla w(z). \tag{3.7}$$

(iii) *The following inner asymptotic formula holds:*

$$u(x) \approx U(z) + \delta \hat{v}(\frac{x-z}{\delta}) \cdot \nabla U(z) \quad \textit{for } x \textit{ near } z \, . \tag{3.8}$$

The inner asymptotic expansion (3.8) uniquely characterizes the shape and the conductivity of the anomaly. In fact, suppose for two Lipschitz domains $B$ and $B'$ and two conductivities $k$ and $k'$ that $\hat{v}(B,k) = \hat{v}(B',k')$ in a domain englobing $B$ and $B'$ then using the jump conditions satisfied by $\hat{v}(B,k)$ and $\hat{v}(B',k')$ we can easily prove that $B = B'$ and $k = k'$.

The asymptotic expansion (3.5) expresses the fact that the conductivity anomaly can be modeled by a dipole far away from $z$. It does not hold uniformly in $\Omega$. It shows that, from an imaging point of view, the location $z$ and the polarization tensor $M$ of the anomaly are the only quantities that can be determined from boundary measurements of the voltage potential, assuming that the noise level is of order $\delta^{d+1}$. It is then important to precisely characterize the polarization tensor and derive some of its properties, such as symmetry, positivity, and isoperimetric inequalities satisfied by its elements, in order to develop efficient algorithms for reconstructing conductivity anomalies of small volume.

We list in the next theorem important properties of the PT.

**Theorem 3.2 (Properties of the polarization tensor).** *For* $0 < k \neq 1 < +\infty$, *let* $M(k,B) = (m_{pq})_{p,q=1}^d$ *be the PT associated with the bounded domain* $B$ *in* $\mathbb{R}^d$ *and the conductivity* $k$. *Then*

(i) $M$ *is symmetric.*
(ii) *If* $k > 1$, *then* $M$ *is positive definite, and it is negative definite if* $0 < k < 1$.
(iii) *The following isoperimetric inequalities for the PT*

$$\begin{cases} \dfrac{1}{k-1} \operatorname{trace}(M) \leq (d-1+\dfrac{1}{k})|B|, \\ (k-1)\operatorname{trace}(M^{-1}) \leq \dfrac{d-1+k}{|B|}, \end{cases} \tag{3.9}$$

*hold, where* trace *denotes the trace of a matrix.*

The polarization tensor $M$ can be explicitly computed for disks and ellipses in the plane and balls and ellipsoids in three-dimensional space. See [18, pp. 81–89]. The formula of the PT for ellipses will be useful here. Let $B$ be an ellipse whose semi-axes are on the $x_1$- and $x_2$-axes and of length $a$ and $b$, respectively. Then, we recall that $M(k,B)$ takes the form

$$M(k,B) = (k-1)|B| \begin{pmatrix} \dfrac{a+b}{a+kb} & 0 \\ 0 & \dfrac{a+b}{b+ka} \end{pmatrix}, \tag{3.10}$$

where $|B|$ denotes the volume of $B$.

Formula (3.5) shows that from boundary measurements we can always represent and visualize an arbitrary shaped anomaly by means of an equivalent ellipse of center $z$ with the same polarization tensor. Further, it is impossible to extract the conductivity from the polarization tensor. The information contained in the polarization tensor is a mixture of the conductivity and the volume. A small anomaly with high conductivity and larger anomaly with lower conductivity can have the same polarization tensor.

The bounds (3.9) are known as the Hashin–Shtrikman bounds. By making use of these bounds, a size estimation of $B$ can be obtained.

## 3.3 Wave Equation

With the notation of Sect. 3.2, consider the initial boundary value problem for the (scalar) wave equation

$$\begin{cases} \partial_t^2 u - \nabla \cdot \Big( \chi(\Omega \setminus \overline{D}) + k\chi(D) \Big) \nabla u = 0 \quad \text{in} \quad \Omega_T, \\ u(x,0) = u_0(x), \quad \partial_t u(x,0) = u_1(x) \quad \text{for} \quad x \in \Omega\,, \\ \dfrac{\partial u}{\partial \nu} = g \quad \text{on} \quad \partial\Omega_T\,, \end{cases} \tag{3.11}$$

where $T < +\infty$ is a final observation time, $\Omega_T = \Omega \times ]0,T[, \partial\Omega_T = \partial\Omega \times ]0,T[$. The initial data $u_0, u_1 \in \mathcal{C}^\infty(\overline{\Omega})$, and the Neumann boundary data $g \in \mathcal{C}^\infty(0,T;\mathcal{C}^\infty(\partial\Omega))$ are subject to compatibility conditions.

Define the background solution $U$ to be the solution of the wave equation in the absence of any anomalies. Thus $U$ satisfies

$$\begin{cases} \partial_t^2 U - \Delta U = 0 \quad \text{in} \quad \Omega_T\,, \\ U(x,0) = u_0(x), \quad \partial_t U(x,0) = u_1(x) \quad \text{for} \quad x \in \Omega\,, \\ \dfrac{\partial U}{\partial \nu} = g \quad \text{on} \quad \partial\Omega_T. \end{cases}$$

The following asymptotic expansion holds as $\delta \to 0$.

**Theorem 3.3 (Perturbations of weighted boundary measurements).** *Set $\Omega_T = \Omega \times ]0,T[$ and $\partial\Omega_T = \partial\Omega \times ]0,T[$. Let $w \in C^\infty(\overline{\Omega}_T)$ satisfy $(\partial_t^2 - \Delta)w(x,t) = 0$ in $\Omega_T$ with $\partial_t w(x,T) = w(x,T) = 0$ for $x \in \Omega$. Define the weighted boundary measurements*

$$I_w(T) := \int_{\partial\Omega_T} (u-U)(x,t) \frac{\partial w}{\partial \nu}(x,t)\, d\sigma(x)\, dt\,.$$

*Then, for any fixed $T > diam(\Omega)$, the following asymptotic expansion for $I_w(T)$ holds as $\delta \to 0$:*

$$I_w(T) \approx \delta^d \int_0^T \nabla U(z,t) M(k,B) \nabla w(z,t)\, dt, \tag{3.12}$$

*where $M(k,B)$ is defined by (3.6).*

Expansion (3.12) is a weighted expansion. Pointwise expansions similar to those in Theorem 3.1 which is for the steady-state model can also be obtained.

Let $y \in \mathbb{R}^3$ be such that $|y - z| \gg \delta$. Choose

$$U(x,t) := U_y(x,t) := \frac{\delta_{t=|x-y|}}{4\pi|x-y|} \quad \text{for } x \neq y\,. \tag{3.13}$$

It is easy to check that $U_y$ is the outgoing Green function to the wave equation:

$$(\partial_t^2 - \Delta) U_y(x,t) = \delta_{x=y}\delta_{t=0} \quad \text{in } \mathbb{R}^3\times]0,+\infty[\,.$$

Moreover, $U_y$ satisfies the initial conditions: $U_y(x,0) = \partial_t U_y(x,0) = 0$ for $x \neq y$. Consider now for the sake of simplicity the wave equation in the whole three-dimensional space with appropriate initial conditions:

$$\begin{cases} \partial_t^2 u - \nabla \cdot \Big( \chi(\mathbb{R}^3 \setminus \overline{D}) + k\chi(D) \Big) \nabla u = \delta_{x=y}\delta_{t=0} & \text{in} \quad \mathbb{R}^3\times]0,+\infty[, \\ u(x,0) = 0, \quad \partial_t u(x,0) = 0 \quad \text{for} \quad x \in \mathbb{R}^3, x \neq y\,. \end{cases} \tag{3.14}$$

For $\rho > 0$, define the operator $P_\rho$ on tempered distributions by

$$P_\rho[\psi](t) = \int_{|\omega|\leq\rho} e^{-\sqrt{-1}\omega t} \hat{\psi}(\omega)\, d\omega,$$

where $\hat{\psi}$ denotes the Fourier transform of $\psi$. Clearly, the operator $P_\rho$ truncates the high-frequency component of $\psi$.

**Theorem 3.4 (Pointwise perturbations).** *Let $u$ be the solution to (3.14). Set $U_y$ to be the background solution. Suppose that $\rho = O(\delta^{-\alpha})$ for some $\alpha < \frac{1}{2}$.*

(i) *The following outer expansion holds*

$$P_\rho[u - U_y](x,t) \approx -\delta^3 \int_{\mathbb{R}} \nabla P_\rho[U_z](x, t-\tau) \cdot M(k,B) \nabla P_\rho[U_y](z,\tau)\, d\tau, \tag{3.15}$$

*for $x$ away from $z$, where $M(k,B)$ is defined by (3.6) and $U_y$ and $U_z$ by (3.13).*

(ii) *The following inner approximation holds:*

$$P_\rho[u - U_y](x,t) \approx \delta \hat{v}\left(\frac{x-z}{\delta}\right) \cdot \nabla P_\rho[U_y](x,t) \quad \text{for } x \text{ near } z\,, \tag{3.16}$$

*where $\hat{v}$ is given by (3.2) and $U_y$ by (3.13).*

Formula (3.15) shows that the perturbation due to the anomaly is in the time-domain a wavefront emitted by a dipolar source located at the point $z$.

Taking the Fourier transform of (3.15) in the time variable gives an expansion of the perturbations resulting from the presence of a small anomaly for solutions to the Helmholtz equation at low frequencies (at wavelengths large compared to the size of the anomaly).

## 3.4 Heat Equation

Suppose that the background $\Omega$ is homogeneous with thermal conductivity 1 and that the anomaly $D = \delta B + z$ has thermal conductivity $0 < k \neq 1 < +\infty$. In this section we consider the following transmission problem for the heat equation:

$$\begin{cases} \partial_t u - \nabla \cdot \Big(\chi(\Omega \setminus \overline{D}) + k\chi(D)\Big)\nabla u = 0 & \text{in} \quad \Omega_T, \\ u(x,0) = u_0(x) \quad \text{for} \quad x \in \Omega\,, \\ \dfrac{\partial u}{\partial \nu} = g \quad \text{on} \quad \partial\Omega_T\,, \end{cases} \tag{3.17}$$

where the Neumann boundary data $g$ and the initial data $u_0$ are subject to a compatibility condition. Let $U$ be the background solution defined as the solution of

$$\begin{cases} \partial_t U - \Delta U = 0 \quad \text{in} \quad \Omega_T\,, \\ U(x,0) = u_0(x) \quad \text{for} \quad x \in \Omega\,, \\ \dfrac{\partial U}{\partial \nu} = g \quad \text{on} \quad \partial\Omega_T. \end{cases}$$

The following asymptotic expansion holds as $\delta \to 0$.

**Theorem 3.5 (Perturbations of weighted boundary measurements).** *Let $w \in C^\infty(\overline{\Omega}_T)$ be a solution to the adjoint problem, namely, satisfy $(\partial_t + \Delta)w(x,t) = 0$ in $\Omega_T$ with $w(x,T) = 0$ for $x \in \Omega$. Define the weighted boundary measurements*

$$I_w(T) := \int_{\partial\Omega_T} (u - U)(x,t)\frac{\partial w}{\partial \nu}(x,t)\, d\sigma(x)\, dt\ .$$

*Then, for any fixed $T > 0$, the following asymptotic expansion for $I_w(T)$ holds as $\delta \to 0$:*

$$I_w(T) \approx -\delta^d \int_0^T \nabla U(z,t) \cdot M(k,B) \nabla w(z,t)\, dt, \tag{3.18}$$

*where $M(k,B)$ is defined by (3.6).*

Note that (3.18) holds for any fixed positive final time $T$ while (3.12) holds only for $T > \operatorname{diam}(\Omega)$. This difference comes from the finite speed propagation property for the wave equation compared to the infinite one for the heat equation.

Consider now the background solution to be the Green function of the heat equation at $y$:

$$U(x,t) := U_y(x,t) := \begin{cases} \dfrac{e^{-\frac{|x-y|^2}{4t}}}{(4\pi t)^{d/2}} & \text{for } t > 0, \\ 0 \quad \text{for } t < 0\,. \end{cases} \tag{3.19}$$

Let $u$ be the solution to the following heat equation with an appropriate initial condition:

$$\begin{cases} \partial_t u - \nabla \cdot \Big( \chi(\mathbb{R}^d \setminus \overline{D}) + k\chi(D) \Big) \nabla u = 0 \quad \text{in} \quad \mathbb{R}^d \times ]0, +\infty[, \\ u(x,0) = U_y(x,0) \quad \text{for} \quad x \in \mathbb{R}^d. \end{cases} \tag{3.20}$$

Proceeding as in the derivation of (3.15), we can prove that $\delta u(x,t) := u - U$ is approximated by

$$-(k-1) \int_0^t \frac{1}{(4\pi(t-\tau))^{d/2}} \int_{\partial D} e^{-\frac{|x-x'|^2}{4(t-\tau)}} \frac{\partial \hat{v}}{\partial \nu}\bigg|_- \Big(\frac{x'-z}{\delta}\Big) \cdot \nabla U_y(x',\tau)\, d\sigma(x')\, d\tau, \tag{3.21}$$

for $x$ near $z$. Therefore, analogously to Theorem 3.4, the following pointwise expansion follows from the approximation (3.21).

**Theorem 3.6 (Pointwise perturbations).** *Let $y \in \mathbb{R}^3$ be such that $|y - z| \gg \delta$. Let $u$ be the solution to (3.20). The following expansion holds*

$$(u-U)(x,t) \approx -\delta^d \int_0^t \nabla U_z(x, t-\tau) M(k,B) \nabla U_y(z,\tau)\, d\tau \text{ for } |x-z| \gg O(\delta)\,, \tag{3.22}$$

*where $M(k,B)$ is defined by (3.6) and $U_y$ and $U_z$ by (3.19).*

When comparing (3.22) and (3.15), we shall point out that for the heat equation the perturbation due to the anomaly is accumulated over time.

An asymptotic formalism for the realistic half space model for thermal imaging well suited for the design of anomaly reconstruction algorithms has been developed in [22].

## 3.5 Modified Stokes System

Consider the modified Stokes system, i.e., the problem of determining $\mathbf{v}$ and $q$ in a domain $\Omega$ from the conditions:

$$\begin{cases} (\Delta + \kappa^2)\mathbf{v} - \nabla q = 0, \\ \nabla \cdot \mathbf{v} = 0, \\ \mathbf{v}|_{\partial\Omega} = \mathbf{g} \, . \end{cases} \tag{3.23}$$

The problem (3.23) governs elastic wave propagation in nearly-incompressible media. In biological media, the compression modulus is 4–6 orders higher than the shear modulus. We can prove that the Lamé system converges to (3.23) as the compression modulus goes to $+\infty$.

Let $(G_{il})_{i,l=1}^d$ be the Dirichlet Green's function for the operator in (3.23), i.e., for $y \in \Omega$,

$$\begin{cases} (\Delta_x + \kappa^2) G_{il}(x,y) - \dfrac{\partial F_i(x-y)}{\partial x_l} = \delta_{il}\delta_y(x) \quad \text{in } \Omega \, , \\ \displaystyle\sum_{l=1}^d \frac{\partial}{\partial x_l} G_{il}(x,y) = 0 \quad \text{in } \Omega, \\ G_{il}(x,y) = 0 \quad \text{on } \partial\Omega \, . \end{cases} \tag{3.24}$$

Denote by $(\mathbf{e}_1, \ldots, \mathbf{e}_d)$ an orthonormal basis of $\mathbb{R}^d$. Let $d(\xi) := (1/d) \sum_k \xi_k \mathbf{e}_k$ and $\hat{\mathbf{v}}_{pq}$, for $p, q = 1, \ldots, d$, be the solution to

$$\begin{cases} \mu \Delta \hat{\mathbf{v}}_{pq} + \nabla \hat{p} = 0 \quad \text{in } \mathbb{R}^d \setminus \overline{B}, \\ \widetilde{\mu} \Delta \hat{\mathbf{v}}_{pq} + \nabla \hat{p} = 0 \quad \text{in } B, \\ \hat{\mathbf{v}}_{pq}|_- - \hat{\mathbf{v}}_{pq}|_+ = 0 \quad \text{on } \partial B, \\ (\hat{p}\mathbf{N} + \widetilde{\mu} \dfrac{\partial \hat{\mathbf{v}}_{pq}}{\partial \mathbf{N}})|_- - (\hat{p}\mathbf{N} + \mu \dfrac{\partial \hat{\mathbf{v}}_{pq}}{\partial \mathbf{N}})|_+ = 0 \quad \text{on } \partial B, \\ \nabla \cdot \hat{\mathbf{v}}_{pq} = 0 \quad \text{in } \mathbb{R}^d, \\ \hat{\mathbf{v}}_{pq}(\xi) \to \xi_p \mathbf{e}_q - \delta_{pq} d(\xi) \quad \text{as } |\xi| \to \infty, \\ \hat{p}(\xi) \to 0 \quad \text{as } |\xi| \to +\infty \, . \end{cases} \tag{3.25}$$

Here $\partial \mathbf{v}/\partial \mathbf{N} = (\nabla \mathbf{v} + (\nabla \mathbf{v})^T) \cdot \mathbf{N}$ and $T$ denotes the transpose.

Define the viscous moment tensor (VMT) $(V_{ijpq})_{i,j,p,q=1,\ldots,d}$ by

$$V_{ijpq} := (\widetilde{\mu} - \mu) \int_B \nabla \hat{\mathbf{v}}_{pq} \cdot (\nabla(\xi_i \mathbf{e}_j) + \nabla(\xi_i \mathbf{e}_j)^T)\, d\xi \,. \tag{3.26}$$

Consider an elastic anomaly $D$ inside a nearly-compressible medium $\Omega$. The anomaly $D$ has a shear modulus $\widetilde{\mu}$ different from that of $\Omega$, $\mu$. The displacement field $\mathbf{u}$ solves the following transmission problem for the modified Stokes problem:

$$\begin{cases} (\mu\Delta + \omega^2)\mathbf{u} + \nabla p = 0 \quad \text{in } \Omega \setminus \overline{D}, \\ (\widetilde{\mu}\Delta + \omega^2)\mathbf{u} + \nabla p = 0 \quad \text{in } D, \\ \mathbf{u}\big|_- = \mathbf{u}\big|_+ \quad \text{on } \partial D \,, \\ (p|_+ - p|_-)\mathbf{N} + \mu \dfrac{\partial \mathbf{u}}{\partial \mathbf{N}}\Big|_+ - \widetilde{\mu} \dfrac{\partial \mathbf{u}}{\partial \mathbf{N}}\Big|_- = 0 \quad \text{on } \partial D, \\ \nabla \cdot \mathbf{u} = 0 \quad \text{in } \Omega \,, \\ \mathbf{u} = \mathbf{g} \quad \text{on } \partial\Omega, \\ \displaystyle\int_\Omega p = 0 \,, \end{cases} \tag{3.27}$$

where $\mathbf{g} \in L^2(\partial\Omega)$ satisfies the compatibility condition $\displaystyle\int_{\partial\Omega} \mathbf{g} \cdot \mathbf{N} = 0$.

Let $(\mathbf{U}, q)$ denote the background solution to the modified Stokes system in the absence of any anomalies, that is, the solution to

$$\begin{cases} (\mu\Delta + \omega^2)\mathbf{U} + \nabla q = 0 \quad \text{in } \Omega, \\ \nabla \cdot \mathbf{U} = 0 \quad \text{in } \Omega \,, \\ \mathbf{U} = \mathbf{g} \quad \text{on } \partial\Omega, \\ \displaystyle\int_\Omega q = 0 \,. \end{cases} \tag{3.28}$$

The following asymptotic expansions hold.

**Theorem 3.7 (Expansions of the displacement field).** *Suppose that $D = \delta B + z$, and let $u$ be the solution of (3.27), where $0 < \widetilde{\mu} \neq \mu < +\infty$.*

(i) *The following inner expansion holds:*

$$\mathbf{u}(x) \approx \mathbf{U}(z) + \delta \sum_{p,q=1}^d \partial_q \mathbf{U}(z)_p \hat{\mathbf{v}}_{pq}(\frac{x-z}{\delta}) \quad \textit{for } x \textit{ near } z \,, \tag{3.29}$$

*where $\hat{\mathbf{v}}_{pq}$ is defined by (3.25)*

(ii) *Let* $(V_{ijpq})$ *be the VMT defined by (3.26). The following outer expansion holds uniformly for* $x \in \partial\Omega$*:*

$$(\mathbf{u} - \mathbf{U})(x) \approx \delta^d \left[ \sum_{i,j,p,q,\ell=1}^{d} \mathbf{e}_\ell \partial_j G_{\ell i}(x,z) \partial_q \mathbf{U}(z)_p V_{ijpq} \right], \tag{3.30}$$

*where* $V_{ijpq}$ *is given by (3.26), and the Green function* $(G_{il})_{i,l=1}^d$ *is defined by (3.24) with* $\kappa^2 = \omega^2/\mu$*,* $\mu$ *being the shear modulus of the background medium.*

The notion of a viscous moment tensor extends the notion of a polarization tensor to quasi-incompressible elasticity. The VMT $V$ characterizes all the information about the elastic anomaly that can be learned from the leading-order term of the outer expansion (3.30). It can be explicitly computed for disks and ellipses in the plane and balls and ellipsoids in three-dimensional space. If $B$ is a two dimensional disk, then

$$V = 4\,|B|\,\mu \frac{(\tilde{\mu} - \mu)}{\tilde{\mu} + \mu} P\,,$$

where $P = (P_{ijpq})$ is the orthogonal projection from the space of symmetric matrices onto the space of symmetric matrices of trace zero, i.e.,

$$P_{ijpq} = \frac{1}{2}(\delta_{ip}\delta_{jq} + \delta_{iq}\delta_{jp}) - \frac{1}{d}\delta_{ij}\delta_{pq}\,.$$

If $B$ is an ellipse of the form

$$\frac{x_1^2}{a^2} + \frac{x_2^2}{b^2} = 1, \quad a \geq b > 0\,, \tag{3.31}$$

then the VMT for $B$ is given by

$$\begin{cases} V_{1111} = V_{2222} = -V_{1122} = -V_{2211} = |B| \dfrac{2\mu(\tilde{\mu} - \mu)}{\mu + \tilde{\mu} - (\tilde{\mu} - \mu)m^2}, \\ V_{1212} = V_{2112} = V_{1221} = V_{2121} = |B| \dfrac{2\mu(\tilde{\mu} - \mu)}{\mu + \tilde{\mu} + (\tilde{\mu} - \mu)m^2}, \\ \text{the remaining terms are zero,} \end{cases} \tag{3.32}$$

where $m = (a - b)/(a + b)$.

If $B$ is a ball in three dimensions, the VMT associated with $B$ and an arbitrary $\tilde{\mu}$ is given by

$$\begin{cases} V_{iiii} = \dfrac{20\mu|B|}{3} \dfrac{\widetilde{\mu} - \mu}{2\widetilde{\mu} + 3\mu} , \quad V_{iijj} = -\dfrac{10\mu|B|}{3} \dfrac{\widetilde{\mu} - \mu}{2\widetilde{\mu} + 3\mu} \ (i \neq j) , \\ V_{ijij} = V_{ijji} = 5\mu|B| \dfrac{\widetilde{\mu} - \mu}{2\widetilde{\mu} + 3\mu}, \ (i \neq j) , \\ \text{the remaining terms are zero.} \end{cases} \tag{3.33}$$

**Theorem 3.8 (Properties of the viscous moment tensor).** *For $0 < \widetilde{\mu} \neq \mu < +\infty$, let $V = (V_{ijpq})_{i,p,q=1}^{d}$ be the VMT associated with the bounded domain $B$ in $\mathbb{R}^d$ and the pair of shear modulus $(\widetilde{\mu}, \mu)$. Then*

(i) *For $i, j, p, q = 1, \ldots, d$,*

$$V_{ijpq} = V_{jipq}, \quad V_{ijpq} = V_{ijqp}, \quad V_{ijpq} = V_{pqij} . \tag{3.34}$$

(ii) *We have*

$$\sum_p V_{ijpp} = 0 \quad \text{for all } i, j \quad \text{and} \quad \sum_i V_{iipq} = 0 \quad \text{for all } p, q ,$$

*or equivalently, $V = PVP$.*

(iii) *The tensor $V$ is positive (negative, resp.) definite on the space of symmetric matrices of trace zero if $\widetilde{\mu} > \mu$ ($\widetilde{\mu} < \mu$, resp.).*

(iv) *The tensor $(1/(2\mu))V$ satisfies the following bounds*

$$\mathrm{Tr}\Big(\frac{1}{2\mu}V\Big) \leq |B|\Big(\frac{\widetilde{\mu}}{\mu} - 1\Big)\Big((d-1)\frac{\mu}{\widetilde{\mu}} + \frac{d(d-1)}{2}\Big) , \tag{3.35}$$

$$\mathrm{Tr}\Big(\frac{1}{2\mu}V\Big)^{-1} \leq \frac{1}{|B|(\frac{\widetilde{\mu}}{\mu} - 1)}\Big((d-1)\frac{\widetilde{\mu}}{\mu} + \frac{d(d-1)}{2}\Big) , \tag{3.36}$$

*where for $C = (C_{ijpq})$, $\mathrm{Tr}(C) := \sum_{i,j=1}^{d} C_{ijij}$.*

Note that the VMT $V$ is a 4-tensor and can be regarded, because of its symmetry, as a linear transformation on the space of symmetric matrices. Note also that, in view of Theorem 3.2, the right-hand sides of (3.35) and (3.36) are exactly in the two-dimensional case ($d = 2$) the Hashin–Shtrikman bounds (3.9) for the PT associated with the same domain $B$ and the conductivity contrast $k = \widetilde{\mu}/\mu$.

## 3.6 Electrical Impedance Imaging

In this section we apply the asymptotic formula (3.5) for the purpose of identifying the location and certain properties of the shape of the conductivity anomalies. We single out two simple fundamental algorithms that take advantage of the smallness of the anomalies: projection-type algorithms and multiple signal classification (MUSIC)-type algorithms. These algorithms are fast, stable, and efficient.

We refer to Chap. 1 for basic mathematical and physical concepts of electrical impedance tomography.

### 3.6.1 Detection of a Single Anomaly: A Projection-Type Algorithm

We briefly discuss a simple algorithm for detecting a single anomaly. We refer to Chap. 1 for further details. The projection-type location search algorithm makes use of constant current sources. We want to apply a special type of current that makes $\partial U$ constant in $D$. Injection current $g = a \cdot \nu$ for a fixed unit vector $a \in \mathbb{R}^d$ yields $\nabla U = a$ in $\Omega$.

Assume for the sake of simplicity that $d = 2$ and $D$ is a disk. Set

$$w(y) = -(1/2\pi) \log|x - y| \quad \text{for } x \in \mathbb{R}^2 \setminus \overline{\Omega}, y \in \Omega \, .$$

Since $w$ is harmonic in $\Omega$, then from (3.10) and (3.7), it follows that

$$I_w[a] \approx \frac{(k-1)|D|}{\pi(k+1)} \frac{(x-z) \cdot a}{|x-z|^2} \, , \qquad x \in \mathbb{R}^2 \setminus \overline{\Omega} \, . \tag{3.37}$$

The first step for the reconstruction procedure is to locate the anomaly. The location search algorithm is as follows. Take two observation lines $\Sigma_1$ and $\Sigma_2$ contained in $\mathbb{R}^2 \setminus \overline{\Omega}$ given by

$$\Sigma_1 := \text{ a line parallel to } a,$$

$$\Sigma_2 := \text{ a line normal to } a \, .$$

Find two points $P_i \in \Sigma_i, i = 1, 2$, so that

$$I_w[a](P_1) = 0, \quad I_w[a](P_2) = \max_{x \in \Sigma_2} |I_w[a](x)| \, .$$

From (3.37), we can see that the intersecting point $P$ of the two lines

$$\Pi_1(P_1) := \{x \mid a \cdot (x - P_1) = 0\}, \tag{3.38}$$

$$\Pi_2(P_2) := \{x \mid (x - P_2) \text{ is parallel to } a\} \tag{3.39}$$

is close to the center $z$ of the anomaly $D$: $|P - z| = O(\delta^2)$.

Once we locate the anomaly, the factor $|D|(k-1)/(k+1)$ can be estimated. As we said before, this information is a mixture of the conductivity and the volume.

### 3.6.2 *Detection of Multiple Anomalies: A MUSIC-Type Algorithm*

Consider $m$ anomalies $D_s = \delta B_s + z_s$, $s = 1, \ldots, m$. Suppose for the sake of simplicity that all the domains $B_s$ are disks. Let $y_l \in \mathbb{R}^2 \setminus \Omega$ for $l = 1, \ldots, n$ denote the source points. Set

$$U_{y_l} = w_{y_l} := -(1/2\pi) \log |x - y_l| \quad \text{for } x \in \Omega, \quad l = 1, \ldots, n .$$

The MUSIC-type location search algorithm for detecting multiple anomalies is as follows. For $n \in \mathbb{N}$ sufficiently large, define the matrix $A = [A_{ll'}]_{l,l'=1}^n$ by

$$A_{ll'} = I_{w_{y_l}}[y_{l'}] := \int_{\partial\Omega} (u - U_{y_{l'}})(x) \frac{\partial w_{y_l}}{\partial \nu}(x) \, d\sigma(x) .$$

Expansion (3.7) yields

$$A_{ll'} \approx -\delta^d \sum_{s=1}^{m} \frac{2(k_s - 1)|B_s|}{k_s + 1} \nabla U_{y_{l'}}(z_s) \nabla U_{y_l}(z_s) .$$

Introduce

$$g_z = \Big( U_{y_1}(z), \ldots, U_{y_n}(z) \Big)^* ,$$

where $v^*$ denotes the transpose of the vector $v$.

**Lemma 3.9.** *Suppose that $n > m$. The following characterization of the location of the anomalies in terms of the range of the matrix $A$ holds:*

$$g_z \in \text{Range}(A) \ \textit{iff} \ z \in \{z_1, \ldots, z_m\} . \tag{3.40}$$

The MUSIC-type algorithm to determine the location of the anomalies is as follows. Let $P_{\text{noise}} = I - P$, where $P$ is the orthogonal projection onto the range of $A$. Given any point $z \in \Omega$, form the vector $g_z$. The point $z$ coincides with the location of an anomaly if and only if $P_{\text{noise}} g_z = 0$. Thus we can form an image of the anomalies by plotting, at each point $z$, the cost function $1/||P_{\text{noise}} g_z||$. The resulting plot will have large peaks at the locations of the anomalies.

Once we locate the anomalies, the factors $|D_s|(k_s - 1)/(k_s + 1)$ can be estimated from the significant singular values of $A$.

## 3.7 Impediography

The core idea of impediography is to couple electric measurements to localized elastic perturbations. A body (a domain $\Omega \subset \mathbb{R}^2$) is electrically probed: One or several currents are imposed on the surface and the induced potentials are measured on the boundary. At the same time, a circular region of a few millimeters in the interior of $\Omega$ is mechanically excited by ultrasonic waves, which dilate this region. The measurements are made as the focus of the ultrasounds scans the entire domain. Several sets of measurements can be obtained by varying amplitudes of the ultrasound waves and the applied currents.

Within each disk of (small) volume, the conductivity is assumed to be constant per volume unit. At a point $x \in \Omega$, within a disk $B$ of volume $V_B$, the electric conductivity $\gamma$ is defined in terms of a density $\rho$ as $\gamma(x) = \rho(x) V_B$.

The ultrasonic waves induce a small elastic deformation of the disk $B$. If this deformation is isotropic, the material points of $B$ occupy a volume $V_B^p$ in the perturbed configuration, which at first order is equal to

$$V_B^p = V_B(1 + 2\frac{\Delta r}{r}) ,$$

where $r$ is the radius of the disk $B$ and $\Delta r$ is the variation of the radius due to the elastic perturbation. As $\Delta r$ is proportional to the amplitude of the ultrasonic wave, we obtain a proportional change of the deformation. Using two different ultrasonic waves with different amplitudes but with the same spot, it is therefore easy to compute the ratio $V_B^p / V_B$. As a consequence, the perturbed electrical conductivity $\gamma^p$ satisfies

$$\forall\, x \in \Omega, \quad \gamma^p(x) = \rho(x) V_B^p \; = \; \gamma(x)\nu(x) ,$$

where $\nu(x) = V_B^p / V_B$ is a known function. We make the following realistic assumptions: (1) the ultrasonic wave expands the zone it impacts, and changes its conductivity: $\forall x \in \Omega$, $\nu(x) > 1$, and (2) the perturbation is not too small: $\nu(x) - 1 \gg V_B$.

### 3.7.1 A Mathematical Model

Let $u$ be the voltage potential induced by a current $g$, in the absence of ultrasonic perturbations. It is given by

$$\begin{cases} \nabla_x \cdot (\gamma(x) \nabla_x u) = 0 & \text{in } \Omega, \\ \gamma(x) \dfrac{\partial u}{\partial \nu} = g & \text{on } \partial\Omega , \end{cases} \tag{3.41}$$

with the convention that $\int_{\partial\Omega} u = 0$. We suppose that the conductivity $\gamma$ of the region close to the boundary of the domain is known, so that ultrasonic probing is limited to interior points. We denote the region (open set) by $\Omega_1$.

Let $u_\delta$ be the voltage potential induced by a current $g$, in the presence of ultrasonic perturbations localized in a disk-shaped domain $D := z + \delta B$ of volume $|D| = O(\delta^2)$. The voltage potential $u_\delta$ is a solution to

$$\begin{cases} \nabla_x \cdot (\gamma_\delta(x) \nabla_x u_\delta(x)) = 0 & \text{in } \Omega\,, \\ \gamma(x) \dfrac{\partial u_\delta}{\partial \nu} = g & \text{on } \partial\Omega\,, \end{cases} \tag{3.42}$$

with the notation

$$\gamma_\delta(x) = \gamma(x) \Big[ 1 + \chi(D)(x)\,(\nu(x) - 1) \Big]\,,$$

where $\chi(D)$ is the characteristic function of the domain $D$.

As the zone deformed by the ultrasound wave is small, we can view it as a small volume perturbation of the background conductivity $\gamma$, and seek an asymptotic expansion of the boundary values of $u_\delta - u$. The method of small-volume expansions shows that comparing $u_\delta$ and $u$ on $\partial\Omega$ provides information about the conductivity. Indeed, we can prove that

$$\begin{aligned} \int_{\partial\Omega} (u_\delta - u) g \, d\sigma &= \int_D \gamma(x) \frac{(\nu(x)-1)^2}{\nu(x)+1} \nabla u \cdot \nabla u \, dx + o(|D|) \\ &= |\nabla u(z)|^2 \int_D \gamma(x) \frac{(\nu(x)-1)^2}{\nu(x)+1} \, dx + o(|D|)\,. \end{aligned}$$

Therefore, we have

$$\gamma(z) \, |\nabla u(z)|^2 = \mathcal{E}(z) + o(1)\,, \tag{3.43}$$

where the function $\mathcal{E}(z)$ is defined by

$$\mathcal{E}(z) = \left( \int_D \frac{(\nu(x)-1)^2}{\nu(x)+1} dx \right)^{-1} \int_{\partial\Omega} (u_\delta - u) g \, d\sigma\,. \tag{3.44}$$

By scanning the interior of the body with ultrasound waves, given an applied current $g$, we then obtain data from which we can compute

$$\mathcal{E}(z) := \gamma(z) |\nabla u(z)|^2$$

in an interior sub-region of $\Omega$. The new inverse problem is now to reconstruct $\gamma$ knowing $\mathcal{E}$.

### 3.7.2 A Substitution Algorithm

The use of $\mathcal{E}$ leads us to transform (3.41), having two unknowns $\gamma$ and $u$ with highly nonlinear dependency on $\gamma$, into the following nonlinear PDE (the 0-Laplacian)

$$\begin{cases} \nabla_x \cdot \left( \dfrac{\mathcal{E}}{|\nabla u|^2} \nabla u \right) = 0 & \text{in } \Omega \, , \\ \dfrac{\mathcal{E}}{|\nabla u|^2} \dfrac{\partial u}{\partial \nu} = g & \text{on } \partial \Omega \, . \end{cases} \tag{3.45}$$

We emphasize that $\mathcal{E}$ is a known function, constructed from the measured data (3.44). Consequently, all the parameters entering in (3.45) are known. So, the ill-posed inverse problem of EIT model is converted into less complicated direct problem (3.45).

The E-substitution algorithm, which will be explained below, uses two currents $g_1$ and $g_2$. We choose this pair of current patterns to have $\nabla u_1 \times \nabla u_2 \neq 0$ for all $x \in \Omega$, where $u_i, i = 1, 2$, is the solution to (3.41). We refer to Chap. 1 and the references therein for an evidence of the possibility of such a choice. The E-substitution algorithm is based on an approximation of a linearized version of problem (3.45).

Suppose that $\gamma$ is a small perturbation of conductivity profile $\gamma_0$: $\gamma = \gamma_0 + \delta\gamma$. Let $u_0$ and $u = u_0 + \delta u$ denote the potentials corresponding to $\gamma_0$ and $\gamma$ with the same Neumann boundary data $g$. It is easily seen that $\delta u$ satisfies $\nabla \cdot (\gamma \nabla \delta u) = -\nabla \cdot (\delta\gamma \nabla u_0)$ in $\Omega$ with the homogeneous Dirichlet boundary condition. Moreover, from

$$\mathcal{E} = (\gamma_0 + \delta\gamma)|\nabla(u_0 + \delta u)|^2 \approx \gamma_0 |\nabla u_0|^2 + \delta\gamma |\nabla u_0|^2 + 2\gamma_0 \nabla u_0 \cdot \nabla \delta u \, ,$$

after neglecting the terms $\delta\gamma \nabla u_0 \cdot \nabla \delta u$ and $\delta\gamma |\nabla \delta u|^2$, it follows that

$$\delta\gamma \approx \frac{\mathcal{E}}{|\nabla u_0|^2} - \gamma_0 - 2\gamma_0 \frac{\nabla \delta u \cdot \nabla u_0}{|\nabla u_0|^2} \, .$$

The E-substitution algorithm is as follows. We start from an initial guess for the conductivity $\gamma$, and solve the corresponding Dirichlet conductivity problem

$$\begin{cases} \nabla \cdot (\gamma \nabla u_0) = 0 \text{ in } \Omega, \\ u_0 = \psi \text{ on } \partial\Omega \, . \end{cases}$$

The data $\psi$ is the Dirichlet data measured as a response to the current $g$ (say $g = g_1$) in absence of elastic deformation. The discrepancy between the data and our guessed solution is

$$\epsilon_0 := \frac{\mathcal{E}}{|\nabla u_0|^2} - \gamma \, . \tag{3.46}$$

We then introduce a corrector, $\delta u$, computed as the solution to

$$\begin{cases} \nabla \cdot (\gamma \nabla \delta u) = -\nabla \cdot (\varepsilon_0 \nabla u_0) & \text{in } \Omega, \\ \delta u = 0 & \text{on } \partial\Omega, \end{cases}$$

and update the conductivity

$$\gamma := \frac{\mathcal{E} - 2\gamma \nabla \delta u \cdot \nabla u_0}{|\nabla u_0|^2}.$$

We iteratively update the conductivity, alternating directions of currents (i.e., with $g = g_2$).

In the case of incomplete data, that is, if $\mathcal{E}$ is only known on a subset $\omega$ of the domain, we can follow an optimal control approach. See [29].

## 3.8 Magneto-Acoustic Imaging

Denote by $\gamma(x)$ the unknown conductivity and let the voltage potential $v$ be the solution to the conductivity problem (3.41). Suppose that the $\gamma$ is a known constant on a neighborhood of the boundary $\partial\Omega$ and let $\gamma_*$ denote $\gamma|_{\partial\Omega}$.

In magneto-acoustic imaging, ultrasonic waves are focused on regions of small diameter inside a body placed on a static magnetic field. The oscillation of each small region results in frictional forces being applied to the ions, making them move. In the presence of a magnetic field, the ions experience Lorentz force. This gives rise to a localized current density within the medium. The current density is proportional to the local electrical conductivity [59]. In practice, the ultrasounds impact a spherical or ellipsoidal zone, of a few millimeters in diameter. The induced current density should thus be sensitive to conductivity variations at the millimeter scale, which is the precision required for breast cancer diagnostic. The feasibility of this conductivity imaging technique has been demonstrated in [43].

Let $z \in \Omega$ and $D$ be a small impact zone around the point $z$. The created current by the Lorentz force density is given by

$$\mathbf{J}_z(x) = c\chi_D(x)\gamma(x)\mathbf{e}, \tag{3.47}$$

for some constant $c$ and a constant unit vector $\mathbf{e}$ both of which are independent of $z$. Here, $\chi_D$ denotes the characteristic function of $D$. With the induced current $\mathbf{J}_z$ the new voltage potential, denoted by $u_z$, satisfies

$$\begin{cases} \nabla \cdot (\gamma \nabla u_z + \mathbf{J}_z) = 0 & \text{in } \Omega, \\ u_z = g & \text{on } \partial\Omega. \end{cases}$$

According to (3.47), the induced electrical potential $w_z := v - u_z$ satisfies the conductivity equation:

$$\begin{cases} \nabla \cdot \gamma \nabla w_z = c \nabla \cdot (\chi_D \gamma \mathbf{e}) & \text{for } x \in \Omega, \\ w_z(x) = 0 & \text{for } x \in \partial\Omega. \end{cases} \tag{3.48}$$

The inverse problem for the vibration potential tomography, which is a synonym of Magneto-Acoustic Imaging, is to reconstruct the conductivity profile $\gamma$ from boundary measurements of $\frac{\partial u_z}{\partial \nu}|_{\partial\Omega}$ or equivalently $\frac{\partial w_z}{\partial \nu}|_{\partial\Omega}$ for $z \in \Omega$.

Since $\gamma$ is assumed to be constant in $D$ and $|D|$ is small, we obtain using Green's identity [5]

$$\int_{\partial\Omega} \gamma_* \frac{\partial w_z}{\partial \nu} g d\sigma \approx -c|D| \nabla(\gamma v)(z) \cdot \mathbf{e}. \tag{3.49}$$

The relation (3.49) shows that, by scanning the interior of the body with ultrasound waves, $c\nabla(\gamma v)(z) \cdot \mathbf{e}$ can be computed from the boundary measurements $\frac{\partial w_z}{\partial \nu}|_{\partial\Omega}$ in $\Omega$. If we can rotate the subject, then $c\nabla(\gamma v)(z)$ for any $z$ in $\Omega$ can be reconstructed. In practice, the constant $c$ is not known. But, since $\gamma v$ and $\partial(\gamma v)/\partial \nu$ on the boundary of $\Omega$ are known, we can recover $c$ and $\gamma v$ from $c\nabla(\gamma v)$ in a constructive way. See [5].

The new inverse problem is now to reconstruct the contrast profile $\gamma$ knowing

$$\mathcal{E}(z) := \gamma(z) v(z) \tag{3.50}$$

for a given boundary potential $g$, where $v$ is the solution to (3.41).

In view of (3.50), $v$ satisfies

$$\begin{cases} \nabla \cdot \dfrac{\mathcal{E}}{v} \nabla v = 0 & \text{in } \Omega, \\ v = g & \text{on } \partial\Omega. \end{cases} \tag{3.51}$$

If we solve (3.51) for $v$, then (3.50) yields the conductivity contrast $\gamma$. Note that to be able to solve (3.51) we need to know the coefficient $\mathcal{E}(z)$ for all $z$, which amounts to scanning all the points $z \in \Omega$ by the ultrasonic beam.

Observe that solving (3.51) is quite easy mathematically: If we put $w = \ln v$, then $w$ is the solution to

$$\begin{cases} \nabla \cdot \mathcal{E} \nabla w = 0 & \text{in } \Omega, \\ w = \ln g & \text{on } \partial\Omega, \end{cases} \tag{3.52}$$

as long as $g > 0$. Thus if we solve (3.52) for $w$, then $v := e^w$ is the solution to (3.51). However, taking an exponent may amplify the error which already exists in the computed data $\mathcal{E}$. In order to avoid this numerical instability, we solve (3.51) iteratively. To do so, we can adopt an iterative scheme similar to the one proposed in the previous section.

Start with $\gamma_0$ and let $v_0$ be the solution of

$$\begin{cases} \nabla \cdot \gamma_0 \nabla v_0 = 0 & \text{in } \Omega, \\ v_0 = g & \text{on } \partial\Omega. \end{cases} \tag{3.53}$$

According to (3.50), our updates, $\gamma_0 + \delta\gamma$ and $v_0 + \delta v$, should satisfy

$$\gamma_0 + \delta\gamma = \frac{\mathcal{E}}{v_0 + \delta v}, \tag{3.54}$$

where

$$\begin{cases} \nabla \cdot (\gamma_0 + \delta\gamma)\nabla(v_0 + \delta v) = 0 & \text{in } \Omega, \\ \delta v = 0 & \text{on } \partial\Omega, \end{cases}$$

or equivalently

$$\begin{cases} \nabla \cdot \gamma_0 \nabla \delta v + \nabla \cdot \delta\gamma \nabla v_0 = 0 & \text{in } \Omega, \\ \delta v = 0 & \text{on } \partial\Omega. \end{cases} \tag{3.55}$$

We then linearize (3.54) to have

$$\gamma_0 + \delta\gamma = \frac{\mathcal{E}}{v_0(1 + \delta v / v_0)} \approx \frac{\mathcal{E}}{v_0}\left(1 - \frac{\delta v}{v_0}\right). \tag{3.56}$$

Thus

$$\delta\gamma = -\frac{\mathcal{E}\delta v}{v_0^2} - \delta, \quad \delta = -\frac{\mathcal{E}}{v_0} + \gamma_0. \tag{3.57}$$

We then find $\delta v$ by solving

$$\begin{cases} \nabla \cdot \gamma_0 \nabla \delta v - \nabla \cdot \left(\frac{\mathcal{E}\delta v}{v_0^2} + \delta\right) \nabla v_0 = 0 & \text{in } \Omega, \\ \delta v = 0 & \text{on } \partial\Omega, \end{cases}$$

or equivalently

$$\begin{cases} \nabla \cdot \gamma_0 \nabla \delta v - \nabla \cdot \left(\frac{\mathcal{E}\nabla v_0}{v_0^2} \delta v\right) = \nabla \cdot \delta \nabla v_0 & \text{in } \Omega, \\ \delta v = 0 & \text{on } \partial\Omega. \end{cases} \tag{3.58}$$

In the case of incomplete data, that is, if $\mathcal{E}$ is only known on a subset $\omega$ of the domain, we can follow an optimal control approach. See [5].

## 3.9 Magnetic Resonance Elastography

Let $\mathbf{u}$ be the solution to the modified Stokes system (3.27). The inverse problem in the magnetic resonance elastography is to reconstruct the shape and the shear modulus of the anomaly $D$ from internal measurements of $\mathbf{u}$.

Based on the inner asymptotic expansion (3.29) of $\delta\mathbf{u}$ ($:= \mathbf{u} - \mathbf{U}$) of the perturbations in the displacement field that are due to the presence of the anomaly, a reconstruction method of binary level set type can be designed.

The first step for the reconstruction procedure is to locate the anomaly. This can be done using the outer expansion of $\delta\mathbf{u}$, i.e., an expansion far away from the elastic anomaly.

Suppose that $z$ is reconstructed. Since the representation $D = z + \delta B$ is not unique, we can fix $\delta$. We use a binary level set representation $f$ of the scaled domain $B$:

$$f(x) = \begin{cases} 1, & x \in B, \\ -1, & x \in \mathbb{R}^3 \setminus \overline{B}. \end{cases} \tag{3.59}$$

Let

$$2h(x) = \widetilde{\mu}\left(f(\frac{x-z}{\delta}) + 1\right) - \mu\left(f(\frac{x-z}{\delta}) - 1\right), \tag{3.60}$$

and let $\beta$ be a regularization parameter. Then the second step is to fix a window $W$ (for instance a sphere containing $z$) and solve the following constrained minimization problem

$$\begin{aligned} \min_{\widetilde{\mu}, f} L(f, \widetilde{\mu}) &= \frac{1}{2}\left\| \delta\mathbf{u}(x) - \delta \sum_{p,q=1}^{d} \partial_q \mathbf{U}(z)_p \hat{\mathbf{v}}_{pq}(\frac{x-z}{\delta}) + \nabla\mathbf{U}(z)(x-z) \right\|^2_{L^2(W)} \\ &\quad + \beta \int_W |\nabla h(x)| \, dx \,, \end{aligned} \tag{3.61}$$

subject to (3.25). Here, $\int_W |\nabla h| \, dx$ is the total variation of the shear modulus and $|\nabla h|$ is understood as a measure:

$$\int_W |\nabla h| = \sup \left\{ \int_W h \nabla \cdot \mathbf{v} \, dx, \quad \mathbf{v} \in \mathcal{C}^1_0(W) \text{ and } |\mathbf{v}| \leq 1 \text{ in } W \right\}.$$

This regularization indirectly controls both the length of the level curves and the jumps in the coefficients.

The local character of the method is due to the decay of

$$\delta \sum_{p,q=1}^{d} \partial_q \mathbf{U}(z)_p \hat{\mathbf{v}}_{pq}(\frac{\cdot - z}{\delta}) - \nabla\mathbf{U}(z)(\cdot - z)$$

away from $z$. This is one of the main features of the method. In the presence of noise, because of a trade-off between accuracy and stability, we have to choose carefully the size of $W$. The size of $W$ should not be so small to preserve some stability and not so big so that we can gain some accuracy. See [8].

The minimization problem (3.61) corresponds to a minimization with respect to $\widetilde{\mu}$ followed by a step of minimization with respect to $f$. The minimization steps are over the set of $\widetilde{\mu}$ and $f$, and can be performed using a gradient based method with a line search. Of importance to us are the optimal bounds satisfied by the viscous moment tensor $V$. We should check for each step whether the bounds (3.35) and (3.36) on $V$ are satisfied or not. In the case they are not, we have to restate the value of $\widetilde{\mu}$. Another way to deal with (3.35) and (3.36) is to introduce them into the minimization problem (3.61) as a constraint. Set $\alpha = \mathrm{Tr}(V)$ and $\beta = \mathrm{Tr}(V^{-1})$ and suppose for simplicity that $\widetilde{\mu} > \mu$. Then, (3.35) and (3.36) can be rewritten (when $d = 3$) as follows

$$\begin{cases} \alpha \leq 2(\widetilde{\mu} - \mu)(3 + \dfrac{2\mu}{\widetilde{\mu}})|D|, \\ \dfrac{2\mu(\widetilde{\mu} - \mu)}{3\mu + 2\widetilde{\mu}}|D| \leq \beta^{-1} . \end{cases} \tag{3.62}$$

## 3.10 Imaging by the Acoustic Radiation Force

A model problem for the acoustic radiation force imaging is (3.14), where $y$ is the location of the pushing ultrasonic beam. The transient wave $u(x,t)$ is the induced wave. The inverse problem is to reconstruct the shape and the conductivity of the small anomaly $D$ from measurements of $u$ on $\mathbb{R}^3 \times ]0, +\infty[$. It is easy to detect $T = |y - z|$ and the location $z$ of the anomaly from measurements of $u(x,t) - U_y(x,t)$.

Suppose that the wavefield in a window $W$ containing the anomaly can be acquired. In view of Theorem 3.4, the shape and the conductivity of $D$ can be approximately reconstructed, analogously to MRE, by minimizing over $k$ and $f$ the following functional:

$$L(f,k) = \frac{1}{2\Delta T} \int_{T-\frac{\Delta T}{2}}^{T+\frac{\Delta T}{2}} \left\| P_\rho[u - U_y] - \delta \hat{v}\left(\frac{x-z}{\delta}\right) \cdot \nabla P_\rho[U_y] \right\|_{L^2(W)}^2 dt$$
$$+\beta \int_W |\nabla h(x)|\, dx \,, \tag{3.63}$$

subject to (3.2). Here $\Delta T = O(\delta/\sqrt{k})$ is small, $2h(x) = k(f(\frac{x-z}{\delta}) + 1) - (f(\frac{x-z}{\delta}) - 1)$, and $f$ given by (3.59).

To detect the anomaly from measurements of the wavefield away from the anomaly one can use a time-reversal technique. As shown in Chap. 2, the main idea of time-reversal is to take advantage of the reversibility of the wave equation in a non-dissipative unknown medium in order to back-propagate signals to the sources that emitted them. In the context of anomaly detection, one measures the perturbation of the wave on a closed surface surrounding the anomaly, and retransmits it through the background medium in a time-reversed chronology. Then the perturbation will travel back to the location of the anomaly. We can show that the time-reversal perturbation focuses on the location $z$ of the anomaly with a focal spot size limited to one-half the wavelength which is in agreement with the Rayleigh resolution limit.

In mathematical words, suppose that we are able to measure the perturbation $w := u - U_y$ and its normal derivative at any point $x$ on a sphere $S$ englobing the anomaly $D$. The time-reversal operation is described by the transform $t \mapsto t_0 - t$. Both the perturbation $w$ and its normal derivative on $S$ are time-reversed and emitted from $S$. Then a time-reversed perturbation, denoted by $w_{\mathrm{tr}}$, propagates inside the volume $\Omega$ surrounded by $S$. Taking into account the definition of the outgoing fundamental solution (3.13) to the wave equation, spatial reciprocity and time reversal invariance of the wave equation, the time-reversed perturbation $w_{\mathrm{tr}}$ due to the anomaly $D$ in $\Omega$ should be defined by

$$w_{\mathrm{tr}}(x,t) = \int_{\mathbb{R}} \int_S \left[ U_x(x',t-s)\frac{\partial w}{\partial \nu}(x',t_0-s) - \frac{\partial U_x}{\partial \nu}(x',t-s) w(x',t_0-s) \right] d\sigma(x')\, ds\,,$$

where

$$U_x(x',t-\tau) = \frac{\delta(t-\tau-|x-x'|)}{4\pi|x-x'|}\,.$$

However, with the high frequency component of $w$ truncated as in Theorem 3.4, we take the following definition:

$$\begin{aligned} w_{\mathrm{tr}}(x,t) = \int_{\mathbb{R}} \int_S & \left[ U_x(x',t-s)\frac{\partial P_\rho[u-U_{\bar{y}}]}{\partial \nu}(x',t_0-s) \right. \\ & \left. - \frac{\partial U_x}{\partial \nu}(x',t-s) P_\rho[u-U_{\bar{y}}](x',t_0-s) \right] d\sigma(x')\,. \end{aligned}$$

According to Theorem 3.4, we have

$$P_\rho[u-U_y](x,t) \approx -\delta^3 \int_{\mathbb{R}} \nabla P_\rho[U_z](x,t-\tau)\cdot p(z,\tau)\, d\tau\,,$$

where

$$p(z,\tau) = M(k,B)\nabla P_\rho[U_y](z,\tau)\,.$$

Thus, since

$$\int_{\mathbb{R}} \int_S \left[ U_x(x', t-s) \frac{\partial P_\rho[U_z]}{\partial \nu}(x', t_0 - s - \tau) \right. \\ \left. - \frac{\partial U_x}{\partial \nu}(x', t-s) P_\rho[U_z](x', t_0 - s - \tau) \right] d\sigma(x')\, ds \\ = P_\rho[U_z](x, t_0 - \tau - t) - P_\rho[U_z](x, t - t_0 + \tau)\,,$$

we arrive at

$$w_{\mathrm{tr}}(x,t) \approx -\delta^3 \int_{\mathbb{R}} p(z,\tau) \cdot \nabla_z \big[ P_\rho[U_z](x, t_0 - \tau - t) - P_\rho[U_z](x, t - t_0 + \tau) \big] d\tau\,.$$

Formula (3.64) can be interpreted as the superposition of incoming and outgoing waves, centered on the location $z$ of the anomaly. Suppose that $p(z,\tau)$ is concentrated at the travel time $\tau = T$. Formula (3.64) takes therefore the form

$$w_{\mathrm{tr}}(x,t) \approx -\delta^3 p \cdot \nabla_z \big[ P_\rho[U_z](x, t_0 - T - t) - P_\rho[U_z](x, t - t_0 + T) \big]\,,$$

where $p = p(z,T)$. The wave $w_{\mathrm{tr}}$ is clearly sum of incoming and outgoing spherical waves.

Formula (3.64) has an important physical interpretation. By changing the origin of time, $T$ can be set to 0 without loss of generality. By taking Fourier transform of (3.64) over the time variable $t$, we obtain that

$$\hat{w}_{\mathrm{tr}}(x,\omega) \propto \delta^3 p \cdot \nabla \Big( \frac{\sin(\omega |x-z|)}{|x-z|} \Big)\,,$$

where $\hat{w}_{\mathrm{tr}}$ denotes the Fourier transform of $w_{\mathrm{tr}}$ and $\omega$ is the wavenumber and, which shows that the time-reversal perturbation $w_{\mathrm{tr}}$ focuses on the location $z$ of the anomaly with a focal spot size limited to one-half the wavelength. An identity parallel to (3.64) can be rigorously derived in the frequency domain. It plays a key role in the resolution limit analysis. See [7].

## 3.11 Infrared Thermal Imaging

In this section we apply (3.18) (with an appropriate choice of test functions $w$ and background solutions $U$) for the purpose of identifying the location of the anomaly $D$. The first algorithm makes use of constant heat flux and, not surprisingly, it is limited in its ability to effectively locate multiple anomalies.

Using many heat sources, we then describe an efficient method to locate multiple anomalies and illustrate its feasibility. For the sake of simplicity we consider only the two-dimensional case.

### 3.11.1 Detection of a Single Anomaly

For $y \in \mathbb{R}^2 \setminus \overline{\Omega}$, let

$$w(x,t) = w_y(x,t) := \frac{1}{4\pi(T-t)} e^{-\frac{|x-y|^2}{4(T-t)}} . \tag{3.64}$$

The function $w$ satisfies $(\partial_t + \Delta)w = 0$ in $\Omega_T$ and the final condition $w|_{t=T} = 0$ in $\Omega$.

Suppose that there is only one anomaly $D = z + \delta B$ with thermal conductivity $k$. For simplicity assume that $B$ is a disk. Choose the background solution $U(x,t)$ to be a harmonic (time-independent) function in $\Omega_T$. We compute

$$\nabla w_y(z,t) = \frac{y-z}{8\pi(T-t)^2} e^{-\frac{|z-y|^2}{4(T-t)}} ,$$

$$M(k,B)\nabla w_y(z,t) = \frac{(k-1)|B|}{k+1} \frac{y-z}{4\pi(T-t)^2} e^{-\frac{|z-y|^2}{4(T-t)}} ,$$

and

$$\int_0^T M(k,B)\nabla w_y(z,t)\, dt = \frac{(k-1)|B|}{k+1} \frac{y-z}{4\pi} \int_0^T \frac{e^{-\frac{|z-y|^2}{4(T-t)}}}{(T-t)^2}\, dt .$$

But

$$\frac{d}{dt} e^{-\frac{|z-y|^2}{4(T-t)}} = \frac{-|z-y|^2}{4} \frac{e^{-\frac{|z-y|^2}{4(T-t)}}}{(T-t)^2}$$

and therefore

$$\int_0^T M(k,B)\nabla w_y(z,t)\, dt = \frac{(k-1)|B|}{k+1} \frac{y-z}{\pi|z-y|^2} e^{-\frac{|z-y|^2}{4(T-t)}} .$$

Then the asymptotic expansion (3.18) yields

$$I_w(T)(y) \approx \delta^2 \frac{k-1}{k+1} |B| \frac{\nabla U(z)\cdot(y-z)}{\pi|y-z|^2} e^{-\frac{|y-z|^2}{4T}} . \tag{3.65}$$

Now we are in a position to present our projection-type location search algorithm for detecting a single anomaly. We prescribe the initial condition $u_0(x) = a \cdot x$ for some fixed unit constant vector $a$ and choose $g = a \cdot \nu$ as an applied time-independent heat flux on $\partial\Omega_T$, where $a$ is taken to be a coordinate unit vector. Take two observation lines $\Sigma_1$ and $\Sigma_2$ contained in $\mathbb{R}^2 \setminus \overline{\Omega}$ such that

$$\Sigma_1 := \text{a line parallel to } a, \qquad \Sigma_2 := \text{a line normal to } a\,.$$

Next we find two points $P_i \in \Sigma_i$ $(i = 1, 2)$ so that $I_w(T)(P_1) = 0$ and

$$I_w(T)(P_2) = \begin{cases} \min\limits_{x \in \Sigma_2} I_w(T)(x) & \text{if } k - 1 < 0\,, \\ \max\limits_{x \in \Sigma_2} I_w(T)(x) & \text{if } k - 1 > 0. \end{cases}$$

Finally, we draw the corresponding lines $\Pi_1(P_1)$ and $\Pi_2(P_2)$ given by (3.38). Then the intersecting point $P$ of $\Pi_1(P_1) \cap \Pi_2(P_2)$ is close to the anomaly $D$: $|P - z| = O(\delta\,|\log \delta|)$ for $\delta$ small enough.

### *3.11.2 Detection of Multiple Anomalies: A MUSIC-Type Algorithm*

Consider $m$ anomalies $D_s = \delta B_s + z_s$, $s = 1, \ldots, m$, whose heat conductivity is $k_s$. Choose

$$U(x,t) = U_{y'}(x,t) := \frac{1}{4\pi t} e^{-\frac{|x-y'|^2}{4t}} \quad \text{for } y' \in \mathbb{R}^2 \setminus \overline{\Omega}$$

or, equivalently, $g$ to be the heat flux corresponding to a heat source placed at the point source $y'$ and the initial condition $u_0(x) = 0$ in $\Omega$, to obtain that

$$I_w(T) \approx -\delta^2 \sum_{s=1}^{m} \frac{(1-k_s)}{64\pi^2} (y' - z_s) M^{(s)} (y - z_s) \\ \times \int_0^T \frac{1}{t^2 (T-t)^2} \exp\left( -\frac{|y - z_s|^2}{4(T-t)} - \frac{|y' - z_s|^2}{4t} \right) dt,$$

where $w$ is given by (3.64) and $M^{(s)}$ is the polarization tensor of $D_s$.

Suppose for the sake of simplicity that all the domains $B_s$ are disks. Then it follows from (3.10) that $M^{(s)} = m^{(s)} I_2$, where $m^{(s)} = 2(k_s - 1)|B_s|/(k_s + 1)$ and $I_2$ is the $2 \times 2$ identity matrix. Let $y_l \in \mathbb{R}^2 \setminus \overline{\Omega}$ for $l \in \mathbb{N}$ be the source points. We assume that the countable set $\{y_l\}_{l \in \mathbb{N}}$ has the property that any analytic function which vanishes in $\{y_l\}_{l \in \mathbb{N}}$ vanishes identically.

The MUSIC-type location search algorithm for detecting multiple anomalies is as follows. For $n \in \mathbb{N}$ sufficiently large, define the matrix $A = [A_{ll'}]_{l,l'=1}^n$ by

$$A_{ll'} := -\delta^2 \sum_{s=1}^m \frac{(1-k_s)}{64\pi^2} m^{(s)} (y_{l'} - z_s) \cdot (y_l - z_s) \\ \times \int_0^T \frac{1}{t^2(T-t)^2} \exp\Big( -\frac{|y_l - z_s|^2}{4(T-t)} - \frac{|y_{l'} - z_s|^2}{4t} \Big) \, dt \, .$$

For $z \in \Omega$, we decompose the symmetric real matrix $C$ defined by

$$C := \Big[ \int_0^T \frac{1}{t^2(T-t)^2} \exp\Big( -\frac{|y_l - z|^2}{4(T-t)} - \frac{|y_{l'} - z|^2}{4t} \Big) \, dt \Big]_{l,l'=1,\ldots,n}$$

as follows:

$$C = \sum_{l=1}^p v_l(z) v_l(z)^* \tag{3.66}$$

for some $p \leq n$, where $v_l \in \mathbb{R}^n$ and $v_l^*$ denotes the transpose of $v_l$. Define the vector $g_z^{(l)} \in \mathbb{R}^{n \times 2}$ for $z \in \Omega$ by

$$g_z^{(l)} = \Big( (y_1 - z) v_{l1}(z), \ldots, (y_n - z) v_{ln}(z) \Big)^*, \quad l = 1, \ldots, p. \tag{3.67}$$

Here $v_{l1}, \ldots, v_{ln}$ are the components of the vector $v_l$, $l = 1, \ldots, p$. Let $y_l = (y_{lx}, y_{ly})$ for $l = 1, \ldots, n$, $z = (z_x, z_y)$, and $z_s = (z_{sx}, z_{sy})$. We also introduce

$$g_{zx}^{(l)} = \Big( (y_{1x} - z_x) v_{l1}(z), \ldots, (y_{nx} - z_x) v_{ln}(z) \Big)^*$$

and

$$g_{zy}^{(l)} = \Big( (y_{1y} - z_y) v_{l1}(z), \ldots, (y_{ny} - z_y) v_{ln}(z) \Big)^* .$$

**Lemma 3.10.** *The following characterization of the location of the anomalies in terms of the range of the matrix $A$ holds:*

$$g_{zx}^{(l)} \text{ and } g_{zy}^{(l)} \subset \text{Range}(A) \quad \forall l \in \{1, \ldots, p\} \quad \text{iff} \quad z \in \{z_1, \ldots, z_m\} \, . \tag{3.68}$$

Note that the smallest number $n$ which is sufficient to efficiently recover the anomalies depends on the (unknown) number $m$. This is the main reason to take $n$ sufficiently large. As for the electrical impedance imaging, the MUSIC-type algorithm for the thermal imaging is as follows. Compute $P_{\text{noise}}$, the projection onto the noise space, by the singular value decomposition of the matrix $A$. Compute the vectors $v_l$ by (3.66). Form an image of the locations,

$z_1, \ldots, z_m$, by plotting, at each point $z$, the quantity $\|g_z^{(l)} \cdot a\| / \|P_{\text{noise}}(g_z^{(l)} \cdot a)\|$ for $l = 1, \ldots, p$, where $g_z^{(l)}$ is given by (3.67) and $a$ is a unit constant vector. The resulting plot will have large peaks at the locations of $z_s$, $s = 1, \ldots, m$.

The algorithms described for reconstructing thermal anomalies can be extended to the realistic half-space model. See [22].

## 3.12 Bibliography and Concluding Remarks

In this chapter, applications of the method of small-volume expansions in emerging medical imaging are outlined. This method leads to very effective and robust reconstruction algorithms in many imaging problems [15]. Of particular interest are emerging multi-physics or hybrid imaging approaches. These approaches allow to overcome the severe ill-posedness character of image reconstruction.

Part (i) in Theorem 3.1 was proven in [14, 33, 40] and in a more general form in [31]. The proof in [14] is based on a decomposition formula of the solution into a harmonic part and a refraction part first derived in [48]. In this connection, see [49–51, 54, 56]. Part (iii) is from [21]. The Hashin-Shtrikman bounds for the polarization tensor were proved in [32, 57]. Theorem 3.7 and the results on the viscous moment tensor in Theorem 3.8 are from [6]. The initial boundary-value problems for the wave equation in the presence of anomalies of small volume have been considered in [1, 17]. See [16] for the time-harmonic regime. Theorem 3.4 is from [7]. See also [19,20,30] for similar results in the case of compressible elasticity. In that paper, a time-reversal approach was designed for locating the anomaly from the outer expansion (3.15). We refer to Chap. 2 for basic physical principles of time reversal. See also [38, 39].

The projection algorithm was introduced in [23, 24, 55, 63]. The MUSIC-type algorithm for locating small electromagnetic anomalies from the response matrix was first developed in [28]. See also [9, 11–13, 34]. It is worth mentioning that the MUSIC-type algorithm is related to time reversal [61,62].

Impediography was proposed in [3] and the substitution algorithm proposed there. An optimal control approach for solving the inverse problem in impediography has been described in [29]. The inversion was considered as a minimization problem, and it was performed in two or three dimensions.

Magnetic resonance elastography was first proposed in [60]. The results provided on this technique are from [6]. For physical principles of radiation force imaging we refer to [26, 27]. Thermal imaging of small anomalies has been considered in [10]. See also [22] where a realistic half space model for thermal imaging was considered and accurate and robust reconstruction algorithms are designed.

To conclude this chapter, it is worth mentioning that the inner expansions derived for the heat equation can be used to improve reconstruction in ultrasonic temperature imaging. The idea behind ultrasonic temperature imaging hinges on measuring local temperature near anomalies. The aim is to reconstruct anomalies with higher spatial and contrast resolution as compared to those obtained from boundary measurements alone.

We would also like to mention that our approach for the magneto-acoustic tomography can be used in photo-acoustic imaging. The photo-acoustic effect is the physical basis for photo-acoustic imaging; it refers to the generation of acoustic waves by the absorption of optical energy. In [4], a new method for reconstructing absorbing regions inside a bounded domain from boundary measurements of the induced acoustic signal has been developed. There, the focusing property of the time-reversed acoustic signal has been shown.

## References

1. H. Ammari, An inverse initial boundary value problem for the wave equation in the presence of imperfections of small volume, SIAM J. Control Optim., 41 (2002), 1194–1211.
2. H. Ammari, *An Introduction to Mathematics of Emerging Biomedical Imaging*, Mathématiques & Applications, Vol. 62, Springer-Verlag, Berlin, 2008.
3. H. Ammari, E. Bonnetier, Y. Capdeboscq, M. Tanter, and M. Fink, Electrical impedance tomography by elastic deformation, SIAM J. Appl. Math., 68 (2008), 1557–1573.
4. H. Ammari, E. Bossy, V. Jugnon, and H. Kang, Mathematical modelling in photo-acoustic imaging, SIAM Rev., to appear.
5. H. Ammari, Y. Capdeboscq, H. Kang, and A. Kozhemyak, Mathematical models and reconstruction methods in magneto-acoustic imaging, Europ. J. Appl. Math., 20 (2009), 303–317.
6. H. Ammari, P. Garapon, H. Kang, and H. Lee, A method of biological tissues elasticity reconstruction using magnetic resonance elastography measurements, Quart. Appl. Math., 66 (2008), 139–175.
7. H. Ammari, P. Garapon, L. Guadarrama Bustos, and H. Kang, Transient anomaly imaging by the acoustic radiation force, J. Diff. Equat., to appear.
8. H. Ammari, P. Garapon, and F. Jouve, Separation of scales in elasticity imaging: A numerical study, J. Comput. Math, to appear.
9. H. Ammari, R. Griesmaier, and M. Hanke, Identification of small inhomogeneities: asymptotic factorization, Math. of Comp., 76 (2007), 1425–1448.
10. H. Ammari, E. Iakovleva, H. Kang, and K. Kim, Direct algorithms for thermal imaging of small inclusions, Multiscale Modeling and Simulation: A SIAM Interdisciplinary Journal, 4 (2005), 1116–1136.
11. H. Ammari, E. Iakovleva, and D. Lesselier, Two numerical methods for recovering small electromagnetic inclusions from scattering amplitude at a fixed frequency, SIAM J. Sci. Comput., 27 (2005), 130–158.
12. H. Ammari, E. Iakovleva, and D. Lesselier, A MUSIC algorithm for locating small inclusions buried in a half-space from the scattering amplitude at a fixed frequency, Multiscale Model. Simul., 3 (2005), 597–628.
13. H. Ammari, E. Iakovleva, D. Lesselier, and G. Perrusson, A MUSIC-type electromagnetic imaging of a collection of small three-dimensional inclusions, SIAM J. Sci. Comput., 29 (2007), 674–709.

14. H. Ammari and H. Kang, High-order terms in the asymptotic expansions of the steady-state voltage potentials in the presence of conductivity inhomogeneities of small diameter, SIAM J. Math. Anal., 34 (2003), 1152–1166.
15. H. Ammari and H. Kang, *Reconstruction of Small Inhomogeneities from Boundary Measurements*, Lecture Notes in Mathematics, Vol. 1846, Springer-Verlag, Berlin, 2004.
16. H. Ammari and H. Kang, Boundary layer techniques for solving the Helmholtz equation in the presence of small inhomogeneities, J. Math. Anal. Appl., 296 (2004), 190–208.
17. H. Ammari and H. Kang, Reconstruction of elastic inclusions of small volume via dynamic measurements, Appl. Math. Opt., 54 (2006), 223–235.
18. H. Ammari and H. Kang, *Polarization and Moment Tensors: with Applications to Inverse Problems and Effective Medium Theory*, Applied Mathematical Sciences, Vol. 162, Springer-Verlag, New York, 2007.
19. H. Ammari, H. Kang, and H. Lee, A boundary integral method for computing elastic moment tensors for ellipses and ellipsoids, J. Comp. Math., 25 (2007), 2–12.
20. H. Ammari, H. Kang, G. Nakamura, and K. Tanuma, Complete asymptotic expansions of solutions of the system of elastostatics in the presence of an inclusion of small diameter and detection of an inclusion, J. Elasticity, 67 (2002), 97–129.
21. H. Ammari and A. Khelifi, Electromagnetic scattering by small dielectric inhomogeneities, J. Math. Pures Appl., 82 (2003), 749–842.
22. H. Ammari, A. Kozhemyak, and D. Volkov, Asymptotic formulas for thermography based recovery of anomalies, Numer. Math.: TMA, 2 (2009), 18–42.
23. H. Ammari, O. Kwon, J.K. Seo, and E.J. Woo, Anomaly detection in T-scan trans-admittance imaging system, SIAM J. Appl. Math., 65 (2004), 252–266.
24. H. Ammari and J.K. Seo, An accurate formula for the reconstruction of conductivity inhomogeneities, Adv. Appl. Math., 30 (2003), 679–705.
25. M. Assenheimer, O. Laver-Moskovitz, D. Malonek, D. Manor, U. Nahliel, R. Nitzan, and A. Saad, The T-scan technology: Electrical impedance as a diagnostic tool for breast cancer detection, Physiol. Meas., 22 (2001), 1–8.
26. J. Bercoff, M. Tanter, and M. Fink, Supersonic shear imaging: a new technique for soft tissue elasticity mapping, IEEE Trans. Ultrasonics, Ferro., Freq. Control, 51 (2004), 396–409.
27. J. Bercoff, M. Tanter, and M. Fink, The role of viscosity in the impulse diffraction field of elastic waves induced by the acoustic radiation force, IEEE Trans. Ultrasonics, Ferro., Freq. Control, 51 (2004), 1523–1536.
28. M. Brühl, M. Hanke, and M.S. Vogelius, A direct impedance tomography algorithm for locating small inhomogeneities, Numer. Math., 93 (2003), 635–654.
29. Y. Capdeboscq, F. De Gournay, J. Fehrenbach, and O. Kavian, An optimal control approach to imaging by modification, preprint.
30. Y. Capdeboscq and H. Kang, Improved Hashin-Shtrikman bounds for elastic moment tensors and an application, Appl. Math. Opt., 57 (2008), 263–288.
31. Y. Capdeboscq and M.S. Vogelius, A general representation formula for the boundary voltage perturbations caused by internal conductivity inhomogeneities of low volume fraction, Math. Modelling Num. Anal., 37 (2003), 159–173.
32. Y. Capdeboscq and M.S. Vogelius, Optimal asymptotic estimates for the volume of internal inhomogeneities in terms of multiple boundary measurements, Math. Modelling Num. Anal., 37 (2003), 227–240.
33. D.J. Cedio-Fengya, S. Moskow, and M.S. Vogelius, Identification of conductivity imperfections of small diameter by boundary measurements: Continuous dependence and computational reconstruction, Inverse Problems, 14 (1998), 553–595.
34. M. Cheney, The linear sampling method and the MUSIC algorithm, Inverse Problems, 17 (2001), 591–595.
35. M. Cheney and D. Isaacson, Distinguishability in impedance imaging, IEEE Trans. Biomed. Engr., 39 (1992), 852–860.
36. M. Cheney, D. Isaacson, and J.C. Newell, Electrical impedance tomography, SIAM Rev., 41 (1999), 85–101.

37. M. Cheney, D. Isaacson, J.C. Newell, S. Simske, and J. Goble, NOSER: an algorithm for solving the inverse conductivity problem, Int. J. Imag. Syst. Technol., 22 (1990), 66–75.
38. A.J. Devaney, Time reversal imaging of obscured targets from multistatic data, IEEE Trans. Antennas Propagat., 523 (2005), 1600–1610.
39. M. Fink, Time-reversal acoustics, Contemp. Math., 408 (2006), 151–179.
40. A. Friedman and M.S. Vogelius, Identification of small inhomogeneities of extreme conductivity by boundary measurements: a theorem on continuous dependence, Arch. Rat. Mech. Anal., 105 (1989), 299–326.
41. D. Gisser, D. Isaacson, and J.C. Newell, Electric current tomography and eigenvalues, SIAM J. Appl. Math., 50 (1990), 1623–1634.
42. J.F. Greenleaf, M. Fatemi, and M. Insana, Selected methods for imaging elastic properties of biological tissues, Annu. Rev. Biomed. Eng., 5 (2003), 57–78.
43. S. Haider, A. Hrbek, and Y. Xu, Magneto-acousto-electrical tomography, preprint.
44. D. Isaacson, Distinguishability of conductivities by electric current computed tomography, IEEE Trans. Medical Imag., 5 (1986), 91–95.
45. D. Isaacson and M. Cheney, Effects of measurements precision and finite numbers of electrodes on linear impedance imaging algorithms, SIAM J. Appl. Math., 51 (1991), 1705–1731.
46. D. Isaacson and E.L. Isaacson, Comments on Calderón's paper: "On an inverse boundary value problem", Math. Compt., 52 (1989), 553–559.
47. V. Isakov, *Inverse Problems for Partial Differential Equations*, Applied Mathematical Sciences, Vol. 127, Springer-Verlag, New York, 1998.
48. H. Kang and J.K. Seo, Layer potential technique for the inverse conductivity problem, Inverse Problems, 12 (1996), 267–278.
49. H. Kang and J.K. Seo, Identification of domains with near-extreme conductivity: Global stability and error estimates, Inverse Problems, 15 (1999), 851–867.
50. H. Kang and J.K. Seo, Inverse conductivity problem with one measurement: Uniqueness of balls in $R^3$, SIAM J. Appl. Math., 59 (1999), 1533–1539.
51. H. Kang and J.K. Seo, Recent progress in the inverse conductivity problem with single measurement, in *Inverse Problems and Related Fields*, CRC Press, Boca Raton, FL, 2000, 69–80.
52. S. Kim, O. Kwon, J.K. Seo, and J.R. Yoon, On a nonlinear partial differential equation arising in magnetic resonance electrical impedance imaging, SIAM J. Math. Anal., 34 (2002), 511–526.
53. Y.J. Kim, O. Kwon, J.K. Seo, and E.J. Woo, Uniqueness and convergence of conductivity image reconstruction in magnetic resonance electrical impedance tomography, Inverse Problems, 19 (2003), 1213–1225.
54. O. Kwon and J.K. Seo, Total size estimation and identification of multiple anomalies in the inverse electrical impedance tomography, Inverse Problems, 17 (2001), 59–75.
55. O. Kwon, J.K. Seo, and J.R. Yoon, A real-time algorithm for the location search of discontinuous conductivities with one measurement, Comm. Pure Appl. Math., 55 (2002), 1–29.
56. O. Kwon, J.R. Yoon, J.K. Seo, E.J. Woo, and Y.G. Cho, Estimation of anomaly location and size using impedance tomography, IEEE Trans. Biomed. Engr., 50 (2003), 89–96.
57. R. Lipton, Inequalities for electric and elastic polarization tensors with applications to random composites. J. Mech. Phys. Solids, 41 (1993), 809–833.
58. A. Manduca, T.E. Oliphant, M.A. Dresner, J.L. Mahowald, S.A. Kruse, E. Amromin, J.P. Felmlee, J.F. Greenleaf, and R.L. Ehman, Magnetic resonance elastography: non-invasive mapping of tissue elasticity, Medical Image Analysis, 5 (2001), 237–254.
59. A. Montalibet, J. Jossinet, A. Matias, and D. Cathignol, Electric current generated by ultrasonically induced Lorentz force in biological media, Medical Biol. Eng. Comput., 39 (2001), 15–20.

60. R. Muthupillai, D.J. Lomas, P.J. Rossman, J.F. Greenleaf, A. Manduca, and R.L. Ehman, Magnetic resonance elastography by direct visualization of propagating acoustic strain waves, Science, 269 (1995), 1854–1857.
61. T.D. Mast, A. Nachman, and R.C. Waag, Focusing and imagining using eigenfunctions of the scattering operator, J. Acoust. Soc. Am., 102 (1997), 715–725.
62. C. Prada, J.-L. Thomas, and M. Fink, The iterative time reversal process: Analysis of the convergence, J. Acoust. Soc. Amer., 97 (1995), 62–71.
63. J.K. Seo, O. Kwon, H. Ammari, and E.J. Woo, Mathematical framework and anomaly estimation algorithm for breast cancer detection using TS2000 configuration, IEEE Trans. Biomedical Engineering, 51 (2004), 1898–1906.
64. R. Sinkus, M. Tanter, S. Catheline, J. Lorenzen, C. Kuhl, E. Sondermann, and M. Fink, Imaging anisotropic and viscous properties of breast tissue by magnetic resonance-elastography, Mag. Res. Med., 53 (2005), 372–387.
65. R. Sinkus, M. Tanter, T. Xydeas, S. Catheline, J. Bercoff, and M. Fink, Viscoelastic shear properties of in vivo breast lesions measured by MR elastography, Mag. Res. Imag., 23 (2005), 159–165.

# Chapter 4
# Electric and Magnetic Activity of the Brain in Spherical and Ellipsoidal Geometry

**George Dassios***

## 4.1 Introduction

Understanding the functional brain is one of the top challenges of contemporary science. The challenge is connected with the fact that we are trying to understand how an organized structure works and the only means available for this task is the structure itself. Therefore an extremely complicated scientific problem is combined with a hard philosophical problem.

Under the given conditions it comes as no surprise that so many, apparently simple, physical and mathematical problems in neuroscience are not generally solved today. One of this problems is the electromagnetic problem of a current field inside an arbitrary conductor. We do understand the physics of this problem, but it is very hard to solve the corresponding mathematical problem if the geometry of the conducting medium diverts from the spherical one. Mathematically, the human brain is an approximately 1.5 L of conductive material in the shape of an ellipsoid with average semiaxes of 6, 6.5 and 9 cm [47]. On the outermost layer of the brain, known as the cerebral cortex, most of the $10^{11}$ neurons contained within the brain are distributed. The neurons are the basic elements of this complicated network and each one of them possesses $10^4$ interconnections with neighboring neurons. At each interconnection, also known as a synapse, neurons communicate via the transfer of particular ions, the neurotransmitters [38]. Neurons are electrochemically excited and they are able to fire instantaneous currents giving rise to very weak magnetic fields which can be measured with the SQUID (Superconducting QUantum Interference Device). The SQUID is the most sensitive apparatus ever built. It can measure magnetic fields as small as $10^{-14}$ T, a sensitivity that is necessary to measure the $10^{-15}$ to $10^{-13}$ T fields resulting

G. Dassios (✉)
Department of Applied Mathematics and Theoretical Physics, University of Cambridge, Cambridge CB3 0WA, UK
e-mail: G.Dassios@damtp.cam.ac.uk

*On leave from the University of Partras and ICE-HT/FORTH, Greece.

H. Ammari, *Mathematical Modeling in Biomedical Imaging I*,
Lecture Notes in Mathematics 1983, DOI 10.1007/978-3-642-03444-2_4,

from brain activity. For comparison we mention that the magnetic fields due to brain activity are about $10^{-9}$ of the Earth's average magnetic field, $10^{-5}$ of the fluctuations of the Earth's magnetic field, and about $10^{-3}$ of the maximum magnetic field generated by the beating heart.

Since the physiological parameters of the brain provide a wavelength of approximately 400 m for the generated electromagnetic field [45], i.e. 2,000 times larger than the human head, it is obvious that the quasi-static theory of electromagnetism [38] is well fitted for modeling the electric and the magnetic activity of the brain. In the quasi-static theory we allow the existence of stationary currents but the electric displacement and the magnetic induction fields are assumed to vary so slowly that their time derivatives can be set equal to zero. Consequently, the electric and the magnetic fields are to a large extent separated, leading to Electroencephalography (EEG) and to Magnetoencephalography (MEG). In EEG a set of electrodes, placed on the scalp, measures the electric potential on the surface of the head, while in MEG a set of coils, adjusted within the SQUID helmet at distances 3–6 cm from the surface of the head, measures the components of the magnetic induction field along the axes of the coils.

My lectures will be focused on some mathematical aspects of the electric and the magnetic activity of the brain, using the simple geometries of the sphere and of the ellipsoid. As it is always the case, our discussion will suffer from the classical discrepancies between mathematical modeling and reality. On the other hand, mathematical modeling provides the appropriate tool for the understanding of the underlying phenomenon, for discovering the basic governing laws, and for identifying the limitations of the algorithms we try to develop. In my lectures, I will be commuting between four central poles: electroencephalography, magnetoencephalography, the sphere, and the ellipsoid. These four poles represent: electricity, magnetism, and their behaviour in isotropic (sphere) and in anisotropic (ellipsoid) environments.

Given the current and the geometry of the brain the *direct* EEG and the *direct* MEG problems consist of finding the electric potential and the magnetic induction, i.e. what can be measured in each case, respectively. The much more interesting and much harder problems of the *inverse* EEG and the *inverse* MEG consist of finding the current from the given EEG and MEG measurements once we know the geometry of the brain. In their general formulation, the corresponding direct mathematical problems are well-posed, while the inverse problems are ill-posed, mainly due to lack of uniqueness [1, 4, 5, 17, 20, 21] a property that was first recognized by Helmholtz.

The source of the mathematical difficulties, in dealing with EEG and MEG problems, is the conductive medium that surrounds the primary neuronal current. The fields generated by the neuronal current give rise to an induction current within the conductor, which in turn generates a magnetic induction field. What we actually measure outside the head is the superposition of both of these fields, and from these measurements we want to identify only the neuronal component of the current which is hidden in this superposition.

The magnetic induction field depends on the conductivity profile and on the geometry of the conductor, and in order to calculate it we need to solve the boundary value problem that determines the electric potential generated by the primary current. Once the electric potential is obtained the exterior magnetic field is uniquely determined via the Geselowitz [25] integral representation.

The simplest mathematical model for the electric and the magnetic activity of the brain is that of a localized dipolar current within a homogeneous conductive sphere. Actually, this simplified model is the only one that is completely solved [7, 16, 22, 46, 49] and forms the basis of all medical applications of EEG and MEG. But the brain is not a sphere. Mathematically, the brain is best approximated by an ellipsoid for which, as we indicated before, the aspect ratio of the most eccentric principal ellipse is equal to 1.5. Hence, the realistic geometry of the brain is far from being a spherical one. On the other hand, the mathematical complexity associated with the ellipsoidal geometry is much higher than the one needed for the sphere. A compromise is achieved if one uses the spheroidal geometry [3, 19, 35, 50] which fits better the shape of the brain but it is still a two-dimensional approximation of the realistic ellipsoidal brain.

As of today no closed form results exist, either for EEG or for MEG, for any other shape besides the sphere. Considerable efforts have been dedicated to the solutions of the direct EEG and the direct MEG problem in ellipsoidal geometry [6,8,10–13,15,18,26–28,32–34,36] and although many partial results are available, the problem is far for its complete mathematical understanding. Even for the cases where a complete series solution for EEG [36], or for MEG [6] has been obtain, the solutions are not easy to handle. Hence, there is a lot to be done until we arrive at practical applications of the ellipsoidal model. As it has been shown, the ellipsoidal geometry is responsible for drastic variations in the behaviour of EEG and MEG compared with the corresponding behaviour in the case of the sphere. For example, in ellipsoidal geometry the "silent" sources are not radial [10, 12] and shells of different conductivity are "visible" by the MEG measurements [11, 13, 18]. Furthermore, in the case where the inversion algorithms use the spherical model to interpret measurements that come from an actual ellipsoid, it is necessary to estimate the error of our approximate interpretation. It is therefore important to further understand the ellipsoidal model and to incorporate its realistic characteristics in the inversion algorithms for EEG and MEG.

An excellent source for bioelectromagnetism is reference [38] while the review article [29] provides a thorough exposition of MEG theory and applications. Effective techniques for EEG lead field theory can be found in [42,43] and for the corresponding MEG case in [40]. Techniques based on surface perturbations can be found in [41, 44], while in [31, 39] the theory of multipole expansions is utilized. Elements of the theory of ellipsoidal harmonics can be found in [23, 30, 37, 48]. Finally, Chap. 8 in [14] contains closed form solutions of potential problems of physical interest in ellipsoidal coordinates.

This chapter is organized as follows. Section 4.2 provides a formulation of the mathematical problems of EEG and MEG. Section 4.3 summarizes the existing results for the spherical model. In Sect. 4.4 we include a compact introduction to the ellipsoidal system and ready-to-use results from the theory of ellipsoidal harmonics. The theory of EEG in ellipsoidal geometry is then presented in Sect. 4.5 and that for MEG in Sect. 4.6. Basic aspects of the inverse EEG and MEG problems are discussed in Sect. 4.7. Finally, a short Sect. 4.8 enumerates some open mathematical problems within the framework described in the present lectures.

## 4.2 Mathematical Formulation

Quasi-static theory of electromagnetism is governed by Maxwell's equations under the assumption that the electric and the magnetic fields do not very in time. Then Faraday's law is written as

$$\nabla \times \boldsymbol{E} = \boldsymbol{0} \tag{4.1}$$

and the Maxwell–Ampere equation reduces to

$$\nabla \times \boldsymbol{B} = \mu_0 \boldsymbol{J} \tag{4.2}$$

where $\mathbf{E}$ is the electric field, $\mathbf{B}$ is the magnetic induction field, $\mathbf{J}$ is the total current and $\mu_0$ is the magnetic permeability of the medium which, for our purpose, it is taken to be a constant everywhere. The two basic equations (4.1) and (4.2) are complemented by the solenoidal property

$$\nabla \cdot \boldsymbol{B} = \boldsymbol{0} \tag{4.3}$$

of the magnetic induction which represents the lack of magnetic monopoles.

For a conductive medium with conductivity $\sigma$, the total current is the sum

$$\boldsymbol{J} = \boldsymbol{J}^p + \boldsymbol{J}^i \tag{4.4}$$

of the primary current $\boldsymbol{J}^p$ plus the induction current defined by

$$\boldsymbol{J}^i = \sigma \boldsymbol{E}. \tag{4.5}$$

Equation (4.1) allows the introduction of the electric potential $u$, such that

$$\boldsymbol{E} = -\nabla u. \tag{4.6}$$

Then taking the divergence of (4.2) we obtain the conservation law

$$\nabla \cdot \boldsymbol{J} = 0 \tag{4.7}$$

which in view of (4.4) and (4.6) implies the differential equation

$$\nabla \cdot [\sigma \nabla u] = \nabla \cdot \boldsymbol{J}^p \tag{4.8}$$

that u has to satisfy. In particular, for homogeneous conductors (4.8) assumes the form

$$\sigma \Delta u(\boldsymbol{r}) = \begin{cases} \nabla \cdot \boldsymbol{J}^p(\boldsymbol{r}), & \boldsymbol{r} \in \operatorname{supp} \boldsymbol{J}^p \\ 0, & \boldsymbol{r} \notin \operatorname{supp} \boldsymbol{J}^p. \end{cases} \tag{4.9}$$

Actually, (4.1) governs electrostatics, (4.2), (4.3) govern magnetostatics, and the two theories are coupled by the equation

$$\boldsymbol{J} = \boldsymbol{J}^p + \sigma \boldsymbol{E} \tag{4.10}$$

only when $\sigma \neq 0$.

Across an interface S separating two regions $V_1, V_2$ with different conductivities $\sigma_1, \sigma_2$, and in the absence of primary currents in the vicinity of S, the *transmission* conditions read

$$u_1 = u_2, \qquad \text{on } S \tag{4.11}$$

$$\sigma_1 \partial_n u_1 = \sigma_2 \partial_n u_2, \qquad \text{on } S \tag{4.12}$$

where $\partial_n$ denotes normal differentiation. Hence, the electric potential should remain continuous while its normal derivative should jump in such a way as to preserves the continuity of the normal components of the induction current.

If S separates a conductive from a nonconductive region and if $u$ denotes the electric potential in the conductive region, then (4.12) becomes

$$\partial_n u = 0, \qquad \text{on } S. \tag{4.13}$$

The direct EEG problem now is formulated as follows.

Let the region $V_c$ occupied by the cerebrum with conductivity $\sigma_c$ and boundary $S_c$. $V_c$ is surrounded by the region $V_f$ occupied by the cerebrospinal fluid whose conductivity is $\sigma_f$ and it is bounded from the outside by $S_f$. In a similar fashion we have the region $V_b$ occupied by the surrounding bone (the skull) with conductivity $\sigma_b$ and outer boundary $S_b$. Finally, comes the region $V_s$ occupied by the skin (the scalp) with conductivity $\sigma_s$ and outer boundary $S_s$ which separates the head from the infinite exterior region $V$ where $\sigma = 0$. Figure 4.1 depicts the appropriate geometry as a set of four nested confocal ellipsoids where it is to be understood that the diameter of

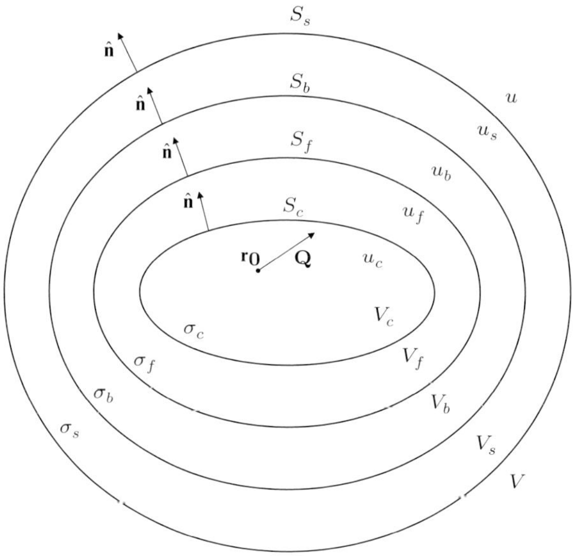


**Fig. 4.1** Geometry of the brain-head system

$S_c$ is approximately 18 cm and that of $S_s$ is about 22–23 cm. Indicative values for the conductivities in $(\Omega\,\mathrm{m})^{-1}$ are: 0.33 for the cerebrum and the scalp, 1.00 for the cerebrospinal fluid and 0.0042 for the skull.

We assume that

$$\boldsymbol{J}^P(\boldsymbol{r}) = \boldsymbol{Q}\delta(\boldsymbol{r} - \boldsymbol{r}_0) \tag{4.14}$$

where $\delta$ is the Dirac measure. In other words, the primary current is a localized dipole at $\boldsymbol{r}_0 \in V_c$ with moment $\mathbf{Q}$. Then we have to solve the following problem

$$\sigma_c \Delta u_c(\mathbf{r}) = \mathbf{Q} \cdot \nabla\delta(\mathbf{r} - \mathbf{r}_0), \qquad \mathbf{r} \in V_c \tag{4.15}$$

$$\Delta u_f(\mathbf{r}) = 0, \qquad \mathbf{r} \in V_f \tag{4.16}$$

$$\Delta u_b(\mathbf{r}) = 0, \qquad \mathbf{r} \in V_b \tag{4.17}$$

$$\Delta u_s(\mathbf{r}) = 0, \qquad \mathbf{r} \in V_s \tag{4.18}$$

$$\Delta u(\mathbf{r}) = 0, \qquad \mathbf{r} \in V \tag{4.19}$$

with

$$u_c(\boldsymbol{r}) = u_f(\boldsymbol{r}), \qquad \sigma_c \partial_n u_c(\boldsymbol{r}) = \sigma_f \partial_n u_f(\boldsymbol{r}), \qquad \boldsymbol{r} \in S_c \tag{4.20}$$

$$u_f(\boldsymbol{r}) = u_b(\boldsymbol{r}), \qquad \sigma_f \partial_n u_f(\boldsymbol{r}) = \sigma_b \partial_n u_b(\boldsymbol{r}), \qquad \boldsymbol{r} \in S_f \tag{4.21}$$

$$u_b(\boldsymbol{r}) = u_s(\boldsymbol{r}), \qquad \sigma_b \partial_n u_b(\boldsymbol{r}) = \sigma_s \partial_n u_s(\boldsymbol{r}), \qquad \boldsymbol{r} \in S_b \tag{4.22}$$

$$u_s(\boldsymbol{r}) = u(\boldsymbol{r}), \qquad \partial_n u_s(\boldsymbol{r}) = 0, \qquad \boldsymbol{r} \in S_s \tag{4.23}$$

and

$$u(\boldsymbol{r}) = O\left(\frac{1}{r^2}\right), r \to \infty \tag{4.24}$$

where $u_c, u_f, u_b, u_s$ and $u$ are the electric potentials in $V_c, V_f, V_b, V_s$ and $V$ respectively. The asymptotic condition (4.24) is a consequence of the fact that the sources (actually the dipole at $\boldsymbol{r}_0$) are compactly supported.

If the primary current is distributed over a compact subset of $V_c$, then we solve problem (4.15)–(4.24) and we integrate the solution with respect to $\boldsymbol{r}_0 \in \operatorname{supp} \boldsymbol{J}^p$ keeping in mind that $\mathbf{Q}$ now is a function of $\boldsymbol{r}_0$. From the point of view of EEG applications what we want are the values of $u_s$, or of $u$ on $S_s$.

Next we turn to the mathematical formulation of the MEG problem. Again we will assume the primary current (4.14) and if we have to deal with a distributed current then we integrate over the support of $\boldsymbol{J}^p$. In both cases, the EEG or the MEG solution that corresponds to the point excitation current (4.14) provides the EEG or the MEG *kernel function* for the given geometry.

The integral form of the magnetostatic equations (4.2), (4.3) is the Biot–Savart law

$$\boldsymbol{B}(\boldsymbol{r}) = \frac{\mu_0}{4\pi}\int_{\mathbb{R}^3} \boldsymbol{J}(\boldsymbol{r}') \times \frac{\boldsymbol{r}-\boldsymbol{r}'}{|\boldsymbol{r}-\boldsymbol{r}'|^3} dv(\boldsymbol{r}') \tag{4.25}$$

where the integration is restricted to the support of $\boldsymbol{J}$, which is the union of the support of $\boldsymbol{J}^p$and the support of $\sigma$. In our brain-head model (4.25) is written as

$$\begin{aligned}\boldsymbol{B}(\boldsymbol{r}) = {} & \frac{\mu_0}{4\pi}\boldsymbol{Q} \times \frac{\boldsymbol{r}-\boldsymbol{r}_0}{|\boldsymbol{r}-\boldsymbol{r}_0|^3} \\ & -\frac{\mu_0}{4\pi}\int_{V^c} \nabla_{\boldsymbol{r}'}[\sigma_c u_c(\boldsymbol{r}')\mathcal{X}(V_c) + \sigma_f u_f(\boldsymbol{r}')\mathcal{X}(V_f) \\ & +\sigma_b u_b(\boldsymbol{r}')\mathcal{X}(V_b) + \sigma_s u_s(\boldsymbol{r}')\mathcal{X}(V_s)\Big] \times \frac{\boldsymbol{r}-\boldsymbol{r}'}{|\boldsymbol{r}-\boldsymbol{r}'|^3} dv(\boldsymbol{r}')\end{aligned} \tag{4.26}$$

where $\mathcal{X}(V_i)$ is the characteristic function of the set $V_i, i = c, f, b, s$.

An appropriate application of integral theorems transforms (4.26) to the Geselowitz formula [9, 25]

$$\begin{aligned}\mathbf{B}(\mathbf{r}) = {} & \frac{\mu_0}{4\pi}\mathbf{Q} \times \frac{\mathbf{r}-\mathbf{r}_0}{|\mathbf{r}-\mathbf{r}_0|^3} \\ & -\frac{\mu_0}{4\pi}(\sigma_c - \sigma_f)\oint_{S_c} u_c(\mathbf{r}')\hat{\mathbf{n}}' \times \frac{\mathbf{r}-\mathbf{r}'}{|\mathbf{r}-\mathbf{r}'|^3} ds(\mathbf{r}')\end{aligned}$$

$$
\begin{aligned}
&-\frac{\mu_0}{4\pi}(\sigma_f-\sigma_b)\oint_{S_f} u_f(\mathbf{r}')\hat{\mathbf{n}}'\times\frac{\mathbf{r}-\mathbf{r}'}{|\mathbf{r}-\mathbf{r}'|^3}ds(\mathbf{r}')\\
&-\frac{\mu_0}{4\pi}(\sigma_b-\sigma_s)\oint_{S_b} u_b(\mathbf{r}')\hat{\mathbf{n}}'\times\frac{\mathbf{r}-\mathbf{r}'}{|\mathbf{r}-\mathbf{r}'|^3}ds(\mathbf{r}')\\
&-\frac{\mu_0}{4\pi}\sigma_s\oint_{S_s} u_s(\mathbf{r}')\hat{\mathbf{n}}'\times\frac{\mathbf{r}-\mathbf{r}'}{|\mathbf{r}-\mathbf{r}'|^3}ds(\mathbf{r}')
\end{aligned}
\tag{4.27}
$$

which holds for $\boldsymbol{r}\in\mathbb{R}^3-S_c\cup S_f\cup S_b\cup S_s$.

From the physical point of view the Geselowitz formula (4.27) replaces the volume distribution of current dipoles $-\sigma\nabla u(\boldsymbol{r}')$ at each point $\boldsymbol{r}'$ in the appropriate region, with a surface distribution of dipoles $-(\sigma_{\text{interior}}-\sigma_{\text{exterior}})u(\boldsymbol{r}')\hat{\boldsymbol{n}}'$ normal to the appropriate interface. Hence, interfaces separating regions of different conductivities are read as surface distributions of normal dipoles, with moments that are proportional to the jumps of the conductivity and the local value of the electric potential.

The corresponding Geselowitz formula for the EEG problem [24] is written as

$$
\begin{aligned}
u(\boldsymbol{r})=&\frac{1}{4\pi\sigma_c}\boldsymbol{Q}\cdot\frac{\boldsymbol{r}-\boldsymbol{r}_0}{|\boldsymbol{r}-\boldsymbol{r}_0|^3}\\
&-\frac{\sigma_c-\sigma_f}{4\pi\sigma_c}\oint_{S_c} u_c(\boldsymbol{r}')\hat{\boldsymbol{n}}'\cdot\frac{\boldsymbol{r}-\boldsymbol{r}'}{|\boldsymbol{r}-\boldsymbol{r}'|^3}ds(\boldsymbol{r}')\\
&-\frac{\sigma_f-\sigma_b}{4\pi\sigma_c}\oint_{S_f} u_f(\boldsymbol{r}')\hat{\boldsymbol{n}}'\cdot\frac{\boldsymbol{r}-\boldsymbol{r}'}{|\boldsymbol{r}-\boldsymbol{r}'|^3}ds(\boldsymbol{r}')\\
&-\frac{\sigma_b-\sigma_s}{4\pi\sigma_c}\oint_{S_b} u_b(\boldsymbol{r}')\hat{\boldsymbol{n}}'\cdot\frac{\boldsymbol{r}-\boldsymbol{r}'}{|\boldsymbol{r}-\boldsymbol{r}'|^3}ds(\boldsymbol{r}')\\
&-\frac{\sigma_s}{4\pi\sigma_c}\oint_{S_s} u_s(\boldsymbol{r}')\hat{\boldsymbol{n}}'\cdot\frac{\boldsymbol{r}-\boldsymbol{r}'}{|\boldsymbol{r}-\boldsymbol{r}'|^3}ds(\boldsymbol{r}')
\end{aligned}
\tag{4.28}
$$

which also holds for every $\boldsymbol{r}$ that does not belong to any one of the interfaces $S_c, S_f, S_b$ or $S_s$. We remark here that (4.27) and (4.28) are actually the vector and the scalar invariants respectively of a related dyadic equation. Indeed, if we consider for simplicity the case of a homogeneous conductor where

$$\sigma_c=\sigma_f=\sigma_b=\sigma_s=\sigma \tag{4.29}$$

and define the dyadic

$$\tilde{\boldsymbol{D}}(\boldsymbol{r})=\boldsymbol{Q}\otimes\frac{\boldsymbol{r}-\boldsymbol{r}_0}{|\boldsymbol{r}-\boldsymbol{r}_0|^3}-\sigma\oint_S u(\boldsymbol{r}')\hat{\boldsymbol{n}}'\otimes\frac{\boldsymbol{r}-\boldsymbol{r}'}{|\boldsymbol{r}-\boldsymbol{r}|^3}ds(\boldsymbol{r}') \tag{4.30}$$

with scalar invariant $S_D$ and vector invariant $\boldsymbol{V}_D$ then, under (4.29), (4.28) coincides with

$$u(\boldsymbol{r}) = \frac{1}{4\pi\sigma} S_D(\boldsymbol{r}) \tag{4.31}$$

and (4.27) coincides with

$$\boldsymbol{B}(\boldsymbol{r}) = \frac{\mu_0}{4\pi} \boldsymbol{V}_D(\boldsymbol{r}). \tag{4.32}$$

For the homogeneous conductor we have

$$u(\boldsymbol{r}) = \frac{1}{4\pi\sigma} \boldsymbol{Q} \cdot \frac{\boldsymbol{r} - \boldsymbol{r}_0}{|\boldsymbol{r} - \boldsymbol{r}_0|^3} - \frac{1}{4\pi} \oint_S u(\boldsymbol{r}') \hat{\boldsymbol{n}}' \cdot \frac{\boldsymbol{r} - \boldsymbol{r}'}{|\boldsymbol{r} - \boldsymbol{r}'|^3} ds(\boldsymbol{r}') \tag{4.33}$$

and

$$\boldsymbol{B}(\boldsymbol{r}) = \frac{\mu_0}{4\pi} \boldsymbol{Q} \times \frac{\boldsymbol{r} - \boldsymbol{r}_0}{|\boldsymbol{r} - \boldsymbol{r}_0|^3} - \frac{\mu_0 \sigma}{4\pi} \oint_S u(\boldsymbol{r}') \hat{\boldsymbol{n}}' \times \frac{\boldsymbol{r} - \boldsymbol{r}'}{|\boldsymbol{r} - \boldsymbol{r}'|^3} ds(\boldsymbol{r}') \tag{4.34}$$

where in both formulae $\boldsymbol{r} \notin S$.

Obviously, the first term on the RHS of (4.27) and of (4.28) describes the contribution of the primary current, while the integrals describe the contribution of the induction current. These integrals carry all the information about the geometry of the conductive regions. Hence, in order to calculate the magnetic field we need first to find the traces of the electric potential on the interfaces supporting the conductivity jumps. This behavior is a consequence of the coupling between the electric and the magnetic field imposed by the conductivity of the medium.

## 4.3 The Spherical Brain Model

Suppose now that the brain-head system is a spherical homogeneous conductor with radius $\alpha$ and conductivity $\sigma$. Then the interior electric potential $u^-$ solves the interior Neumann problem

$$\sigma \Delta u^-(\boldsymbol{r}) = \boldsymbol{Q} \cdot \nabla \delta(\boldsymbol{r} - \boldsymbol{r}_0), \qquad |\boldsymbol{r}| < \alpha \tag{4.35}$$

$$\partial_n u^-(\boldsymbol{r}) = 0, \qquad |\boldsymbol{r}| = \alpha. \tag{4.36}$$

Once (4.35), (4.36) is solved the trace of $u^-$ on $|\boldsymbol{r}| = \alpha$ is used to solve the exterior Dirichlet problem

$$\Delta u^+(\mathbf{r}) = 0, \qquad |\mathbf{r}| > \alpha \tag{4.37}$$

$$u^+(\mathbf{r}) = u^-(\mathbf{r}), \qquad |\mathbf{r}| = \alpha \tag{4.38}$$

$$u^+(\mathbf{r}) = O\left(\frac{1}{r^2}\right), \qquad \mathbf{r} \to +\infty. \tag{4.39}$$

Furthermore, the trace of either $u^-$ or $u^+$ on $|\boldsymbol{r}| = \alpha$ is used to calculate the $\boldsymbol{B}$ field from (4.34). Of course the above program can be applied to any homogeneous conductor, but as we will see in the sequel the sphere is an extremely special case.

It is easy to see that $u^-$ assumes the form

$$u^-(\boldsymbol{r}) = \frac{1}{4\pi\sigma}\boldsymbol{Q}\cdot\frac{\boldsymbol{r}-\boldsymbol{r}_0}{|\boldsymbol{r}-\boldsymbol{r}_0|^3} + w(\boldsymbol{r}) \tag{4.40}$$

where $w(\boldsymbol{r})$ is a harmonic function satisfying the condition

$$\partial_n w(\boldsymbol{r}) = -\frac{1}{4\pi\sigma}\partial_n\boldsymbol{Q}\cdot\frac{\boldsymbol{r}-\boldsymbol{r}_0}{|\boldsymbol{r}-\boldsymbol{r}_0|^3} \tag{4.41}$$

on the boundary $|\boldsymbol{r}| = \alpha$.

Straightforward expansion in spherical harmonics leads to the solution

$$u^-(\boldsymbol{r}) = \frac{1}{4\pi\sigma}\boldsymbol{Q}\cdot\nabla_{\boldsymbol{r}_0}\left[\frac{1}{|\boldsymbol{r}-\boldsymbol{r}_0|} + 4\pi\sum_{n=1}^{\infty}\sum_{m=-n}^{n}\frac{n+1}{n(2n+1)}\frac{r_0^n r^n}{\alpha^{2n+1}}Y_n^{m^*}(\hat{\boldsymbol{r}}_0)Y_n^m(\hat{\boldsymbol{r}})\right] \tag{4.42}$$

for $|\boldsymbol{r}| < \alpha$, where $\hat{\boldsymbol{r}}$ denotes the unit vector and the directional derivative $\boldsymbol{Q}\cdot\nabla_{\boldsymbol{r}_0}$ depends only on the dipolar source. The complex spherical harmonics used in the expansion (4.42) satisfy the condition

$$\oint_{S^2} Y_n^{m^*}(\hat{\boldsymbol{r}})Y_{n'}^{m'}(\hat{\boldsymbol{r}})ds(\hat{\boldsymbol{r}}) = \delta_{\mathrm{nn'}}\delta_{\mathrm{mm'}}. \tag{4.43}$$

with $S^2$ denoting the unit sphere and $\delta_{\mathrm{nn'}}$ being the Kronecker symbol.

The series (4.42) can be summed as follows. First we use the addition theorem

$$\frac{4\pi}{2n+1}\sum_{m=-n}^{n} Y_n^{m^*}(\hat{\boldsymbol{r}}_0)Y_n^m(\hat{\boldsymbol{r}}) = P_n(\hat{\boldsymbol{r}}\cdot\hat{\boldsymbol{r}}_0) \tag{4.44}$$

where $P_n$ is the Legendre polynomial of degree $n$, to rewrite (4.42) as

$$u^-(\boldsymbol{r}) = \frac{1}{4\pi\sigma}\boldsymbol{Q}\cdot\nabla_{\boldsymbol{r}_0}\left[\frac{1}{|\boldsymbol{r}-\boldsymbol{r}_0|} + \sum_{n=1}^{\infty}\left(1+\frac{1}{n}\right)\frac{r_0^n r^n}{\alpha^{2n+1}}P_n(\hat{\boldsymbol{r}}\cdot\hat{\boldsymbol{r}}_0)\right]. \tag{4.45}$$

Then we show that the function

$$f(\rho) = \sum_{n=1}^{\infty}\frac{\rho^n}{n}P_n(\cos\theta), \rho < 1 \tag{4.46}$$

solves the initial value problem

$$\begin{aligned} \rho f'(\rho) &= (1 - 2\rho\cos\theta + \rho^2)^{-1/2} - 1 \\ f(0) &= 0 \end{aligned} \tag{4.47}$$

which has the solution

$$f(\rho) = -\ln\frac{1 - \rho\cos\theta + \sqrt{1 - 2\rho\cos\theta + \rho^2}}{2}. \tag{4.48}$$

Finally, we can use the expansion

$$\frac{1}{|\boldsymbol{r} - \boldsymbol{r}_0|} = \sum_{n=0}^{\infty}\sum_{m=-n}^{n}\frac{4\pi}{2n+1}\frac{r_0^n}{r^{n+1}}Y_n^{m^*}(\hat{\boldsymbol{r}_0})Y_n^m(\hat{\boldsymbol{r}}). \tag{4.49}$$

Putting everything together we arrive at the closed form expression

$$u^-(\boldsymbol{r}) = \frac{1}{4\pi\sigma}\boldsymbol{Q}\cdot\nabla_{\boldsymbol{r}_0}\left[\frac{1}{P} + \frac{\alpha}{r}\frac{1}{R} - \frac{1}{\alpha}\ln\frac{rR + \boldsymbol{r}\cdot\boldsymbol{R}}{2\alpha^2}\right] \tag{4.50}$$

where

$$\boldsymbol{P} = \boldsymbol{r} - \boldsymbol{r}_0 \tag{4.51}$$

$$\boldsymbol{R} = \frac{\alpha^2}{r^2}\boldsymbol{r} - \boldsymbol{r}_0 \tag{4.52}$$

which gives the boundary values

$$u^-(\alpha\hat{\boldsymbol{r}}) = \frac{1}{4\pi\sigma}\boldsymbol{Q}\cdot\nabla_{\boldsymbol{r}_0}\left[\frac{2}{|\alpha\hat{\boldsymbol{r}} - \boldsymbol{r}_0|} - \frac{1}{\alpha}\ln\frac{|\alpha\hat{\boldsymbol{r}} - \boldsymbol{r}_0| + \hat{\boldsymbol{r}}\cdot(\alpha\hat{\boldsymbol{r}} - \boldsymbol{r}_0)}{2\alpha}\right]. \tag{4.53}$$

Applying $\boldsymbol{Q}\cdot\nabla_{\boldsymbol{r}_0}$ in (4.50) and (4.53) we also obtain the expressions

$$u^-(\boldsymbol{r}) = \frac{\boldsymbol{Q}}{4\pi\sigma}\cdot\left[\frac{\boldsymbol{P}}{P^3} + \frac{\alpha}{r}\frac{\boldsymbol{R}}{R^3} + \frac{1}{\alpha R}\frac{R\boldsymbol{r} + r\boldsymbol{R}}{Rr + \boldsymbol{r}\cdot\boldsymbol{R}}\right], |\boldsymbol{r}| < \alpha \tag{4.54}$$

and

$$u^-(\alpha\hat{\boldsymbol{r}}) = \frac{\boldsymbol{Q}}{4\pi\sigma}\cdot\left[2\frac{\alpha\hat{\boldsymbol{r}} - \boldsymbol{r}_0}{|\alpha\hat{\boldsymbol{r}} - \boldsymbol{r}_0|^3} + \frac{1}{\alpha|\alpha\hat{\boldsymbol{r}} - \boldsymbol{r}_0|}\frac{|\alpha\hat{\boldsymbol{r}} - \boldsymbol{r}_0|\hat{\boldsymbol{r}} + (\alpha\hat{\boldsymbol{r}} - \boldsymbol{r}_0)}{|\alpha\hat{\boldsymbol{r}} - \boldsymbol{r}_0| + \hat{\boldsymbol{r}}\cdot(\alpha\hat{\boldsymbol{r}} - \boldsymbol{r}_0)}\right]. \tag{4.55}$$

Obviously, the solution of the Neumann problem (4.35), (4.36) is unique up to an additive constant, which we can take to be zero since the EEG measurements are always potential differences from some reference point.

Using the data (4.53) in (4.38) and following similar arguments as with the internal potential $u^-$ we can solve (4.37)–(4.39) and obtain

$$u^+(\boldsymbol{r}) = \frac{\boldsymbol{Q}}{4\pi\sigma} \cdot \left[ 2\frac{\boldsymbol{P}}{P^3} + \frac{1}{rP} \frac{P\boldsymbol{r} + r\boldsymbol{P}}{Pr + \boldsymbol{r} \cdot \boldsymbol{P}} \right] \tag{4.56}$$

for $|\boldsymbol{r}| > \alpha$. Obviously

$$u^+(\alpha\hat{\boldsymbol{r}}) = u^-(\alpha\hat{\boldsymbol{r}}). \tag{4.57}$$

An inspection of the solutions (4.54) and (4.56) reveals that besides the common term

$$\frac{\boldsymbol{Q}}{4\pi\sigma} \cdot \frac{\boldsymbol{P}}{P^3}$$

which comes from the particular solution of Poisson's equation for $u^-$, and from the boundary condition for $u^+$, the other two terms of $u^-$ and $u^+$ are connected by the Kelvin's transformation [7]

$$\boldsymbol{r} \longmapsto \frac{\alpha^2}{r^2}\boldsymbol{r}. \tag{4.58}$$

Indeed, Kelvin's theorem

$$\Delta f(\boldsymbol{r}) \longmapsto \left(\frac{r}{\alpha}\right)^5 \Delta \frac{\alpha}{r} f\left(\frac{\alpha^2}{r^2}\boldsymbol{r}\right) \tag{4.59}$$

shows that if $f(\boldsymbol{r})$ is a harmonic function then the function $\frac{\alpha}{r} f(\frac{\alpha^2}{r^2}\boldsymbol{r})$ is also harmonic. Hence, if we apply the harmonicity preserving transformation (4.58) we see immediately that the interior harmonic function

$$u^-(\boldsymbol{r}) - \frac{1}{4\pi\sigma}\boldsymbol{Q} \cdot \frac{\boldsymbol{P}}{P^3} = \frac{\boldsymbol{Q}}{4\pi\sigma} \cdot \left[ \frac{\alpha}{r} \frac{\boldsymbol{R}}{R^3} + \frac{1}{\alpha R} \frac{R\boldsymbol{r} + r\boldsymbol{R}}{Rr + \boldsymbol{r} \cdot \boldsymbol{R}} \right] \tag{4.60}$$

is transformed into the exterior harmonic function

$$u^+(\boldsymbol{r}) - \frac{1}{4\pi\sigma}\boldsymbol{Q} \cdot \frac{\boldsymbol{P}}{P^3} = \frac{\boldsymbol{Q}}{4\pi\sigma} \cdot \left[ \frac{\boldsymbol{P}}{P^3} + \frac{1}{rP} \frac{P\boldsymbol{r} + r\boldsymbol{P}}{Pr + \boldsymbol{r} \cdot \boldsymbol{P}} \right] \tag{4.61}$$

and vice versa. That indicates that the above solutions can be interpreted in terms of images as it was originally approached by two medical doctors (!) [49].

If we introduce the function

$$\Psi(\boldsymbol{r}) = \frac{1}{4\pi\sigma}\boldsymbol{Q} \cdot \frac{\boldsymbol{r} - \boldsymbol{r}_0}{|\boldsymbol{r} - \boldsymbol{r}_0|^3} = \frac{1}{4\pi\sigma}(\boldsymbol{Q} \cdot \nabla_{\boldsymbol{r}_0}) \frac{1}{|\boldsymbol{r} - \boldsymbol{r}_0|} \tag{4.62}$$

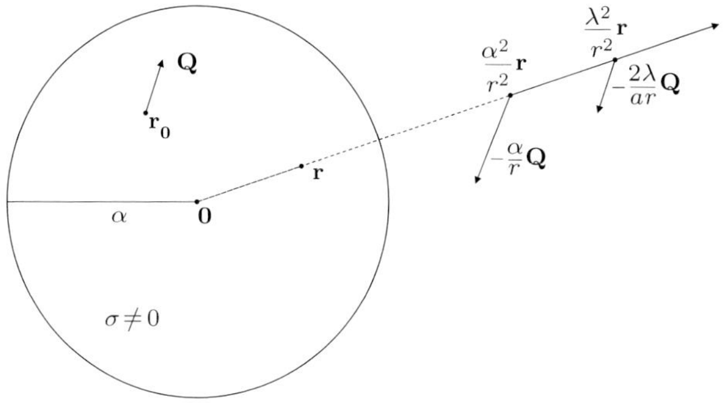


**Fig. 4.2** The spherical image system

which represents the electric potential due to a dipole at the point $\boldsymbol{r}_0$ with moment $\boldsymbol{Q}$, then we can easily justify that

$$u^{-}(\boldsymbol{r}) = \Psi(\boldsymbol{r}) + \frac{\alpha}{r}\Psi\left(\frac{\alpha^2}{r^2}\boldsymbol{r}\right) + \frac{2}{\alpha}\int_{\alpha}^{+\infty}\frac{\lambda}{r}\Psi\left(\frac{\lambda^2}{r^2}\boldsymbol{r}\right)d\lambda \tag{4.63}$$

for $|\boldsymbol{r}| < \alpha$ (Fig. 4.2).

By virtue of reciprocity, formula (4.63) interprets the contribution of the conductive medium, to the electric potential at the interior point $\boldsymbol{r}$, as the electric potential at $\boldsymbol{r}_0$ due to a dipole with moment $-\frac{\alpha}{r}\boldsymbol{Q}$ at the image point $\frac{\alpha^2}{r^2}\boldsymbol{r}$, plus a continuous distribution of dipoles with moment $-\frac{2\lambda}{\alpha r}\boldsymbol{Q}$ at the image point $\frac{\lambda^2}{r^2}\boldsymbol{r}$ for every $\lambda \in [\alpha, +\infty), r < \alpha$. A straightforward integration in (4.63) leads to (4.54).

Similarly, applying Kelvin's theorem to the expression $(u^{-} - \Psi)$ we obtain the following representation for the exterior potential in terms of images

$$u^{+}(\boldsymbol{r}) = 2\Psi(\boldsymbol{r}) + \frac{2}{\alpha^2}\int_{\alpha}^{+\infty}\lambda\Psi\left(\frac{\lambda^2}{\alpha^2}\boldsymbol{r}\right)d\lambda. \tag{4.64}$$

for $|\boldsymbol{r}| > \alpha$.

Consequently, the contribution of the conductive medium to the electric potential at the exterior point $\boldsymbol{r}$ is equal to the electric potential at $\boldsymbol{r}_0$ due to a dipole with moment $-\boldsymbol{Q}$ at $\boldsymbol{r}$, plus a continuous distribution of dipoles with moments $-\frac{2\lambda}{\alpha^2}\boldsymbol{Q}$ at the point $\frac{\lambda^2}{\alpha^2}\boldsymbol{r}$ for every $\lambda \in [\alpha, +\infty), r > \alpha$. Performing the integration in (4.64) we recover (4.56).

Next we turn to the exterior magnetic induction for the sphere which represents magnetoencephalography. The behavior of MEG in spherical geometry is very unique for the following simple reason. If we consider the radial component of $\boldsymbol{B}$, given by (4.34), and we note that for $\hat{\boldsymbol{n}}' = \hat{\boldsymbol{r}}'$

$$\hat{\boldsymbol{n}}' \times (\boldsymbol{r} - \boldsymbol{r}') \cdot \hat{\boldsymbol{r}} = 0 \tag{4.65}$$

we obtain

$$- \hat{\boldsymbol{r}} \cdot \boldsymbol{B}(\boldsymbol{r}) = \frac{\mu_0}{4\pi} \frac{\boldsymbol{Q} \times \boldsymbol{r}_0 \cdot \hat{\boldsymbol{r}}}{|\boldsymbol{r} - \boldsymbol{r}_0|^3}. \tag{4.66}$$

In other words, the radial component of the magnetic induction is independent of the electric potential and therefore independent of the radius of the conductive sphere. This observation allowed Sarvas [46] to calculate $\boldsymbol{B}$ outside the sphere in closed form, in the following simple way. Outside the sphere $\boldsymbol{B}$ is both irrotational (since $\boldsymbol{J} = \boldsymbol{0}$ there) and solenoidal [by (4.3)]. Hence it is the gradient of a harmonic function $\frac{\mu_0}{4\pi} U$ which we will call the scalar magnetic potential

$$\boldsymbol{B}(\boldsymbol{r}) = \frac{\mu_0}{4\pi} \nabla U(\boldsymbol{r}), |\boldsymbol{r}| > \alpha. \tag{4.67}$$

Using integration along a ray from $\boldsymbol{r}$ to infinity and the fact that the magnetic potential vanishes at infinity, we can write $U$ as follows :

$$\begin{aligned} U(\boldsymbol{r}) &= - \int_r^{+\infty} \frac{\partial}{\partial r'} U(\boldsymbol{r}') dr' = - \int_r^{+\infty} \hat{\boldsymbol{r}} \cdot \nabla_{\boldsymbol{r}'} U(\boldsymbol{r}') dr' \\ &= - \frac{4\pi}{\mu_0} \int_{\boldsymbol{r}}^{+\infty} \hat{\boldsymbol{r}}' \cdot \boldsymbol{B}(\boldsymbol{r}') dr' = \boldsymbol{Q} \times \boldsymbol{r}_0 \cdot \hat{\boldsymbol{r}} \int_r^{+\infty} \frac{dr'}{|r'\hat{\boldsymbol{r}} - \boldsymbol{r}_0|^3}, \end{aligned} \tag{4.68}$$

for $r > \alpha$.

A straightforward calculation gives

$$\int_r^{+\infty} \frac{dr'}{|r'\hat{\boldsymbol{r}} - \boldsymbol{r}_0|^3} = \frac{r}{F(\boldsymbol{r}, \boldsymbol{r}_0)} \tag{4.69}$$

where

$$F(\boldsymbol{r}, \boldsymbol{r}_0) = |\boldsymbol{r} - \boldsymbol{r}_0| \left[ |\boldsymbol{r}| |\boldsymbol{r} - \boldsymbol{r}_0| + \boldsymbol{r} \cdot (\boldsymbol{r} - \boldsymbol{r}_0) \right] \tag{4.70}$$

and therefore the magnetic potential assumes the form

$$U(\boldsymbol{r}) = \frac{\boldsymbol{Q} \times \boldsymbol{r}_0 \cdot \boldsymbol{r}}{F(\boldsymbol{r}, \boldsymbol{r}_0)} \tag{4.71}$$

and the magnetic induction is then given by

$$\begin{aligned} \boldsymbol{B}(\boldsymbol{r}) &= \frac{\mu_0}{4\pi} \frac{\boldsymbol{Q} \times \boldsymbol{r}_0}{F^2(\boldsymbol{r}, \boldsymbol{r}_0)} \cdot [\tilde{\boldsymbol{I}} - \boldsymbol{r} \otimes \nabla] F(\boldsymbol{r}, \boldsymbol{r}_0) \\ &= \frac{\mu_0}{4\pi} \frac{\boldsymbol{Q} \times \boldsymbol{r}_0}{F(\boldsymbol{r}, \boldsymbol{r}_0)} - \frac{\mu_0}{4\pi} \frac{\boldsymbol{Q} \times \boldsymbol{r}_0 \cdot \boldsymbol{r}}{F^2(\boldsymbol{r}, \boldsymbol{r}_0)} \nabla F(\boldsymbol{r}, \boldsymbol{r}_0) \end{aligned} \tag{4.72}$$

where $\tilde{\boldsymbol{I}}$ is the identity dyadic and

$$\begin{aligned} \nabla F(\boldsymbol{r}, \boldsymbol{r}_0) &= |\boldsymbol{r} - \boldsymbol{r}_0| \left( |\boldsymbol{r}| + |\boldsymbol{r} - \boldsymbol{r}_0| \right) \hat{\boldsymbol{r}} \\ &\quad + \left[ |\boldsymbol{r} - \boldsymbol{r}_0|^2 + 2|\boldsymbol{r}||\boldsymbol{r} - \boldsymbol{r}_0| + \boldsymbol{r} \cdot (\boldsymbol{r} - \boldsymbol{r}_0) \right] \frac{\boldsymbol{r} - \boldsymbol{r}_0}{|\boldsymbol{r} - \boldsymbol{r}_0|}. \end{aligned} \tag{4.73}$$

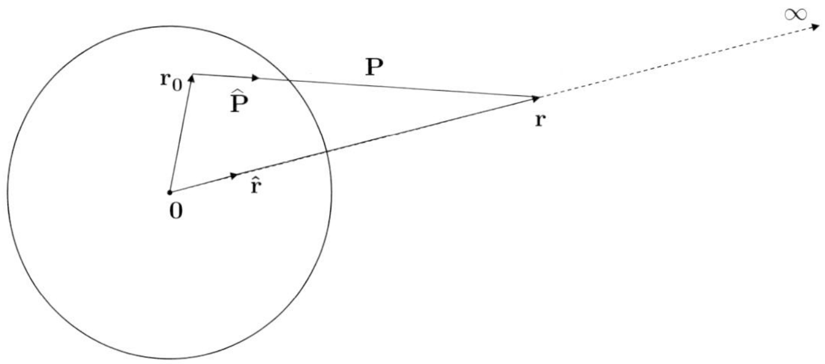


**Fig. 4.3** The path integral for the calculation of the magnetic potential

Note that the only two vectors involved in the expressions for $U$ and $\boldsymbol{B}$ are $\boldsymbol{r}$ and

$$\boldsymbol{P} = \boldsymbol{r} - \boldsymbol{r}_0 \tag{4.74}$$

which specify the observation point with respect to the center of the conductor and to the source point $\boldsymbol{r}_0$, respectively (see Fig. 4.3). In terms of $\boldsymbol{r}$ and $\boldsymbol{P}$ (4.71) and (4.72) are written as

$$U(\boldsymbol{r}) = \boldsymbol{Q} \cdot \frac{\hat{\boldsymbol{r}} \times \hat{\boldsymbol{P}}}{P(1 + \hat{\boldsymbol{r}} \cdot \hat{\boldsymbol{P}})} \tag{4.75}$$

and

$$\begin{aligned} \boldsymbol{B}(\boldsymbol{r}) = & \frac{\mu_0}{4\pi} \frac{\boldsymbol{Q} \times (\boldsymbol{r} - \boldsymbol{P})}{rP^2(1 + \hat{\boldsymbol{r}} \cdot \hat{\boldsymbol{P}})} - \frac{\mu_0}{4\pi} \boldsymbol{Q} \cdot \frac{\hat{\boldsymbol{r}} \times \hat{\boldsymbol{P}}}{P^2(1 + \hat{\boldsymbol{r}} \cdot \hat{\boldsymbol{P}})^2} \\ & \times \left[ \left(1 + \frac{P}{r}\right) \hat{\boldsymbol{r}} + \left(2 + \frac{P}{r} + \hat{\boldsymbol{r}} \cdot \hat{\boldsymbol{P}}\right) \hat{\boldsymbol{P}} \right] \end{aligned} \tag{4.76}$$

which is also written as

$$\boldsymbol{B}(\boldsymbol{r}) = \frac{\mu_0}{4\pi} \frac{\boldsymbol{Q} \times (\boldsymbol{r} - \boldsymbol{P})}{rP^2(1 + \hat{\boldsymbol{r}} \cdot \hat{\boldsymbol{P}})} - \frac{\mu_0}{4\pi} \frac{U(\boldsymbol{r})}{P} \left[ \hat{\boldsymbol{P}} + \frac{r + P}{r(1 + \hat{\boldsymbol{r}} \cdot \hat{\boldsymbol{P}})} (\hat{\boldsymbol{r}} + \hat{\boldsymbol{P}}) \right] \tag{4.77}$$

where we remind that the wedge on a vector means that the vector has unit length. The magnetic potential $U$, as well as its expansion in exterior vector spherical harmonics, was first obtained by Bronzan [2], but it was rediscovered in the framework of MEG by Sarvas [46]. All of the mathematical theory of magnetoencephalography for the spherical model of the brain, for the last 20 years, is founded on the Sarvas solution (4.71), (4.72).

Closing this section for the sphere we expand the magnetic potential in terms of exterior spherical harmonic functions. In fact, it is not easy to expand the form (4.71) of U, but we can rewrite it as

$$
\begin{aligned}
U(\boldsymbol{r}) &= \boldsymbol{Q} \times \boldsymbol{r}_0 \cdot \hat{\boldsymbol{r}} \int_r^{+\infty} \frac{dr'}{|r'\hat{\boldsymbol{r}} - \boldsymbol{r}_0|^3} \\
&= \boldsymbol{Q} \times \boldsymbol{r}_0 \cdot \int_r^{+\infty} \frac{r'\hat{\boldsymbol{r}} - \boldsymbol{r}_0}{|r'\hat{\boldsymbol{r}} - \boldsymbol{r}_0|^3} \frac{dr'}{r'} \\
&= \boldsymbol{Q} \times \boldsymbol{r}_0 \cdot \nabla_{\boldsymbol{r}_0} \int_r^{+\infty} \frac{1}{|r'\hat{\boldsymbol{r}} - \boldsymbol{r}_0|} \frac{dr'}{r'} \\
&= \boldsymbol{Q} \times \boldsymbol{r}_0 \cdot \nabla_{\boldsymbol{r}_0} \int_0^1 \frac{dt}{|\boldsymbol{r} - t\boldsymbol{r}_0|} \\
&= \boldsymbol{Q} \times \boldsymbol{r}_0 \cdot \nabla_{\boldsymbol{r}_0} \int_0^1 \sum_{n=1}^{\infty} \frac{(tr_0)^n}{r^{n+1}} P_n(\hat{\boldsymbol{r}} \cdot \hat{\boldsymbol{r}}_0) dt \\
&= \boldsymbol{Q} \times \boldsymbol{r}_0 \cdot \nabla_{\boldsymbol{r}_0} \sum_{n=1}^{\infty} \frac{1}{n+1} \frac{r_0^n}{r^{n+1}} P_n(\hat{\boldsymbol{r}} \cdot \hat{\boldsymbol{r}}_0)
\end{aligned}
\tag{4.78}
$$

where we have used the transformation $r' = r/t$.

In view of the addition theorem, (4.78) is also written as

$$
U(\boldsymbol{r}) = \boldsymbol{Q} \times \boldsymbol{r}_0 \cdot \nabla_{\boldsymbol{r}_0} \sum_{n=1}^{\infty} \sum_{m=-n}^{n} \frac{4\pi}{(n+1)(2n+1)} \frac{r_0^n}{r^{n+1}} Y_n^{m^*}(\hat{\boldsymbol{r}}_0) Y_n^m(\hat{\boldsymbol{r}}) \tag{4.79}
$$

where $Y_n^m$ are the normalized complex spherical harmonics. Note that

$$
U(\boldsymbol{r}) = O\left(\frac{1}{r^2}\right), r \to \infty \tag{4.80}
$$

and the angular differential operator $\boldsymbol{Q} \times \boldsymbol{r}_0 \cdot \nabla_{\boldsymbol{r}_0}$ depends only on the dipolar source. Interchanging the algebraic operations in this triple product operator, and using the fact that

$$
\boldsymbol{r}_0 \times \nabla_{\boldsymbol{r}_0} Y_n^m(\hat{\boldsymbol{r}}_0) = \boldsymbol{C}_n^m(\hat{\boldsymbol{r}}_0) \tag{4.81}
$$

where $\boldsymbol{C}_n^m$ is the indicated vector spherical harmonic we can rewrite (4.79) as

$$
U(\boldsymbol{r}) = \boldsymbol{Q} \cdot \sum_{n=1}^{\infty} \sum_{m=-n}^{n} \frac{4\pi}{(n+1)(2n+1)} \frac{r_0^n}{r^{n+1}} \boldsymbol{C}_n^{m^*}(\hat{\boldsymbol{r}}_0) Y_n^m(\hat{\boldsymbol{r}}). \tag{4.82}
$$

From (4.82) we observe that the source of $U$ is expressible only in terms of the vector harmonics $\boldsymbol{C}_n^m$. Therefore, in accord with the result obtained in [20, 21] for the sphere and in [5, 17] for any star-shaped conductor, the part of the current that is needed to generate U lives only in the subspace spanned by $\{\boldsymbol{C}_n^m\}$, while the part that lives in the orthogonal complement of this subspace is magnetically "silent". In other words, the null space of the MEG sources is "twice" as large as the space that can actually be detected outside the head. As was proved in [17], this result is characteristic of the physics of the problem and has nothing to do with the extreme symmetry of the sphere. Obviously, formulae (4.67) and (4.82) provide the corresponding spherical expansion for the magnetic induction field.

Finally, two important observations are: (a) any dipole with a radial moment $\mathbf{Q}$ is magnetically "silent", since $\boldsymbol{Q} \times \boldsymbol{r}_0$ vanishes and therefore no magnetic field can be detected outside the sphere, and (b) the magnetic potential U and therefore also $\mathbf{B}$, are independent of the radius of the sphere, so that the extent of the conduction region is also "silent" for the case of the sphere.

## 4.4 Elements of Ellipsoidal Harmonics

In an anisotropic space the ellipsoid plays the role that the sphere plays in an isotropic space. This is the reason why the ellipsoid appears so naturally in so many apparently unrelated areas in Science and Mathematics. We just mention here the ellipsoid of inertia, the gyroscopic ellipsoid, the polarization ellipsoid, the diagonal form of any positive definite quadratic form, such as the energy functionals, or even the shape of the celestial bodies, and the brain itself. Besides its role in the study of anisotropic domains, the ellipsoid provides a very good approximation of any convex three-dimensional body. The ellipsoid, being a genuine 3D shape, gives rise to mathematical problems that are much harder than the corresponding 1D problems associated with the sphere. The theory of ellipsoidal harmonics [30, 37, 48] is a fascinating theory with very elegant structure and a plethora of applications, which unfortunately is not so well spread today, perhaps because of its demand for calculational techniques. For this reason we will present here the basic background needed to solve boundary value problems in ellipsoidal geometry. The theory of ellipsoidal harmonics, as well as the theory of curvilinear coordinates, was introduced by Lamé in a series of papers published in the first half of the nineteenth century [30], during his efforts to solve the problem-of temperature distribution within an ellipsoid in thermal equilibrium. Then, during the following 70 years, through the basic contributions of Ferrers, Heine, Niven, Lindemann, Dixon, Liouville, Stieltjes, Whittaker, Hermite, Hobson and others [30, 48] the theory was shaped essentially in the form that is available today. Although many theoretical results on Lamé functions exist in the literature, from the applied point of view, there are still open problems that need to be settled. For example, the separation constants that enter the Lamé functions of degree greater than three are not explicitly know, and a closed form expression for the normalized constants of the surface ellipsoidal harmonics is known only for degree less or equal to three [8, 18, 36]. Even more useful would be the development of general formulae that would allow the transformation of harmonics from ellipsoidal to spherical coordinates and vice versa, although approximate and inconvenient expressions of this kind do exist in the literature.

In order to introduce a spherical system, all we need to do is to pick up a center, from which all radial distances are determined, and to specify the unit sphere, from which all orientations are specified by its points. A confocal ellipsoidal system is defined also by picking up a center, but instead of the unit sphere we now need to specify a fundamental ellipsoid

$$\frac{x_1^2}{\alpha_1^2} + \frac{x_2^2}{\alpha_2^2} + \frac{x_3^2}{\alpha_3^2} = 1 \tag{4.83}$$

with $0 < \alpha_3 < \alpha_2 < \alpha_1 < +\infty$, which introduces the confocal characteristics of the system as we will explain in the sequel.

The ellipsoidal coordinate system is based on the following observation. The quadratic form

$$\frac{x_1^2}{\alpha_1^2 - \lambda} + \frac{x_2^2}{\alpha_2^2 - \lambda} + \frac{x_3^2}{\alpha_3^2 - \lambda} = 1 \tag{4.84}$$

defines a family of:

(i) *ellipsoids*, for $\lambda \in (-\infty, \alpha_3^2)$
(ii) *1-hyperboloids* (hyperboloids of one sheet), for $\lambda \in (\alpha_3^2, \alpha_2^2)$
(iii) *2-hyperboloids* (hyperboloids of two sheets), for $\lambda \in (\alpha_2^2, \alpha_1^2)$.

For $\lambda > \alpha_1^2$ (4.84) defines an imaginary family of surfaces. For $\lambda = \alpha_3^2$ the ellipsoid (4.84) collapses to the *focal ellipse*

$$\frac{x_1^2}{\alpha_1^2 - \alpha_3^2} + \frac{x_2^2}{\alpha_2^2 - \alpha_3^2} = 1 \tag{4.85}$$

while for $\lambda = \alpha_2^2$ it collapses to the *focal hyperbola*

$$\frac{x_1^2}{\alpha_1^2 - \alpha_2^2} + \frac{x_3^2}{\alpha_3^2 - \alpha_2^2} = 1. \tag{4.86}$$

The focal ellipse and the focal hyperbola provide the "back-bone" of the ellipsoidal system and play the role of the polar axis of the spherical system. It can be proved [30] that the cubic in $\lambda$ polynomial (4.84) has, for every point $(x_1, x_2, x_3)$ with $x_1 x_2 x_3 \neq 0$, three real roots $\lambda_1, \lambda_2, \lambda_3$ such that

$$-\infty < \lambda_3 < \alpha_3^2 < \lambda_2 < \alpha_2^2 < \lambda_1 < \alpha_1^2. \tag{4.87}$$

This 1–1 correspondence between points $(x_1, x_2, x_3) \in \mathbb{R}^3$ and roots $(\lambda_1, \lambda_2, \lambda_3)$ in $\mathbb{R}^3$ defines the ellipsoidal coordinate system, so that from $(x_1, x_2, x_3)$ there pass exactly one ellipsoid, one 1-hyperboloid and one 2-hyperboloid. All these three families of quadrics are confocal and they are based on the six foci

$$(\pm h_2, 0, 0), \qquad (\pm h_3, 0, 0), \qquad (0, \pm h_1, 0)$$

where the semifocal distances are defined as

$$h_1^2 = \alpha_2^2 - \alpha_3^2, \qquad h_2^2 = \alpha_1^2 - \alpha_3^2, \qquad h_3^2 = \alpha_1^2 - \alpha_2^2. \tag{4.88}$$

Obviously, the foci of the focal ellipse are $(\pm h_3, 0, 0)$ and its semiaxes are $h_2$ and $h_1$, where $h_1 < h_2$ and $h_3 < h_2$. Figure 4.4 depicts the six foci, the semiaxes of the fundamental ellipsoid (4.83), the focal ellipse (4.85) and the focal hyperbola (4.86).

In Fig. 4.5 the three coordinate surfaces of the ellipsoidal system are shown separately.

The Cartesian planes $x_1 = 0, x_2 = 0, x_3 = 0$ intersect any one of the confocal quadrics either in an ellipse or in a hyperbola which are called *principal ellipses* and *principal hyperbolas* of the corresponding quadric.

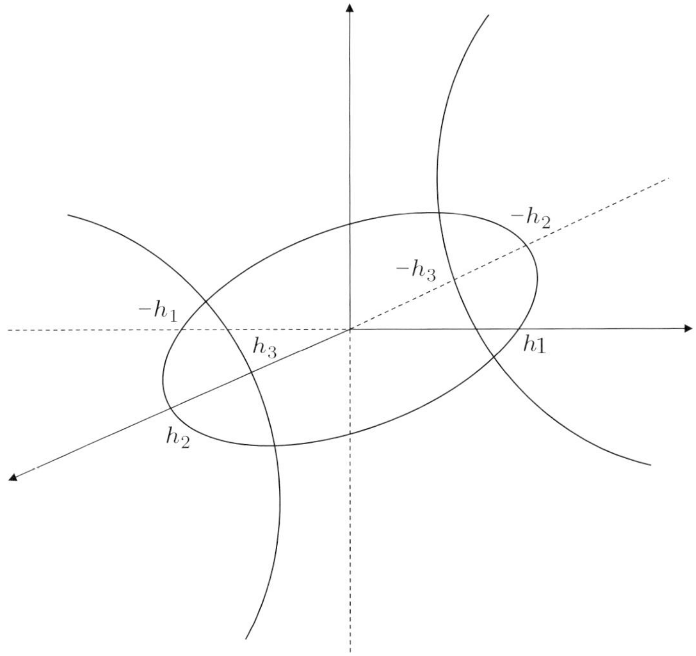


**Fig. 4.4** The focal ellipse and the focal hyperbola

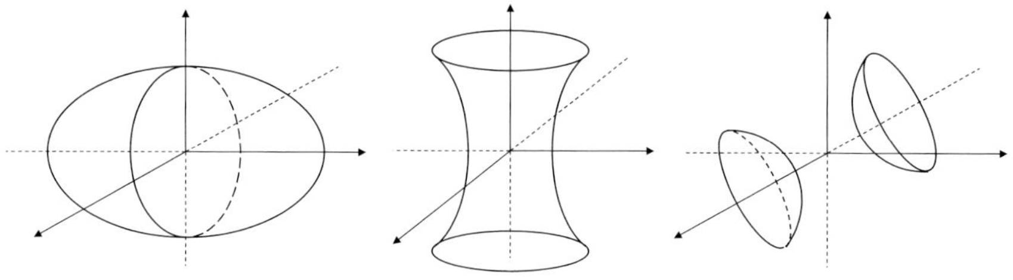


**Fig. 4.5** The ellipsoidal coordinate surfaces

The standard ellipsoidal coordinates are denoted by $(\rho, \mu, \nu)$ where

$$\rho^2 = \alpha_1^2 - \lambda_3, \qquad \mu^2 = \alpha_1^2 - \lambda_2, \qquad \nu^2 = \alpha_1^2 - \lambda_1 \tag{4.89}$$

and they are related with the Cartesian coordinates by

$$\begin{aligned} x_1 &= \frac{1}{h_2 h_3} \rho\mu\nu \\ x_2 &= \frac{1}{h_1 h_3} \sqrt{\rho^2 - h_3^2}\sqrt{\mu^2 - h_3^2}\sqrt{h_3^2 - \nu^2} \\ x_3 &= \frac{1}{h_1 h_2} \sqrt{\rho^2 - h_2^2}\sqrt{h_2^2 - \mu^2}\sqrt{h_2^2 - \nu^2} \end{aligned} \tag{4.90}$$

where

$$0 \leqslant \nu^2 \leqslant h_3^2 \leqslant \mu^2 \leqslant h_2^2 \leqslant \rho^2. \tag{4.91}$$

The coordinate $\rho$ specifies the ellipsoid, the coordinate $\mu$ specifies the 1-hyperboloid and the coordinate $\nu$ specifies the 2-hyperboloid that pass through any generic point in space. The ellipsoidal system is orthogonal and it forms a dextral system in the order $(\rho, \nu, \mu)$. Nevertheless, it is historically established to write the ellipsoidal coordinates in the order $(\rho, \mu, \nu)$, perhaps since this order corresponds to successive intervals of variation of the coordinates $\rho, \mu$ and $\nu$ as it is seen in (4.91). As far as the sign conventions that identify points in different octanes are concerned, we can avoid them if we use the symmetries of the fundamental ellipsoid which allow us to work only with points in the first octant, where all three coordinates are positive.

Inversion of the expressions (4.90) to obtain $\rho, \mu, \nu$ in terms of $x_1, x_2, x_3$ leads to complicated formulae [23] since they are roots of irreducible cubic polynomials.

The Gaussian map on the ellipsoid

$$\frac{x_1^2}{\rho^2} + \frac{x_2^2}{\rho^2 - h_3^2} + \frac{x_3^2}{\rho^2 - h_2^2} = 1 \tag{4.92}$$

gives the outward unit normal

$$\hat{\boldsymbol{\rho}} = \frac{\rho}{h_\rho} \sum_{i=1}^{3} \frac{x_i}{\rho^2 - \alpha_1^2 + \alpha_i^2} \hat{\boldsymbol{x}}_i \tag{4.93}$$

and similarly for the 1-hyperboloid

$$\frac{x_1^2}{\mu^2} + \frac{x_2^2}{\mu^2 - h_3^2} + \frac{x_3^2}{\mu^2 - h_2^2} = 1 \tag{4.94}$$

we obtain

$$\hat{\boldsymbol{\mu}} = \frac{\mu}{h_\mu} \sum_{i=1}^{3} \frac{x_i}{\mu^2 - \alpha_1^2 + \alpha_i^2} \hat{\boldsymbol{x}}_i \tag{4.95}$$

and for the 2-hyperboloid

$$\frac{x_1^2}{\nu^2} + \frac{x_2^2}{\nu^2 - h_3^2} + \frac{x_3^2}{\nu^2 - h_2^2} = 1 \tag{4.96}$$

we obtain

$$\hat{\boldsymbol{\nu}} = \frac{\nu}{h_\nu} \sum_{i=1}^{3} \frac{x_i}{\nu^2 - \alpha_1^2 + \alpha_i^2} \hat{\boldsymbol{x}}_i \tag{4.97}$$

where the metric coefficients $h_\rho, h_\mu, h_\nu$ are given by

$$h_\rho^2 = \frac{(\rho^2 - \mu^2)(\rho^2 - \nu^2)}{(\rho^2 - h_3^2)(\rho^2 - h_2^2)} \tag{4.98}$$

$$h_\mu^2 = \frac{(\mu^2 - \nu^2)(\mu^2 - \rho^2)}{(\mu^2 - h_3^2)(\mu^2 - h_2^2)} \tag{4.99}$$

$$h_\nu^2 = \frac{(\nu^2 - \rho^2)(\nu^2 - \mu^2)}{(\nu^2 - h_3^2)(\nu^2 - h_2^2)}. \tag{4.100}$$

Although simple and smooth as a surface, the ellipsoid is highly non symmetric. In fact, its transformation group contains only eight elements, three rotations around the principal axes by $\pi$, three reflections with respect to the principal planes, one inversion with respect to its center and of course the identity transformation. Compare with the spheroids that have a one-parameter group of transformations and with the sphere that has a two-parameter group of transformations.

Laplace's operator in ellipsoidal coordinates reads

$$\begin{aligned}\Delta = &\frac{1}{(\rho^2-\mu^2)(\rho^2-\nu^2)}\left[(\rho^2-h_3^2)(\rho^2-h_2^2)\frac{\partial^2}{\partial\rho^2}+\rho(2\rho^2-h_3^2-h_2^2)\frac{\partial}{\partial\rho}\right]\\ &+\frac{1}{(\mu^2-\rho^2)(\mu^2-\nu^2)}\left[(\mu^2-h_3^2)(\mu^2-h_2^2)\frac{\partial^2}{\partial\mu^2}+\mu(2\mu^2-h_3^2-h_2^2)\frac{\partial}{\partial\mu}\right]\\ &+\frac{1}{(\nu^2-\rho^2)(\nu^2-\mu^2)}\left[(\nu^2-h_3^2)(\nu^2-h_2^2)\frac{\partial^2}{\partial\nu^2}+\nu(2\nu^2-h_3^2-h_2^2)\frac{\partial}{\partial\nu}\right]\end{aligned} \tag{4.101}$$

and it is remarkable to recognize that, although each one of the metric coefficients $h_\rho, h_\mu, h_\nu$ with respect of which the Laplacian is expressed involves four square roots, in the final expression of the Laplacian no square roots are involved. Instead, all six coefficients are rational functions of $\rho, \mu, \nu$ involving the two independent semifocal distances $h_3, h_2$. The semifocal distance $h_1$ is related to $h_2, h_3$ via

$$h_1^2 = h_2^2 - h_3^2. \tag{4.102}$$

A simple observation shows that, in its standard form, (4.101) is not separable. Nevertheless, Lamé, using an ingenious transformation and some tricky manipulations [30], managed to actually separate Laplace's equation in ellipsoidal coordinates and he showed that the dependence of the solution on each one of the $(\rho, \mu, \nu)$ variables is governed by exactly the same ordinary differential equation

$$\begin{aligned}(x^2-h_3^2)(x^2-h_2^2)y''(x)+x(2x^2-h_3^2-h_2^2)y'(x)\\ +[(h_3^2+h_2^2)p-n(n+1)x^2]y(x)=0\end{aligned} \tag{4.103}$$

which is known today as the Lamé equation. A series of basic arguments leads to the form (4.103), where $n = 0, 1, 2, \ldots$ and $p \in \mathbb{R}$ are the separation constants. The particular choice of $n$ is guided by the demand to obtain polynomial solutions, just as we do with the spherical case, and it can be proved [30] that for each $n$ there are exactly $(2n+1)$ real values of $p$ for which the corresponding solutions of (4.103) are linearly independent. The calculation of these $2n+1$ values of $p$ is the biggest problem in obtaining closed form solutions of ellipsoidal harmonics. Without being technical, we can say that the values of $p$ come as roots of four different polynomials for each $n$ and, of course, their degree increases with $n$. In fact, up to $n = 3$ we only need to solve no higher than quadratic polynomials, up to $n = 7$ we need to solve polynomials up to the fourth degree, and for $n \geqslant 8$ we need to solve polynomials of degree greater of equal to 5. Hence, in principle, the only ellipsoidal harmonics we know in closed form are the 64 harmonics that correspond to $n \leqslant 7$.

Using the Frobenious method to obtain series, and then polynomial, solutions of the Lamé equations we are lead to four different classes of Lamé functions which are known as $K, L, M$ and $N$ classes. Each class contains functions of the following structure:

$$K(x) = x^n + a_1 x^{n-2} + a_2 x^{n-4} + \cdots \tag{4.104}$$

$$L(x) = \sqrt{|x^2 - h_3^2|}(x^{n-1} + b_1 x^{n-3} + \cdots) \tag{4.105}$$

$$M(x) = \sqrt{|x^2 - h_2^2|}(x^{n-1} + c_1 x^{n-3} + \cdots) \tag{4.106}$$

$$N(x) = \sqrt{|x^2 - h_3^2|}\sqrt{|x^2 - h_2^2|}(x^{n-2} + d_1 x^{n-4} + \cdots). \tag{4.107}$$

Note that the above polynomials, depending on $n$, are either even or odd, and that they end either with a linear or with a constant term. So, the only difference of the Lamé functions in the variables $(\rho, \mu, \nu)$ is their domains, which, for solutions of Laplace's equation in the first octant, are

$$h_2 < \rho < +\infty, \qquad h_3 < \mu < h_2, \qquad 0 < \nu < h_3. \tag{4.108}$$

If we enumerate the constants $p_1, p_2, \ldots, p_{2n+1}$ and denote the Lamé function that corresponds to $(n, p_m)$ by $E_n^m(x)$, then we define the *ellipsoidal harmonics of the first kind* (interior harmonics)

$$\mathbb{E}_n^m(\rho, \mu, \nu) = E_n^m(\rho) E_n^m(\mu) E_n^m(\nu) \tag{4.109}$$

for $n = 0, 1, 2, \ldots$ and $m = 1, 2, \ldots, 2n+1$.

The *ellipsoidal harmonics of the second kind* (exterior harmonics) are defined as

$$\mathbb{F}_n^m(\rho, \mu, \nu) = F_n^m(\rho) E_n^m(\mu) E_n^m(\nu) \tag{4.110}$$

where $F_n^m(\rho)$ is the second linearly independent solution of (4.103) which can be obtained by using reduction of order once $E_n^m(\rho)$ is known. This leads to

$$F_n^m(\rho) = (2n+1) E_n^m(\rho) \int_\rho^{+\infty} \frac{dx}{[E_n^m(x)]^2 \sqrt{x^2 - h_3^2}\sqrt{x^2 - h_2^2}} \tag{4.111}$$

so that all exterior ellipsoidal harmonics are given in terms of elliptic integrals [48]. It is of interest to observe that as $\rho \to \infty$

$$F_n^m(\rho) = O\left(\frac{1}{\rho^{n+1}}\right) \tag{4.112}$$

so that as $\rho$ increases and the ellipsoids tend to become spheres, the Lamé functions $F_n^m$ coincide with the radial dependence $r^{-(n+1)}$ of the exterior spherical harmonics. On the other hand, as $\rho \longrightarrow h_2$, i.e. as the ellipsoid degenerates to the focal ellipse, not all of $F_n^m$'s tend to infinity as it can be seen for example by

$$F_0^1(h_2) = \int_{h_2}^{+\infty} \frac{dx}{\sqrt{x^2 - h_3^2}\sqrt{x^2 - h_2^2}} \tag{4.113}$$

which is a nonvanishing convergent improper integral. This peculiarity of the ellipsoidal system is not shared with the spherical system where every exterior harmonic is singular at the origin. It is customary to define the elliptic integrals

$$I_n^m(\rho) = \int_{\rho}^{+\infty} \frac{dx}{[E_n^m(x)]^2\sqrt{x^2 - h_3^2}\sqrt{x^2 - h_2^2}} \tag{4.114}$$

so that

$$F_n^m(\rho) = (2n+1)E_n^m(\rho)I_n^m(\rho). \tag{4.115}$$

If we rewrite the Lamé equation (4.103) in the form

$$\frac{d}{dx}\left[\sqrt{x^2 - h_3^2}\sqrt{x^2 - h_2^2}\frac{d}{dx}y(x)\right] + \left[\frac{h_2^2 + h_3^2}{\sqrt{x^2 - h_3^2}\sqrt{x^2 - h_2^2}}p - \frac{n(n+1)x^2}{\sqrt{x^2 - h_3^2}\sqrt{x^2 - h_2^2}}\right]y(x) = 0 \tag{4.116}$$

where for the case of the ellipsoid $x = \rho$, we see that we have to deal with a singular Sturm–Liouville problem where p is the eigenvalue parameter and

$$w(x) = \frac{h_3^2 + h_2^2}{\sqrt{x^2 - h_3^2}\sqrt{x^2 - h_2^2}} > 0 \tag{4.117}$$

is the weighting function of the associate inner product. Consequently, all the eigenvalues $p$ are real and eigensolutions corresponding to different eigenvalues are orthogonal. Furthermore, the eigenvalues are ordered as

$$0 < p_1 < p_2 < \cdots < p_{2n+1} \underset{n\to\infty}{\longrightarrow} \infty \tag{4.118}$$

and the set of all eigensolutions forms a complete set with respect to the associate inner product.

The theory of ellipsoidal harmonics has also been developed by Niven [30,48] in the framework of the Cartesian coordinate system and the obtained expressions are particularly symmetric and elegant. Nevertheless, the problem of calculating the separation constants remains open in the Cartesian forms too. Note that this problem does not appear in the case of spherical harmonics since there, the second separation constant assumes the values

$$m = -n, -n+1, \ldots, 0, \ldots, n-1, n \tag{4.119}$$

which are imposed by the $2\pi$-periodicity condition in the azimuthal angle. So, moving from the degree $n$ to the degree $(n+1)$ we keep the previous $(2n+1)$ values of $m$ and we add two more, the values $\pm(n+1)$. But, for the ellipsoidal harmonics, as we move from the harmonics of degree $n$ to those of degree $(n+1)$, none of the $(2n+1)$ $p$-eigenvalues for $n$ remains a $p$-eigenvalue for (n+1).

We are not going to discuss here the theory of ellipsoidal harmonics in Cartesian form. Instead we are going to give the ellipsoidal harmonics of degree less or equal to 3, both in ellipsoidal and in Cartesian form, as ready-to-use building blocks to construct solutions of boundary value problems for the Laplace's equation in domains interior or exterior to a fundamental ellipsoid.

It is important to understand here that, in contrast to the spherical case, where given a center there is only one spherical system with this center, for the case of the ellipsoid, there are infinitely many ellipsoidal systems with the given center. This is due to the fact, as we mentioned earlier, that in order to define an ellipsoidal system we need to specify its center as well as the position of its six foci, which for the spherical case they are all located at the center. Hence, in order to be able to solve any boundary value problem we need to choose the ellipsoidal system in such a way as to fit the actual boundary by choosing a particular value of $\rho$. This is secured if we use the boundary of our domain to be the fundamental ellipsoid (4.83) and construct the ellipsoidal system that is based on it.

The ellipsoidal harmonics of degree 0 are:

$$\mathbb{E}_0^1(\rho, \mu, \nu) = 1 \tag{4.120}$$

$$\mathbb{F}_0^1(\rho, \mu, \nu) = I_0^1(\rho). \tag{4.121}$$

The ellipsoidal harmonics of degree 1 are:

$$\mathbb{E}_1^1(\rho, \mu, \nu) = \rho\mu\nu = h_2 h_3 x_1 \tag{4.122}$$

$$\mathbb{E}_1^2(\rho, \mu, \nu) = \sqrt{\rho^2 - h_3^2}\sqrt{\mu^2 - h_3^2}\sqrt{h_3^2 - \nu^2} = h_1 h_3 x_2 \tag{4.123}$$

$$\mathbb{E}_1^3(\rho, \mu, \nu) = \sqrt{\rho^2 - h_2^2}\sqrt{h_2^2 - \mu^2}\sqrt{h_2^2 - \nu^2} = h_1 h_2 x_3 \tag{4.124}$$

$$\mathbb{F}_1^m(\rho, \mu, \nu) = 3\mathbb{E}_1^m(\rho, \mu, \nu) I_1^m(\rho), m = 1, 2, 3. \tag{4.125}$$

The ellipsoidal harmonics of degree 2 are:

$$\begin{aligned}\mathbb{E}_2^1(\rho,\mu,\nu) &= (\rho^2-\alpha_1^2+\Lambda)(\mu^2-\alpha_1^2+\Lambda)(\nu^2-\alpha_1^2+\Lambda)\\ &= (\Lambda-\alpha_1^2)(\Lambda-\alpha_2^2)(\Lambda-\alpha_3^2)\left[\sum_{i=1}^{3}\frac{x_i^2}{\Lambda-\alpha_i^2}+1\right]\end{aligned} \tag{4.126}$$

$$\begin{aligned}\mathbb{E}_2^2(\rho,\mu,\nu) &= (\rho^2-\alpha_1^2+\Lambda')(\mu^2-\alpha_1^2+\Lambda')(\nu^2-\alpha_1^2+\Lambda')\\ &= (\Lambda'-\alpha_1^2)(\Lambda'-\alpha_2^2)(\Lambda'-\alpha_3^2)\left[\sum_{i=1}^{3}\frac{x_i^2}{\Lambda'-\alpha_i^2}+1\right]\end{aligned} \tag{4.127}$$

$$\mathbb{E}_2^3(\rho,\mu,\nu) = \rho\sqrt{\rho^2-h_3^2}\,\mu\sqrt{\mu^2-h_3^2}\,\nu\sqrt{h_3^2-\nu^2} = h_1h_2h_3^2x_1x_2 \tag{4.128}$$

$$\mathbb{E}_2^4(\rho,\mu,\nu) = \rho\sqrt{\rho^2-h_2^2}\,\mu\sqrt{h_2^2-\mu^2}\,\nu\sqrt{h_2^2-\nu^2} = h_1h_2^2h_3x_1x_3 \tag{4.129}$$

$$\begin{aligned}\mathbb{E}_2^5(\rho,\mu,\nu) &= \sqrt{\rho^2-h_3^2}\sqrt{\rho^2-h_2^2}\sqrt{\mu^2-h_3^2}\sqrt{h_2^2-\mu^2}\sqrt{h_3^2-\nu^2}\sqrt{h_2^2-\nu^2}\\ &= h_1^2h_2h_3x_2x_3\end{aligned} \tag{4.130}$$

$$\mathbb{F}_2^m(\rho,\mu,\nu) = 5\mathbb{E}_2^m(\rho,\mu,\nu)I_2^m(\rho),\, m=1,2,\ldots,5 \tag{4.131}$$

where the constants $\Lambda$, $\Lambda'$ are the two roots of the quadratic

$$\sum_{i=1}^{3}\frac{1}{\Lambda-\alpha_i^2} = 0. \tag{4.132}$$

Finally the ellipsoidal harmonics of degree 3 are:

$$\begin{aligned}\mathbb{E}_3^1(\rho,\mu,\nu) &= \rho(\rho^2-\alpha_1^2+\Lambda_1)\mu(\mu^2-\alpha_1^2+\Lambda_1)\nu(\nu^2-\alpha_1^2+\Lambda_1)\\ &= h_2h_3(\Lambda_1-\alpha_1^2)(\Lambda_1-\alpha_2^2)(\Lambda_1-\alpha_3^2)x_1\left[\sum_{i=1}^{3}\frac{x_i^2}{\Lambda_1-\alpha_i^2}+1\right]\end{aligned} \tag{4.133}$$

$$\mathbb{E}_3^2(\rho,\mu,\nu) = \text{replace } \Lambda_1 \text{ by } \Lambda_1' \text{ in } \mathbb{E}_3^1 \tag{4.134}$$

$$\begin{aligned}\mathbb{E}_3^3(\rho,\mu,\nu) &= \sqrt{\rho^2-h_3^2}(\rho^2-\alpha_1^2+\Lambda_2)\sqrt{\mu^2-h_3^2}(\mu^2-\alpha_1^2+\Lambda_2)\\ &\quad\times\sqrt{h_3^2-\nu^2}(\nu^2-\alpha_1^2+\Lambda_2)\\ &= h_1h_3(\Lambda_2-\alpha_1^2)(\Lambda_2-\alpha_2^2)(\Lambda_2-\alpha_3^2)x_2\left[\sum_{i=1}^{3}\frac{x_i^2}{\Lambda_2-\alpha_i^2}+1\right]\end{aligned} \tag{4.135}$$

$$\mathbb{E}_3^4(\rho,\mu,\nu) = \text{replace } \Lambda_2 \text{ by } \Lambda_2' \text{ in } \mathbb{E}_3^3 \tag{4.136}$$

$$\begin{aligned}\mathbb{E}_3^5(\rho,\mu,\nu) &= \sqrt{\rho^2-h_2^2}(\rho^2-\alpha_1^2+\Lambda_3)\sqrt{h_2^2-\mu^2}(\mu^2-\alpha_1^2+\Lambda_3)\\ &\quad\times\sqrt{h_2^2-\nu^2}(\nu^2-\alpha_1^2+\Lambda_3)\\ &= h_1h_2(\Lambda_3-\alpha_1^2)(\Lambda_3-\alpha_2^2)(\Lambda_3-\alpha_3^2)x_3\left[\sum_{i=1}^{3}\frac{x_i^2}{\Lambda_3-\alpha_i^2}+1\right]\end{aligned} \tag{4.137}$$

$$\mathbb{E}_3^6(\rho,\mu,\nu) = \text{replace } \Lambda_3 \text{ by } \Lambda_3' \text{ in } \mathbb{E}_3^5 \tag{4.138}$$

$$\begin{aligned}\mathbb{E}_3^7(\rho,\mu,\nu) &= \rho\sqrt{\rho^2-h_3^2}\sqrt{\rho^2-h_2^2}\mu\sqrt{\mu^2-h_3^2}\sqrt{h_2^2-\mu^2}\nu\sqrt{h_3^2-\nu^2}\sqrt{h_2^2-\nu^2}\\ &= h_1^2h_2^2h_3^2x_1x_2x_3\end{aligned} \tag{4.139}$$

$$\mathbb{F}_3^m(\rho,\mu,\nu) = 7\mathbb{E}_3^m(\rho,\mu,\nu)I_3^m(\rho),\, m=1,2,\ldots,7 \tag{4.140}$$

where $\Lambda_1, \Lambda_1'$ are the roots of

$$\sum_{i=1}^{3}\frac{1}{\Lambda_1-\alpha_i^2}+\frac{2}{\Lambda_1-\alpha_1^2}=0 \tag{4.141}$$

$\Lambda_2, \Lambda_2'$ are the roots of

$$\sum_{i=1}^{3}\frac{1}{\Lambda_2-\alpha_i^2}+\frac{2}{\Lambda_2-\alpha_2^2}=0 \tag{4.142}$$

and $\Lambda_3, \Lambda_3'$ are the roots of

$$\sum_{i=1}^{3}\frac{1}{\Lambda_3-\alpha_i^2}+\frac{2}{\Lambda_3-\alpha_3^2}=0. \tag{4.143}$$

The surface ellipsoidal harmonics $\{E_n^m(\mu)E_n^m(\nu)\}_{n,m}$ form a complete orthogonal system over the surface of every ellipsoid $\rho=\rho_0$ with respect to the inner product

$$(f(\mu,\nu),g(\mu,\nu)) = \oint_{\rho=\rho_0} f(\mu,\nu)g(\mu,\nu)\frac{ds(\mu,\nu)}{\sqrt{\rho_0^2-\mu^2}\sqrt{\rho_0^2-\nu^2}}. \tag{4.144}$$

That is for $n=0,1,\ldots$ and $m=1,2,\ldots,2n+1$

$$\oint_{\rho=\rho_0} E_n^m(\mu)E_n^m(\nu)E_{n'}^{m'}(\mu)E_{n'}^{m'}(\nu)\frac{ds(\mu,\nu)}{\sqrt{\rho_0^2-\mu^2}\sqrt{\rho_0^2-\nu^2}} = \gamma_n^m \delta_{\mathrm{nn'}}\delta_{\mathrm{mm'}} \tag{4.145}$$

where the normalization constants $\gamma_n^m$ are not in general known, but for $n \leqslant 3$ they can be computed via a long and tedious procedure that finally leads to the following values [10]:

$$\gamma_0^1 = 4\pi \tag{4.146}$$
$$\gamma_1^1 = \frac{4\pi}{3}h_2^2h_3^2 \tag{4.147}$$
$$\gamma_1^2 = \frac{4\pi}{3}h_1^2h_3^2 \tag{4.148}$$
$$\gamma_1^3 = \frac{4\pi}{3}h_1^2h_2^2 \tag{4.149}$$
$$\gamma_2^1 = -\frac{8\pi}{5}(\Lambda-\Lambda')(\Lambda-\alpha_1^2)(\Lambda-\alpha_2^2)(\Lambda-\alpha_3^2) \tag{4.150}$$
$$\gamma_2^2 = \frac{8\pi}{5}(\Lambda-\Lambda')(\Lambda'-\alpha_1^2)(\Lambda'-\alpha_2^2)(\Lambda'-\alpha_3^2) \tag{4.151}$$
$$\gamma_2^3 = \frac{4\pi}{15}h_1^2h_2^2h_3^4 \tag{4.152}$$
$$\gamma_2^4 = \frac{4\pi}{15}h_1^2h_2^4h_3^2 \tag{4.153}$$
$$\gamma_2^5 = \frac{4\pi}{15}h_1^4h_2^2h_3^2 \tag{4.154}$$
$$\gamma_3^1 = -\frac{8\pi}{21}h_2^2h_3^2(\Lambda_1-\Lambda_1')(\Lambda_1-\alpha_1^2)(\Lambda_1-\alpha_2^2)(\Lambda_1-\alpha_3^2) \tag{4.155}$$
$$\gamma_3^2 = \frac{8\pi}{21}h_2^2h_3^2(\Lambda_1-\Lambda_1')(\Lambda_1'-\alpha_1^2)(\Lambda_1'-\alpha_2^2)(\Lambda_1'-\alpha_3^2) \tag{4.156}$$
$$\gamma_3^3 = -\frac{8\pi}{21}h_1^2h_3^2(\Lambda_2-\Lambda_2')(\Lambda_2-\alpha_1^2)(\Lambda_2-\alpha_2^2)(\Lambda_2-\alpha_3^2) \tag{4.157}$$
$$\gamma_3^4 = \frac{8\pi}{21}h_1^2h_3^2(\Lambda_2-\Lambda_2')(\Lambda_2'-\alpha_1^2)(\Lambda_2'-\alpha_2^2)(\Lambda_2'-\alpha_3^2) \tag{4.158}$$
$$\gamma_3^5 = -\frac{8\pi}{21}h_1^2h_2^2(\Lambda_3-\Lambda_3')(\Lambda_3-\alpha_1^2)(\Lambda_3-\alpha_2^2)(\Lambda_3-\alpha_3^2) \tag{4.159}$$
$$\gamma_3^6 = \frac{8\pi}{21}h_1^2h_2^2(\Lambda_3-\Lambda_3')(\Lambda_3'-\alpha_1^2)(\Lambda_3'-\alpha_2^2)(\Lambda_3'-\alpha_3^2) \tag{4.160}$$
$$\gamma_3^7 = \frac{4\pi}{105}h_1^4h_2^4h_3^4. \tag{4.161}$$

As we mentioned above these form of ellipsoidal harmonics can be used to solve many boundary value problems, essentially in the same way we solve spherical boundary value problem, as long as no harmonics of degree greater than 3 are necessary.

## 4.5 EEG in Ellipsoidal Geometry

Complete analysis of the content of this section can be found in [11, 16, 27, 32, 33, 36]. The direct electroencephalography problem for the homogeneous ellipsoidal conductor

$$\frac{x_1^2}{\alpha_1^2} + \frac{x_2^2}{\alpha_2^2} + \frac{x_3^2}{\alpha_3^2} = 1 \tag{4.162}$$

where $0 < \alpha_3 < \alpha_2 < \alpha_1 < +\infty$, which is excited by a current dipole of moment $\boldsymbol{Q}$ at the interior point $\boldsymbol{r}_0$ is stated as follows: find the interior electric potential $u^-$ that solves the Neumann problem

$$\sigma \Delta u^-(\boldsymbol{\rho}) = \nabla \cdot \boldsymbol{Q}\delta(\boldsymbol{\rho} - \boldsymbol{\rho}_0), \qquad \rho \in [h_2, \alpha_1) \tag{4.163}$$

$$\partial_\rho u(\boldsymbol{\rho}) = 0, \qquad \rho = \alpha_1 \tag{4.164}$$

where $\boldsymbol{\rho} = (\rho, \mu, \nu)$ is the position point and $\boldsymbol{\rho}_0 = (\rho_0, \mu_0, \nu_0)$ is the location of the dipole expressed in ellipsoidal coordinates. Obviously, $\rho_0 \in [h_2, \alpha_1)$.

Since the fundamental solution

$$G(\boldsymbol{r}; \boldsymbol{r}_0) = -\frac{1}{4\pi|\boldsymbol{r} - \boldsymbol{r}_0|} \tag{4.165}$$

solves the equation

$$\Delta G(\boldsymbol{r}; \boldsymbol{r}_0) = \delta(\boldsymbol{r} - \boldsymbol{r}_0) \tag{4.166}$$

we can easily identify a particular solution of Poisson's equation (4.163) in the form

$$u_0^-(\boldsymbol{\rho}) = \frac{1}{4\pi\sigma}\boldsymbol{Q} \cdot \nabla_{\boldsymbol{r}_0} \frac{1}{|\boldsymbol{r} - \boldsymbol{r}_0|}, \tag{4.167}$$

where $\boldsymbol{\rho}$ is the point $\boldsymbol{r}$ expressed in ellipsoidal coordinates. Note that the directional differentiation $\boldsymbol{Q} \cdot \nabla_{\boldsymbol{r}_0}$ depends only on the dipolar source. Then,

$$u^-(\boldsymbol{\rho}) = \frac{1}{4\pi\sigma}\boldsymbol{Q} \cdot \nabla_{\boldsymbol{r}_0} \frac{1}{|\boldsymbol{r} - \boldsymbol{r}_0|} + w(\boldsymbol{\rho}) \tag{4.168}$$

where $w$ is an interior harmonic function satisfying the boundary condition

$$\partial_\rho w(\boldsymbol{\rho}) = -\frac{1}{4\pi\sigma}\partial_\rho \boldsymbol{Q} \cdot \nabla_{\boldsymbol{r}_0} \frac{1}{|\boldsymbol{r} - \boldsymbol{r}_0|} \tag{4.169}$$

on $\rho = \alpha_1$. Expanding (4.165) in ellipsoidal harmonics we obtain

$$\frac{1}{|\boldsymbol{r} - \boldsymbol{r}_0|} = \sum_{n=0}^{\infty} \sum_{m=1}^{2n+1} \frac{4\pi}{2n+1} \frac{1}{\gamma_n^m} \mathbb{E}_n^m(\boldsymbol{\rho}_0) \mathbb{F}_n^m(\boldsymbol{\rho}) \tag{4.170}$$

which holds for $\rho > \rho_0$. In (4.170) the constants $\gamma_n^m$ are given by (4.145), $\mathbb{E}_n^m(\boldsymbol{\rho}_0)$ are the interior ellipsoidal harmonics evaluated at the position of the dipole and $\mathbb{F}_n^m(\boldsymbol{\rho})$ are the exterior ellipsoidal harmonics evaluated at the field point $\boldsymbol{\rho}$. Note that the particular expansion (4.170) holds all the way to the boundary $\rho = \alpha_1$ and therefore it can be used to calculate the coefficients of the interior expansion

$$w(\boldsymbol{\rho}) = \sum_{n=0}^{\infty} \sum_{m=1}^{2n+1} b_n^m \mathbb{E}_n^m(\boldsymbol{\rho}), h_2 \leqslant \rho < \alpha_1 \tag{4.171}$$

of the harmonic function $w$.

Since

$$\partial_\rho F_n^m(\rho) = (2n+1)(\partial_\rho E_n^m(\rho))I_n^m(\rho) - \frac{(2n+1)}{\sqrt{\rho^2 - h_3^2}\sqrt{\rho^2 - h_2^2}E_n^m(\rho)} \tag{4.172}$$

it follows that

$$b_n^m = \frac{\boldsymbol{Q} \cdot \nabla \mathbb{E}_n^m(\boldsymbol{\rho}_0)}{\gamma_n^m} \left[ \frac{1}{\alpha_2 \alpha_3 E_n^m(\alpha_1) E_n^{m\prime}(\alpha_1)} - I_n^m(\alpha_1) \right] \tag{4.173}$$

for $n \geqslant 1$, while since

$$\boldsymbol{Q} \cdot \nabla \mathbb{E}_0^1(\boldsymbol{\rho}_0) = 0 \tag{4.174}$$

the constant $b_0^1$ is undetermined, in accordance with the fact that the solution of the Neumann problem (4.163), (4.164) is unique up to an additive constant. Since EEG measurements are always potential differences we can take $b_0^1 = 0$.

Finally, the electric potential is given by

$$\begin{aligned} u^-(\boldsymbol{\rho}) = \frac{1}{\sigma} \sum_{n=1}^{\infty} \sum_{m=1}^{2n+1} \frac{1}{\gamma_n^m} (\boldsymbol{Q} \cdot \nabla \mathbb{E}_n^m(\boldsymbol{\rho}_0)) \\ \times \left[ I_n^m(\rho) - I_n^m(\alpha_1) + \frac{1}{\alpha_2 \alpha_3 E_n^m(\alpha_1) E_n^{m\prime}(\alpha_1)} \right] \mathbb{E}_n^m(\boldsymbol{\rho}) \end{aligned} \tag{4.175}$$

which holds near the boundary, $\rho \in (\rho_0, \alpha_1)$, or by

$$\begin{aligned} u^-(\boldsymbol{\rho}) = & \frac{1}{4\pi\sigma} \boldsymbol{Q} \cdot \nabla_{\mathbf{r}_0} \left[ \frac{1}{|\boldsymbol{r} - \boldsymbol{r}_0|} \right. \\ & \left. + 4\pi \sum_{n=1}^{\infty} \sum_{m=1}^{2n+1} \frac{1}{\gamma_n^m} \left( \frac{1}{\alpha_2 \alpha_3 E_n^m(\alpha_1) E_n^{m\prime}(\alpha_1)} \quad I_n^m(\alpha_1) \right) \mathbb{E}_n^m(\boldsymbol{\rho}_0) \mathbb{E}_n^m(\boldsymbol{\rho}) \right] \end{aligned} \tag{4.176}$$

which holds everywhere inside the ellipsoid, $\rho \in [h_2, \alpha_1)$. The elliptic integrals $I_n^m$ are given in (4.114).

It is of interest to write the solution (4.175) in terms of Cartesian coordinates. We do this for the $n = 1$ and the $n = 2$ terms of this expansion. To this end, we first calculate the action of the source dependent operator $\boldsymbol{Q} \cdot \nabla_{\mathbf{r}_0}$ on the source dependent eigenfunctions $\mathbb{E}_n^m, n = 1, 2, \ldots$

$$\boldsymbol{Q} \cdot \nabla \mathbb{E}_1^m(\boldsymbol{\rho}_0) = h_1 h_2 h_3 \frac{Q_m}{h_m}, m = 1, 2, 3 \tag{4.177}$$

$$\boldsymbol{Q} \cdot \nabla \mathbb{E}_2^1(\boldsymbol{\rho}_0) = 2(\Lambda - \alpha_1^2)(\Lambda - \alpha_2^2)(\Lambda - \alpha_3^2) \sum_{i=1}^{3} \frac{Q_i x_{0i}}{\Lambda - \alpha_i^2} \tag{4.178}$$

$$\boldsymbol{Q} \cdot \nabla \mathbb{E}_2^2(\boldsymbol{\rho}_0) = 2(\Lambda' - \alpha_1^2)(\Lambda' - \alpha_2^2)(\Lambda' - \alpha_3^2) \sum_{i=1}^{3} \frac{Q_i x_{0i}}{\Lambda' - \alpha_i^2} \tag{4.179}$$

$$\boldsymbol{Q} \cdot \nabla \mathbb{E}_2^3(\boldsymbol{\rho}_0) = h_1 h_2 h_3^2 (Q_1 x_{02} + Q_2 x_{01}) \tag{4.180}$$

$$\boldsymbol{Q} \cdot \nabla \mathbb{E}_2^4(\boldsymbol{\rho}_0) = h_1 h_2^2 h_3 (Q_1 x_{03} + Q_3 x_{01}) \tag{4.181}$$

$$\boldsymbol{Q} \cdot \nabla \mathbb{E}_2^5(\boldsymbol{\rho}_0) = h_1^2 h_2 h_3 (Q_2 x_{03} + Q_3 x_{02}). \tag{4.182}$$

Furthermore,

$$E_1^m(\alpha_1) = \alpha_m \tag{4.183}$$

$$E_1^{m\prime}(\alpha_1) = \frac{\alpha_1}{\alpha_m} \tag{4.184}$$

$$E_2^1(\alpha_1) = \Lambda \tag{4.185}$$

$$E_2^{1\prime}(\alpha_1) = 2\alpha_1 \tag{4.186}$$

$$E_2^2(\alpha_1) = \Lambda' \tag{4.187}$$

$$E_2^{2\prime}(\alpha_1) = 2\alpha_1 \tag{4.188}$$

$$E_2^{6-m}(\alpha_1) = \frac{\alpha_1 \alpha_2 \alpha_3}{\alpha_m} \tag{4.189}$$

$$E_2^{6-m\prime}(\alpha_1) = \alpha_1 \alpha_2 \alpha_3 \frac{\alpha_1}{\alpha_m} \left[ \frac{1}{\alpha_1^2} + \frac{1}{\alpha_2^2} + \frac{1}{\alpha_3^2} - \frac{1}{\alpha_m^2} \right] \tag{4.190}$$

for $m = 1, 2, 3$.

Finally inserting (4.177)–(4.190), (4.122)–(4.124), (4.126)–(4.130) and (4.147)–(4.154) in (4.175) we obtain

$$u^-(\boldsymbol{r}) = \frac{3}{4\pi\sigma} \sum_{m=1}^{3} \left[ I_1^m(\rho) - I_1^m(\alpha_1) + \frac{1}{\alpha_1 \alpha_2 \alpha_3} \right] Q_m x_m$$

$$- \frac{5}{4\pi\sigma} \frac{(\Lambda - \alpha_1^2)(\Lambda - \alpha_2^2)(\Lambda - \alpha_3^2)}{\Lambda - \Lambda'} \left[ I_2^1(\rho) - I_2^1(\alpha_1) + \frac{1}{2\Lambda \alpha_1 \alpha_2 \alpha_3} \right]$$

$$\times \left[ \sum_{k=1}^{3} \frac{Q_k x_{0k}}{\Lambda - \alpha_k^2} \right] \left[ \sum_{m=1}^{3} \frac{x_m^2}{\Lambda - \alpha_m^2} + 1 \right]$$

$$+ \frac{5}{4\pi\sigma} \frac{(\Lambda' - \alpha_1^2)(\Lambda' - \alpha_2^2)(\Lambda' - \alpha_3^2)}{\Lambda - \Lambda'} \left[ I_2^2(\rho) - I_2^2(\alpha_1) + \frac{1}{2\Lambda' \alpha_1 \alpha_2 \alpha_3} \right]$$

$$\times \left[ \sum_{k=1}^{3} \frac{Q_k x_{0k}}{\Lambda' - \alpha_k^2} \right] \left[ \sum_{m=1}^{3} \frac{x_m^2}{\Lambda' - \alpha_m^2} + 1 \right]$$

$$+ \frac{15}{4\pi\sigma} \left[ I_2^3(\rho) - I_2^3(\alpha_1) + \frac{1}{\alpha_1 \alpha_2 \alpha_3 (\alpha_1^2 + \alpha_2^2)} \right] (Q_1 x_{02} + Q_2 x_{01}) x_1 x_2$$

$$+ \frac{15}{4\pi\sigma} \left[ I_2^4(\rho) - I_2^4(\alpha_1) + \frac{1}{\alpha_1 \alpha_2 \alpha_3 (\alpha_1^2 + \alpha_3^2)} \right] (Q_1 x_{03} + Q_3 x_{01}) x_1 x_3$$

$$+ \frac{15}{4\pi\sigma} \left[ I_2^5(\rho) - I_2^5(\alpha_1) + \frac{1}{\alpha_1 \alpha_2 \alpha_3 (\alpha_2^2 + \alpha_3^2)} \right] (Q_2 x_{03} + Q_3 x_{02}) x_2 x_3$$

$$+ O(\mathrm{el}_3), \rho_0 < \rho < \alpha_1 \tag{4.191}$$

where $O(\mathrm{el}_3)$ stands for ellipsoidal terms of order greater or equal to three.

**Remarks.** Comparing the expression of $u^-$ in the ellipsoidal and in the Cartesian system we observe that although the ellipsoidal system provides a more compact representation it is actually the Cartesian form that reveals the symmetries of the geometry. This behaviour reflects the symmetric way that the Cartesian coordinates enter the defining equation (4.162) as opposed to the ellipsoidal system where the surface (4.162) is defined by $\rho = \alpha_1$, an equation that involves only the $\rho$-component of the system. Furthermore, expression (4.175) is more compact since the $\rho$, the $\mu$ and the $\nu$ dependence enter through the same Lamé functions, which are hidden within the ellipsoidal harmonics $\mathbb{E}_n^m$ together with their dependence on the separation constants.

In order to appreciate the peculiar way in which results in the ellipsoidal geometry reduce to the corresponding results for the sphere, we try to recover the spherical expansion

$$\begin{aligned} u^-(\boldsymbol{r}) = & \frac{1}{4\pi\sigma\alpha^3} \left( 2 + \frac{\alpha^3}{r^3} \right) (\boldsymbol{Q} \cdot \boldsymbol{r}) - \frac{r^2}{4\pi\sigma\alpha^5} \left( \frac{3}{2} + \frac{\alpha^5}{r^5} \right) (\boldsymbol{Q} \cdot \boldsymbol{r}_0) \\ & + \frac{3}{4\pi\sigma\alpha^5} \left( \frac{3}{2} + \frac{\alpha^5}{r^5} \right) (\boldsymbol{Q} \cdot \boldsymbol{r})(\boldsymbol{r}_0 \cdot \boldsymbol{r}) + O(sp_3), \end{aligned} \tag{4.192}$$

where $r_0 < r < \alpha$ and $O(sp_3)$ stands for multipole spherical terms of order higher or equal to the octapole, which comes from the expansion of (4.42). It is trivial to see that as $\alpha_1 \to \alpha, \alpha_2 \to \alpha, \alpha_3 \to \alpha$ we obtain

$$h_1 \to 0,\ h_2 \to 0,\ h_3 \to 0 \tag{4.193}$$

$$\Lambda \to \alpha^2,\ \Lambda' \to \alpha^2 \tag{4.194}$$

$$\rho \to r,\ \ \mu \to 0,\ \ \nu \to 0 \tag{4.195}$$

and

$$I_n^m(\rho) \to \frac{1}{(2n+1)r^{(2n+1)}}, \qquad I_n^m(\alpha_1) \to \frac{1}{(2n+1)\alpha^{(2n+1)}} \tag{4.196}$$

but these do not help to find the limits of the undetermined ratios coming from the $\mathbb{E}_2^1$ and $\mathbb{E}_2^2$ terms in (4.191). The reason is that in the limiting process $(\alpha_1, \alpha_2, \alpha_3) \to (\alpha, \alpha, \alpha)$ we deal with three unspecified rates of convergence which control the limiting values of the undetermined ratios.

In order to obtain a unique limit we need to perform some algebra before we go to analysis. Indeed, by straightforward calculations we obtain

$$\begin{aligned}
&\frac{\mathbb{E}_2^1(\rho,\mu,\nu)}{\Lambda(\Lambda-\Lambda')(\Lambda-\alpha_m^2)} - \frac{\mathbb{E}_2^2(\rho,\mu,\nu)}{\Lambda'(\Lambda-\Lambda')(\Lambda'-\alpha_m^2)}\\
&= \frac{1}{\Lambda\Lambda'}\sum_{\kappa=1}^{3} x_\kappa^2\left[\frac{3}{2}(\Lambda+\Lambda') - \alpha_\kappa^2 - \alpha_m^2 + 3\delta_{\kappa m}(\alpha_\kappa^2 - \Lambda - \Lambda')\right] + 1 - \frac{\alpha_1^2\alpha_2^2\alpha_3^2}{\alpha_m^2\Lambda\Lambda'}
\end{aligned} \tag{4.197}$$

$$\begin{aligned}
&\frac{\mathbb{E}_2^1(\rho,\mu,\nu)I_2^1(\rho)}{(\Lambda-\Lambda')(\Lambda-\alpha_m^2)} - \frac{\mathbb{E}_2^2(\rho,\mu,\nu)I_2^2(\rho)}{(\Lambda-\Lambda')(\Lambda'-\alpha_m^2)}\\
&= \left(r^2 - 3x_m^2 + \alpha_m^2 - \frac{\alpha_1^2+\alpha_2^2+\alpha_3^2}{3}\right) I_2^1(\rho)\\
&+ \frac{\mathbb{E}_2^2(\rho,\mu,\nu)}{\Lambda'-\alpha_m^2}\left[\frac{3}{2}I_3^7(\rho) - \frac{1}{2\rho\sqrt{\rho^2-h_3^2}\sqrt{\rho^2-h_2^2}(\Lambda-\alpha_1^2+\rho^2)(\Lambda'-\alpha_1^2+\rho^2)}\right]
\end{aligned} \tag{4.198}$$

where

$$\frac{\mathbb{E}_2^2(\rho,\mu,\nu)}{\Lambda'-\alpha_1^2} = (\Lambda'-\alpha_2^2)(x_3^2-x_1^2) + (\Lambda'-\alpha_3^2)(x_2^2-x_1^2) + (\Lambda'-\alpha_2^2)(\Lambda'-\alpha_3^2) \tag{4.199}$$

$$\frac{\mathbb{E}_2^2(\rho,\mu,\nu)}{\Lambda'-\alpha_2^2} = (\Lambda'-\alpha_1^2)(x_3^2-x_2^2) + (\Lambda'-\alpha_3^2)(x_1^2-x_2^2) + (\Lambda'-\alpha_1^2)(\Lambda'-\alpha_3^2) \tag{4.200}$$

$$\frac{\mathbb{E}_2^2(\rho,\mu,\nu)}{\Lambda'-\alpha_3^2} = (\Lambda'-\alpha_1^2)(x_2^2-x_3^2)+(\Lambda'-\alpha_2^2)(x_1^2-x_3^2)+(\Lambda'-\alpha_1^2)(\Lambda'-\alpha_2^2) \tag{4.201}$$

and

$$I_3^7(\rho) = \int_\rho^{+\infty} \frac{dt}{t^2(t^2-h_2^2)^{3/2}(t^2-h_3^2)^{3/2}}. \tag{4.202}$$

Since the RHS of (4.197)–(4.201) are now continuous functions of $(\alpha_1,\alpha_2,\alpha_3)$ at the point $(\alpha,\alpha,\alpha)$ the necessary limits are obtained by simple evaluation. Then it is straightforward to check that as $\alpha_i \to \alpha$ expression (4.191) reduces to (4.192).

Note that the result (4.192) involves ordinary inner products among the vectors $\boldsymbol{Q}, \boldsymbol{r}_0$ and $\boldsymbol{r}$. In fact, the same is true for the expression (4.191) but we now have to do with *weighted inner products* with weights that vary from inner product to inner product such as

$$\sum_{k=1}^{3} \frac{Q_k x_{0k}}{\Lambda-\alpha_k^2} \text{ or } \sum_{k=1}^{3} I_1^k(\alpha_1) Q_k x_k.$$

These variable weights bring all the difficulties into the problem, since in every case we have to find the appropriate weight for each particular inner product. The anisotropic character of the ellipsoidal geometry is reflected precisely upon these weighted inner products, which establish the directional characteristics of the space in a non uniform manner.

In a similar way, we can solve the exterior EEG problem

$$\Delta u^+(\boldsymbol{\rho}) = 0, \ \rho > \alpha_1 \tag{4.203}$$

$$u^+(\boldsymbol{\rho}) = f(\mu,\nu), \rho = \alpha_1 \tag{4.204}$$

$$u^+(\boldsymbol{\rho}) = O\left(\frac{1}{\rho^2}\right), \rho \to \infty, \tag{4.205}$$

where the Dirichlet data $f(\mu,\nu)$ are given by (4.175) evaluated at $\rho = \alpha_1$,

$$f(\mu,\nu) = \frac{1}{\sigma} \sum_{n=1}^{\infty} \sum_{m=1}^{2n+1} \frac{\boldsymbol{Q}\cdot\nabla\mathbb{E}_n^m(\boldsymbol{\rho}_0)}{\gamma_n^m \alpha_2\alpha_3 E_n^{m\prime}(\alpha_1)} E_n^m(\mu)E_n^m(\nu). \tag{4.206}$$

The solution of (4.203)–(4.205) is given by

$$u^+(\boldsymbol{\rho}) = \frac{1}{\sigma\alpha_2\alpha_3} \sum_{n=1}^{\infty} \sum_{m=1}^{2n+1} \frac{\boldsymbol{Q}\cdot\nabla\mathbb{E}_n^m(\boldsymbol{\rho}_0)}{\gamma_n^m E_n^{m\prime}(\alpha_1)F_n^m(\alpha_1)} \mathbb{F}_n^m(\boldsymbol{\rho}) \tag{4.207}$$

for every $\rho > \alpha_1$.

For the sake of completeness we give the electric potential for the realistic case of the ellipsoidal model, with three confocal shells of different conductivity, corresponding to the cerebrospinal fluid, to the bone (skull) and to the skin (scalp), surrounding the cerebrum tissue in this order [27]. In fact, we have to solve the multilayered transmission problem (4.15)–(4.24), where $\rho = c_1$ specifies the boundary $S_c$ of the cerebrum, $\rho = f_1$ specifies the outer boundary $S_f$ of the region occupied by the cerebrospinal fluid, $\rho = b_1$ specifies the outer boundary $S_b$ of the skull and $\rho = s_1$ specifies the outer boundary $S_s$ of the scalp.

If $(c_1, c_2, c_3)$, $(f_1, f_2, f_3)$, $(b_1, b_2, b_3)$ and $(s_1, s_2, s_3)$ denote the semiaxes of the ellipsoidal boundaries $S_c, S_f, S_b$ and $S_s$ respectively, then because of confocality we have

$$h_1^2 = c_2^2 - c_3^2 = f_2^2 - f_3^2 = b_2^2 - b_3^2 = s_2^2 - s_3^2 \tag{4.208}$$

$$h_2^2 = c_1^2 - c_3^2 = f_1^2 - f_3^2 = b_1^2 - b_3^2 = s_1^2 - s_3^2 \tag{4.209}$$

$$h_3^2 = c_1^2 - c_2^2 = f_1^2 - f_2^2 = b_1^2 - b_2^2 = s_1^2 - s_2^2. \tag{4.210}$$

We introduce the functions

$$I_n^m(x, y) = I_n^m(x) - I_n^m(y) = \int_x^y \frac{dt}{[E_n^m(t)]^2 \sqrt{t^2 - h_3^2}\sqrt{t^2 - h_2^2}} \tag{4.211}$$

and the geometrical constants

$$\begin{aligned} S_n^m &= E_n^m(s_1) E_n^{m\prime}(s_1) s_2 s_3 \\ B_n^m &= E_n^m(b_1) E_n^{m\prime}(b_1) b_2 b_3 \\ F_n^m &= E_n^m(f_1) E_n^{m\prime}(f_1) f_2 f_3 \\ C_n^m &= E_n^m(c_1) E_n^{m\prime}(c_1) c_2 c_3. \end{aligned} \tag{4.212}$$

Then the potential for the exterior to the head region is given by

$$u(\boldsymbol{r}) = \sum_{n=1}^{\infty} \sum_{m=1}^{2n+1} \frac{I_n^m(\rho)}{I_n^m(s_1)} \frac{1}{S_n^m} \frac{1}{G_{3,n}^m} \frac{\boldsymbol{Q} \cdot \nabla \mathbb{E}_n^m(\boldsymbol{\rho}_0)}{\gamma_n^m} \mathbb{E}_n^m(\rho, \mu, \nu), \rho > s_1 \tag{4.213}$$

for the skin region by

$$u_s(\boldsymbol{r}) = u(s_1) + \sum_{n=1}^{\infty} \sum_{m=1}^{2n+1} I_n^m(\rho, s_1) \frac{1}{G_{3,n}^m} \frac{\boldsymbol{Q} \cdot \nabla \mathbb{E}_n^m(\boldsymbol{\rho}_0)}{\gamma_n^m} \mathbb{E}_n^m(\rho, \mu, \nu), b_1 < \rho < s_1 \tag{4.214}$$

for the scull region by

$$u_b(\boldsymbol{r}) = u_s(b_1) + \sum_{n=1}^{\infty}\sum_{m=1}^{2n+1} I_n^m(\rho, b_1)\frac{1}{\sigma_b}\frac{G_{1,n}^m}{G_{3,n}^m}\frac{\boldsymbol{Q}\cdot\nabla\mathbb{E}_n^m(\boldsymbol{\rho}_0)}{\gamma_n^m}\mathbb{E}_n^m(\rho,\mu,\nu), f_1 < \rho < b_1 \tag{4.215}$$

for the fluid region by

$$u_f(\boldsymbol{r}) = u_b(f_1) + \sum_{n=1}^{\infty}\sum_{m=1}^{2n+1} I_n^m(\rho, f_1)\frac{1}{\sigma_f}\frac{G_{2,n}^m}{G_{3,n}^m}\frac{\boldsymbol{Q}\cdot\nabla\mathbb{E}_n^m(\boldsymbol{\rho}_0)}{\gamma_n^m}\mathbb{E}_n^m(\rho,\mu,\nu), c_1 < \rho < f_1 \tag{4.216}$$

and finally for the region occupied by the cerebrum by

$$u_c(\boldsymbol{r}) = u_f(c_1) + \sum_{n=1}^{\infty}\sum_{m=1}^{2n+1} I_n^m(\rho, c_1)\frac{1}{\sigma_c}\frac{\boldsymbol{Q}\cdot\nabla\mathbb{E}_n^m(\boldsymbol{\rho}_0)}{\gamma_n^m}\mathbb{E}_n^m(\rho,\mu,\nu), \rho < c_1. \tag{4.217}$$

The constants $G_{1,n}^m, G_{2,n}^m$ and $G_{3,n}^m$ are given by

$$G_{1,n}^m = \sigma_b + (\sigma_b - \sigma_s)\left(I_n^m(b_1, s_1) + \frac{1}{S_n^m} - \frac{1}{B_n^m}\right)B_n^m \tag{4.218}$$

$$\begin{aligned} G_{2,n}^m &= \sigma_f \\ &\quad + (\sigma_f - \sigma_b)\left(I_n^m(f_1, s_1) + \frac{1}{S_n^m} - \frac{1}{F_n^m}\right)F_n^m \\ &\quad + (\sigma_b - \sigma_s)\left(I_n^m(b_1, s_1) + \frac{1}{S_n^m} - \frac{1}{B_n^m}\right)B_n^m \\ &\quad + \frac{(\sigma_f - \sigma_b)(\sigma_b - \sigma_s)}{\sigma_b} I_n^m(f_1, b_1)\left(I_n^m(b_1, s_1) + \frac{1}{S_n^m} - \frac{1}{B_n^m}\right)B_n^m F_n^m \end{aligned} \tag{4.219}$$

and

$$\begin{aligned} G_{3,n}^m &= \sigma_c \\ &\quad + (\sigma_c - \sigma_f)\left(I_n^m(c_1, s_1) + \frac{1}{S_n^m} - \frac{1}{C_n^m}\right)C_n^m \\ &\quad + (\sigma_f - \sigma_b)\left(I_n^m(f_1, s_1) + \frac{1}{S_n^m} - \frac{1}{F_n^m}\right)F_n^m \end{aligned}$$

$$
\begin{aligned}
&+(\sigma_b-\sigma_s)\left(I_n^m(b_1,s_1)+\frac{1}{S_n^m}-\frac{1}{B_n^m}\right)B_n^m\\
&+\frac{(\sigma_c-\sigma_f)(\sigma_f-\sigma_b)}{\sigma_f}I_n^m(c_1,f_1)\left(I_n^m(f_1,s_1)+\frac{1}{S_n^m}-\frac{1}{F_n^m}\right)F_n^mC_n^m\\
&+\frac{(\sigma_c-\sigma_f)(\sigma_b-\sigma_s)}{\sigma_b}I_n^m(c_1,b_1)\left(I_n^m(b_1,s_1)+\frac{1}{S_n^m}-\frac{1}{B_n^m}\right)B_n^mC_n^m\\
&+\frac{(\sigma_f-\sigma_b)(\sigma_b-\sigma_s)}{\sigma_b}I_n^m(f_1,b_1)\left(I_n^m(b_1,s_1)+\frac{1}{S_n^m}-\frac{1}{B_n^m}\right)B_n^mF_n^m\\
&+\frac{(\sigma_c-\sigma_f)(\sigma_f-\sigma_b)(\sigma_b-\sigma_s)}{\sigma_f\sigma_b}I_n^m(c_1,f_1)\left(I_n^m(b_1,s_1)+\frac{1}{S_n^m}-\frac{1}{B_n^m}\right)\\
&\times\left(I_n^m(f_1,b_1)-\frac{1}{F_n^m}\right)B_n^mF_n^mC_n^m
\end{aligned}
\tag{4.220}
$$

and the arbitrary constant of the problem has been set equal to zero. We observe that the value of the skin-potential $u_s$ is equal to the trace of the exterior potential on the outer boundary of the skin-shell plus series expansion that depends on the observation point within the skin-shell. Similarly, the bone-potential is equal to the trace that the skin-potential leaves on the outer surface of the bone and an expansion depending on the observation point within the bone. This structure propagates all the way down to the cerebrum as it can be seen in (4.213)–(4.217). All terms of the above potentials are normalized by the constants $G_{3,n}^m$, given in (4.220), which involve the conductivity jumps across the interfaces and geometrical characteristics of the layered region. Note also that the elliptic integrals which enter the expressions (4.214)–(4.217) extend from the observation point $\rho$ to the outer boundary of the particular conductivity shell, and a similar behavior appears in (4.213). If we want to eliminate any one of these layers all we have to do is to equate the conductivities of the appropriate two successive layers which will effect only the values of the constants $G_{1,n}^m, G_{2,n}^m, G_{3,n}^m$. The reduction of the results (4.213)–(4.220) to the spherical shells model is trivial.

## 4.6 MEG in Ellipsoidal Geometry

In this section we first construct the quadrupolic term of the magnetic induction field in closed form and then we show how to obtain the full multipole expansion in terms of elliptic integrals. The quadrupolic solution for the general case where the cerebrum is surrounded by three shells of different conductivities is given in [18], but in order to be able to discuss some aspects of the inhomogeneities in conductivity we will present the results for the single layer model [13].

We note here that although a complete expansion for the trace of the electric potential on the surface of an ellipsoidal conductor can be obtained (with the exception of some constants) relatively easy [10,33], the corresponding solution of the magnetoencephalography problem is much harder to obtain [6].

As we explained in Sect. 4.2 if the electric potential $u^-$ on $\rho = \alpha_1$ is known, then the magnetic induction field is given by

$$\boldsymbol{B}(\boldsymbol{r}) = \frac{\mu_0}{4\pi}\boldsymbol{Q} \times \frac{\boldsymbol{r}-\boldsymbol{r}_0}{|\boldsymbol{r}-\boldsymbol{r}_0|^3} - \frac{\mu_0\sigma}{4\pi}\oint_{\rho=\alpha_1} u^-(\boldsymbol{r}')\hat{\boldsymbol{\rho}}' \times \frac{\boldsymbol{r}-\boldsymbol{r}'}{|\boldsymbol{r}-\boldsymbol{r}'|^3} ds(\boldsymbol{r}') \quad (4.221)$$

where $u^-$ at $\rho = \alpha_1$ can be obtained either from (4.175) or from (4.207) as

$$u^-(\alpha_1, \mu, \nu) = \frac{1}{\sigma\alpha_2\alpha_3}\sum_{n=1}^{\infty}\sum_{m=1}^{2n+1} \frac{\boldsymbol{Q}\cdot\nabla\mathbb{E}_n^m(\boldsymbol{\rho}_0)}{\gamma_n^m E_n^{m\prime}(\alpha_1)} E_n^m(\mu)E_n^m(\nu). \quad (4.222)$$

From (4.93) we see that on $\rho = \alpha_1$ the unit normal vector $\hat{\boldsymbol{\rho}}$ is expanded in surface ellipsoidal harmonics as

$$\begin{aligned}\hat{\boldsymbol{\rho}} &= \frac{\alpha_1\alpha_2\alpha_3}{\sqrt{\alpha_1^2-\mu^2}\sqrt{\alpha_1^2-\nu^2}}\sum_{m=1}^{3}\frac{x_m\hat{\boldsymbol{x}}_m}{\rho^2-\alpha_1^2+\alpha_m^2} \\ &= \frac{\alpha_1\alpha_2\alpha_3}{h_1h_2h_3}\frac{1}{\sqrt{\alpha_1^2-\mu^2}\sqrt{\alpha_1^2-\nu^2}}\sum_{m=1}^{3}\frac{h_m}{\alpha_m}E_1^m(\mu)E_1^m(\nu)\hat{\boldsymbol{x}}_m. \quad (4.223)\end{aligned}$$

Having now $u^-$ and $\hat{\boldsymbol{\rho}}$ expanded in surface ellipsoidal harmonics we also need to expand the dipole field $(\boldsymbol{r}-\boldsymbol{r}')|\boldsymbol{r}-\boldsymbol{r}'|^{-3}$. But this is readily obtained from (4.170) as

$$\frac{\boldsymbol{r}-\boldsymbol{r}'}{|\boldsymbol{r}-\boldsymbol{r}'|^3} = \nabla_{\boldsymbol{r}'}\frac{1}{|\boldsymbol{r}-\boldsymbol{r}'|} = \sum_{n=1}^{\infty}\sum_{m=1}^{2n+1}\frac{4\pi}{2n+1}\frac{1}{\gamma_n^m}(\nabla\mathbb{E}_n^m(\boldsymbol{\rho}'))\mathbb{F}_n^m(\boldsymbol{\rho}) \quad (4.224)$$

which holds for $\rho > \alpha_1$.

The obvious thing to do now is to take the appropriate product of (4.222), (4.223) and (4.224) and integrate over the surface $\rho = \alpha_1$ using the orthogonality of the surface ellipsoidal harmonics, where the weighting function

$$l(\mu, \nu) = (\alpha_1^2 - \mu^2)^{-1/2}(\alpha_1^2 - \nu^2)^{-1/2} \quad (4.225)$$

is provided by the expansion of the unit normal (4.223). Unfortunately, there are no general formulae that express the product of any two surface ellipsoidal harmonics in terms of surface ellipsoidal harmonics. Hence, any such formula has to be derived individually by performing the necessary calculations. This is a long and tedious series of calculations [10] which for terms of degree 1 and 2 leads to

$$\hat{\boldsymbol{\rho}}' \times \left. \frac{\boldsymbol{r} - \boldsymbol{r}'}{|\boldsymbol{r} - \boldsymbol{r}'|^3} \right|_{\rho'=\alpha_1}$$

$$= \alpha_1\alpha_2\alpha_3 l(\mu',\nu') \left[ \frac{1}{3} \sum_{m=1}^{3} \alpha_m^2 \left( \hat{\boldsymbol{x}}_m \cdot \tilde{\boldsymbol{M}}^{-1}(\alpha_1) \times \tilde{\boldsymbol{F}}(\boldsymbol{r}) \cdot \hat{\boldsymbol{x}}_m \right) \right.$$

$$+ \frac{3}{h_1h_2h_3} \sum_{m=1}^{3} \alpha_m h_m \left( \hat{\boldsymbol{x}}_m \cdot \tilde{\boldsymbol{M}}^{-1}(\alpha_1) \times \tilde{\boldsymbol{H}}_1(\rho) \cdot \boldsymbol{r} \right) E_1^m(\mu') F_1^m(\nu')$$

$$- \frac{1}{3(\Lambda - \Lambda')} \sum_{m=1}^{3} \frac{\alpha_m^2}{\Lambda - \alpha_m^2} \left( \hat{\boldsymbol{x}}_m \cdot \tilde{\boldsymbol{M}}^{-1}(\alpha_1) \times \tilde{\boldsymbol{F}}(\boldsymbol{r}) \cdot \hat{\boldsymbol{x}}_m \right) E_2^1(\mu') E_2^1(\nu')$$

$$+ \frac{1}{3(\Lambda - \Lambda')} \sum_{m=1}^{3} \frac{\alpha_m^2}{\Lambda' - \alpha_m^2} \left( \hat{\boldsymbol{x}}_m \cdot \tilde{\boldsymbol{M}}^{-1}(\alpha_1) \times \tilde{\boldsymbol{F}}(\boldsymbol{r}) \cdot \hat{\boldsymbol{x}}_m \right) E_2^2(\mu') E_2^2(\nu')$$

$$\left. + \frac{1}{h_1h_2h_3} \sum_{\substack{i,j=1 \\ i\neq j}}^{3} \frac{\alpha_i\alpha_j}{h_{6-(i+j)}} \left( \hat{\boldsymbol{x}}_i \cdot \tilde{\boldsymbol{M}}^{-1}(\alpha_1) \times \tilde{\boldsymbol{F}}(\boldsymbol{r}) \cdot \hat{\boldsymbol{x}}_j \right) E_2^{i+j}(\mu') E_2^{i+j}(\nu') \right]$$

$$+ O(\mathrm{el}_3') \tag{4.226}$$

where the dyadics $\tilde{\boldsymbol{M}}, \tilde{\boldsymbol{F}}, \tilde{\boldsymbol{H}}_1(\rho)$ are defined by

$$\tilde{\boldsymbol{M}}(\rho) = \sum_{m-1}^{3} (\rho^2 - \alpha_1^2 + \alpha_m^2) \hat{\boldsymbol{x}}_m \otimes \hat{\boldsymbol{x}}_m \tag{4.227}$$

$$\tilde{\boldsymbol{F}}(\boldsymbol{r}) = -\frac{\mathbb{F}_2^1(\boldsymbol{r})}{\Lambda - \Lambda'} \tilde{\boldsymbol{\Lambda}} + \frac{\mathbb{F}_2^2(\boldsymbol{r})}{\Lambda - \Lambda'} \tilde{\boldsymbol{\Lambda}}' + 15 \boldsymbol{r} \otimes \boldsymbol{r} : \tilde{\tilde{\mathbf{H}}}_2(\rho) \tag{4.228}$$

$$\tilde{\boldsymbol{H}}_1(\rho) = \sum_{m=1}^{3} I_1^m(\rho) \hat{\boldsymbol{x}}_m \otimes \hat{\boldsymbol{x}}_m \tag{4.229}$$

with

$$\tilde{\boldsymbol{\Lambda}} = \sum_{m=1}^{3} \frac{\hat{\boldsymbol{x}}_m \otimes \hat{\boldsymbol{x}}_m}{\Lambda - \alpha_m^2}$$

$$\tilde{\boldsymbol{\Lambda}}' = \sum_{m=1}^{3} \frac{\hat{\boldsymbol{x}}_m \otimes \hat{\boldsymbol{x}}_m}{\Lambda' - \alpha_m^2} \tag{4.230}$$

and the tetradic $\tilde{\tilde{\boldsymbol{H}}}_2(\rho)$ is given by

$$\tilde{\tilde{\boldsymbol{H}}}_2(\rho) = \sum_{\substack{i,j=1\\ i\neq j}}^{3} I_2^{i+j}(\rho)\hat{\boldsymbol{x}}_i \otimes \hat{\boldsymbol{x}}_j \otimes \hat{\boldsymbol{x}}_i \otimes \hat{\boldsymbol{x}}_j. \tag{4.231}$$

Then (4.222) and (4.223) implies that

$$\oint_{\rho=\alpha_1} u^-(\boldsymbol{r}')\hat{\boldsymbol{\rho}}' \times \frac{\boldsymbol{r}-\boldsymbol{r}'}{|\boldsymbol{r}-\boldsymbol{r}'|^3} ds(\boldsymbol{r}') = \sum_{m=1}^{3} \zeta_m \boldsymbol{\beta}_m \gamma_1^m + \sum_{m=1}^{5} \theta_m \boldsymbol{\delta}_m \gamma_2^m + O(\mathrm{el}_3) \tag{4.232}$$

where

$$\boldsymbol{\beta}_m = 3\frac{\alpha_1\alpha_2\alpha_3}{h_1h_2h_3}\frac{h_m}{\alpha_m}\left(\hat{\boldsymbol{x}}_m \otimes \boldsymbol{r} \underset{\cdot}{\times} \tilde{\boldsymbol{H}}_1(\rho)\right), m = 1,2,3 \tag{4.233}$$

$$\boldsymbol{\delta}_1 = -\frac{\alpha_1\alpha_2\alpha_3}{3(\Lambda-\Lambda')}\tilde{\boldsymbol{\Lambda}} \underset{\cdot}{\times} \tilde{\boldsymbol{F}}(\boldsymbol{r}) \tag{4.234}$$

$$\boldsymbol{\delta}_2 = \frac{\alpha_1\alpha_2\alpha_3}{3(\Lambda-\Lambda')}\tilde{\boldsymbol{\Lambda}}' \underset{\cdot}{\times} \tilde{\boldsymbol{F}}(\boldsymbol{r}) \tag{4.235}$$

$$\boldsymbol{\delta}_3 = \frac{\alpha_1\alpha_2\alpha_3}{h_1h_2h_3^2}\left[\frac{\alpha_2}{\alpha_1}\hat{\boldsymbol{x}}_1 \otimes \hat{\boldsymbol{x}}_2 + \frac{\alpha_1}{\alpha_2}\hat{\boldsymbol{x}}_2 \otimes \hat{\boldsymbol{x}}_1\right] \underset{\cdot}{\times} \tilde{\boldsymbol{F}}(\boldsymbol{r}) \tag{4.236}$$

$$\boldsymbol{\delta}_4 = \frac{\alpha_1\alpha_2\alpha_3}{h_1h_2^2h_3}\left[\frac{\alpha_3}{\alpha_1}\hat{\boldsymbol{x}}_1 \otimes \hat{\boldsymbol{x}}_3 + \frac{\alpha_1}{\alpha_3}\hat{\boldsymbol{x}}_3 \otimes \hat{\boldsymbol{x}}_1\right] \underset{\cdot}{\times} \tilde{\boldsymbol{F}}(\boldsymbol{r}) \tag{4.237}$$

$$\boldsymbol{\delta}_5 = \frac{\alpha_1\alpha_2\alpha_3}{h_1^2h_2h_3}\left[\frac{\alpha_3}{\alpha_2}\hat{\boldsymbol{x}}_2 \otimes \hat{\boldsymbol{x}}_3 + \frac{\alpha_2}{\alpha_3}\hat{\boldsymbol{x}}_3 \otimes \hat{\boldsymbol{x}}_2\right] \underset{\cdot}{\times} \tilde{\boldsymbol{F}}(\boldsymbol{r}) \tag{4.238}$$

with the cross-dot product defined as

$$(\boldsymbol{a} \otimes \boldsymbol{b}) \underset{\cdot}{\times} (\boldsymbol{c} \otimes \boldsymbol{d}) = (\boldsymbol{a} \times \boldsymbol{c})(\boldsymbol{b} \cdot \boldsymbol{d}) \tag{4.239}$$

and

$$\zeta_m = \frac{\alpha_m h_m}{\sigma \, \mathrm{V} \, h_1h_2h_3}(\boldsymbol{Q} \cdot \hat{\boldsymbol{x}}_m), m = 1,2,3 \tag{4.240}$$

$$\theta_1 = -\frac{5}{6\sigma V(\Lambda-\Lambda')}(\boldsymbol{Q} \otimes \boldsymbol{r}_0 : \tilde{\boldsymbol{\Lambda}}) \tag{4.241}$$

$$\theta_2 - \frac{5}{6\sigma V(\Lambda-\Lambda')}(\boldsymbol{Q} \otimes \boldsymbol{r}_0 : \tilde{\boldsymbol{\Lambda}}') \tag{4.242}$$

$$\theta_3 = \frac{5\alpha_1\alpha_2\alpha_3}{\sigma\,\mathrm{V}\,h_1h_2h_3}\frac{\boldsymbol{Q}\otimes\boldsymbol{r}_0 : (\hat{\boldsymbol{x}}_1\otimes\hat{\boldsymbol{x}}_2+\hat{\boldsymbol{x}}_2\otimes\hat{\boldsymbol{x}}_1)}{\alpha_3h_3(\alpha_1^2+\alpha_2^2)} \tag{4.243}$$

$$\theta_4 = \frac{5\alpha_1\alpha_2\alpha_3}{\sigma\,\mathrm{V}\,h_1h_2h_3}\frac{\boldsymbol{Q}\otimes\boldsymbol{r}_0 : (\hat{\boldsymbol{x}}_1\otimes\hat{\boldsymbol{x}}_3+\hat{\boldsymbol{x}}_3\otimes\hat{\boldsymbol{x}}_1)}{\alpha_2h_2(\alpha_1^2+\alpha_3^2)} \tag{4.244}$$

$$\theta_5 = \frac{5\alpha_1\alpha_2\alpha_3}{\sigma\,\mathrm{V}\,h_1h_2h_3}\frac{\boldsymbol{Q}\otimes\boldsymbol{r}_0 : (\hat{\boldsymbol{x}}_2\otimes\hat{\boldsymbol{x}}_3+\hat{\boldsymbol{x}}_3\otimes\hat{\boldsymbol{x}}_2)}{\alpha_1h_1(\alpha_2^2+\alpha_3^2)} \tag{4.245}$$

with the double dot product defined by

$$(\boldsymbol{a}\otimes\boldsymbol{b}) : (\boldsymbol{c}\otimes\boldsymbol{d}) = (\boldsymbol{a}\cdot\boldsymbol{c})(\boldsymbol{b}\cdot\boldsymbol{d}). \tag{4.246}$$

In fact, if we further calculate the dipole term we arrive at

$$\sum_{m=1}^{3}\zeta_m\boldsymbol{\beta}_m\gamma_1^m = \frac{3}{\sigma}\sum_{m=1}^{3}(\boldsymbol{Q}\cdot\hat{\boldsymbol{x}}_m)\left(\hat{\boldsymbol{x}}_m\otimes\boldsymbol{r}\underset{\cdot}{\times}\tilde{\boldsymbol{H}}_1(\rho)\right) = \frac{3}{\sigma}\boldsymbol{Q}\otimes\boldsymbol{r}\underset{\cdot}{\times}\tilde{\boldsymbol{H}}_1(\rho) \tag{4.247}$$

which, if substituted in (4.221) one can easily check that it cancels the corresponding dipole term of the multipole expansion of the magnetic field due to the dipole at $\boldsymbol{r}_0$. This result is general and is a consequence of the equation $\nabla\cdot\boldsymbol{B}=0$ which prevents the existence of magnetic monopoles.

Therefore the quadrupolic term of $\boldsymbol{B}$ is given by

$$\begin{aligned}\boldsymbol{B}(\boldsymbol{r}) =& \frac{\mu_0}{4\pi}\boldsymbol{Q}\otimes\boldsymbol{r}_0\underset{\cdot}{\times}\tilde{\boldsymbol{F}}(\boldsymbol{r})\\ &+\frac{\mu_0}{12\pi}\frac{(\Lambda-\alpha_1^2)(\Lambda-\alpha_2^2)(\Lambda-\alpha_3^2)}{\Lambda-\Lambda'}\boldsymbol{Q}\otimes\boldsymbol{r}_0 : \tilde{\boldsymbol{\Lambda}}\otimes\tilde{\boldsymbol{\Lambda}}\underset{\cdot}{\times}\tilde{\boldsymbol{F}}(\boldsymbol{r})\\ &-\frac{\mu_0}{12\pi}\frac{(\Lambda'-\alpha_1^2)(\Lambda'-\alpha_2^2)(\Lambda'-\alpha_3^2)}{\Lambda-\Lambda'}\boldsymbol{Q}\otimes\boldsymbol{r}_0 : \tilde{\boldsymbol{\Lambda}}'\otimes\tilde{\boldsymbol{\Lambda}}'\underset{\cdot}{\times}\tilde{\boldsymbol{F}}(\boldsymbol{r})\\ &-\frac{\mu_0}{4\pi}\boldsymbol{Q}\otimes\boldsymbol{r}_0 : \frac{(\hat{\boldsymbol{x}}_1\otimes\hat{\boldsymbol{x}}_2+\hat{\boldsymbol{x}}_2\otimes\hat{\boldsymbol{x}}_1)\otimes(\alpha_2^2\hat{\boldsymbol{x}}_1\otimes\hat{\boldsymbol{x}}_2+\alpha_1^2\hat{\boldsymbol{x}}_2\otimes\hat{\boldsymbol{x}}_1)}{\alpha_1^2+\alpha_2^2}\underset{\cdot}{\times}\tilde{\boldsymbol{F}}(\boldsymbol{r})\\ &-\frac{\mu_0}{4\pi}\boldsymbol{Q}\otimes\boldsymbol{r}_0 : \frac{(\hat{\boldsymbol{x}}_1\otimes\hat{\boldsymbol{x}}_3+\hat{\boldsymbol{x}}_3\otimes\hat{\boldsymbol{x}}_1)\otimes(\alpha_3^2\hat{\boldsymbol{x}}_1\otimes\hat{\boldsymbol{x}}_3+\alpha_1^2\hat{\boldsymbol{x}}_3\otimes\hat{\boldsymbol{x}}_1)}{\alpha_1^2+\alpha_3^2}\underset{\cdot}{\times}\tilde{\boldsymbol{F}}(\boldsymbol{r})\\ &-\frac{\mu_0}{4\pi}\boldsymbol{Q}\otimes\boldsymbol{r}_0 : \frac{(\hat{\boldsymbol{x}}_2\otimes\hat{\boldsymbol{x}}_3+\hat{\boldsymbol{x}}_3\otimes\hat{\boldsymbol{x}}_2)\otimes(\alpha_3^2\hat{\boldsymbol{x}}_2\otimes\hat{\boldsymbol{x}}_3+\alpha_2^2\hat{\boldsymbol{x}}_3\otimes\hat{\boldsymbol{x}}_2)}{\alpha_2^2+\alpha_3^2}\underset{\cdot}{\times}\tilde{\boldsymbol{F}}(\boldsymbol{r})\\ &+O(\mathrm{el}_3).\end{aligned} \tag{4.248}$$

By means of the scalar identities

$$\frac{(\Lambda-\alpha_1^2)(\Lambda-\alpha_2^2)(\Lambda-\alpha_3^2)}{(\Lambda-\alpha_i^2)^2}=(\Lambda-\alpha_i^2)-\sum_{m=1}^{3}(\Lambda-\alpha_m^2),\, i=1,2,3 \quad (4.249)$$

and

$$\frac{(\Lambda'-\alpha_1^2)(\Lambda'-\alpha_2^2)(\Lambda'-\alpha_3^2)}{(\Lambda'-\alpha_i^2)^2}=(\Lambda'-\alpha_i^2)-\sum_{m=1}^{3}(\Lambda'-\alpha_m^2),\, i=1,2,3 \quad (4.250)$$

we can easily prove the tetradic formula

$$\frac{(\Lambda-\alpha_1^2)(\Lambda-\alpha_2^2)(\Lambda-\alpha_3^2)}{3(\Lambda-\Lambda')}\tilde{\boldsymbol{\Lambda}}\otimes\tilde{\boldsymbol{\Lambda}}-\frac{(\Lambda'-\alpha_1^2)(\Lambda'-\alpha_2^2)(\Lambda'-\alpha_3^2)}{3(\Lambda-\Lambda')}\tilde{\boldsymbol{\Lambda}}'\otimes\tilde{\boldsymbol{\Lambda}}'$$
$$=\frac{1}{3}\tilde{\boldsymbol{I}}\otimes\tilde{\boldsymbol{I}}-\sum_{m=1}^{3}\hat{\boldsymbol{x}}_i\otimes\hat{\boldsymbol{x}}_i\otimes\hat{\boldsymbol{x}}_i\otimes\hat{\boldsymbol{x}}_i. \quad (4.251)$$

Furthermore, the identities

$$\frac{(\hat{\boldsymbol{x}}_i\otimes\hat{\boldsymbol{x}}_j+\hat{\boldsymbol{x}}_j\otimes\hat{\boldsymbol{x}}_i)\otimes(\alpha_j^2\hat{\boldsymbol{x}}_i\otimes\hat{\boldsymbol{x}}_j+\alpha_i^2\hat{\boldsymbol{x}}_j\otimes\hat{\boldsymbol{x}}_i)}{\alpha_i^2+\alpha_j^2}$$
$$=\hat{\boldsymbol{x}}_i\otimes\hat{\boldsymbol{x}}_j\otimes\hat{\boldsymbol{x}}_i\otimes\hat{\boldsymbol{x}}_j+\hat{\boldsymbol{x}}_j\otimes\hat{\boldsymbol{x}}_i\otimes\hat{\boldsymbol{x}}_j\otimes\hat{\boldsymbol{x}}_i$$
$$+\frac{(\alpha_i^2\hat{\boldsymbol{x}}_i\otimes\hat{\boldsymbol{x}}_j-\alpha_j^2\hat{\boldsymbol{x}}_j\otimes\hat{\boldsymbol{x}}_i)\otimes(\hat{\boldsymbol{x}}_j\otimes\hat{\boldsymbol{x}}_i-\hat{\boldsymbol{x}}_i\otimes\hat{\boldsymbol{x}}_j)}{\alpha_i^2+\alpha_j^2}, \quad i\neq j \quad (4.252)$$

$$\hat{\boldsymbol{x}}_1\times\tilde{\boldsymbol{I}}=\hat{\boldsymbol{x}}_3\otimes\hat{\boldsymbol{x}}_2-\hat{\boldsymbol{x}}_2\otimes\hat{\boldsymbol{x}}_3 \quad (4.253)$$
$$\hat{\boldsymbol{x}}_2\times\tilde{\boldsymbol{I}}=\hat{\boldsymbol{x}}_1\otimes\hat{\boldsymbol{x}}_3-\hat{\boldsymbol{x}}_3\otimes\hat{\boldsymbol{x}}_1 \quad (4.254)$$
$$\hat{\boldsymbol{x}}_3\times\tilde{\boldsymbol{I}}=\hat{\boldsymbol{x}}_2\otimes\hat{\boldsymbol{x}}_1-\hat{\boldsymbol{x}}_1\otimes\hat{\boldsymbol{x}}_2 \quad (4.255)$$

can be used to write (4.248) in the form

$$\boldsymbol{B}(\boldsymbol{r})=\frac{\mu_0}{4\pi}(\boldsymbol{Q}\cdot\boldsymbol{r}_0)\tilde{\boldsymbol{I}}\underset{\cdot}{\times}\tilde{\boldsymbol{F}}(\boldsymbol{r})-\frac{\mu_0}{4\pi}(\boldsymbol{d}\times\tilde{\boldsymbol{I}})\underset{\cdot}{\times}\tilde{\boldsymbol{F}}(\boldsymbol{r})+O(\mathrm{el}_3) \quad (4.256)$$

where

$$\boldsymbol{d}=(\boldsymbol{Q}\cdot\tilde{\boldsymbol{M}}(\alpha_1)\times\boldsymbol{r}_0)\cdot\tilde{\boldsymbol{N}}_1 \quad (4.257)$$

and

$$\tilde{\boldsymbol{N}}_1=\sum_{m=1}^{3}\frac{\hat{\boldsymbol{x}}_m\otimes\hat{\boldsymbol{x}}_m}{\alpha_1^2+\alpha_2^2+\alpha_3^2-\alpha_m^2}. \quad (4.258)$$

Further manipulations of (4.256) lead finally to the expression

$$\boldsymbol{B}(\boldsymbol{r}) = \frac{\mu_0}{4\pi}\boldsymbol{d}\cdot\left[\frac{\mathbb{F}_2^1(\boldsymbol{r})}{\Lambda-\Lambda'}\tilde{\boldsymbol{\Lambda}} - \frac{\mathbb{F}_2^2(\boldsymbol{r})}{\Lambda-\Lambda'}\tilde{\boldsymbol{\Lambda}}' - \frac{3}{h_1^2h_2^2h_3^2}\sum_{\substack{i,j=1\\ i\neq j}}^{3} h_ih_j\mathbb{F}_2^{i+j}(\boldsymbol{r})\hat{\boldsymbol{x}}_i\otimes\hat{\boldsymbol{x}}_j\right] + O(\mathrm{el}_3) \tag{4.259}$$

where the exterior ellipsoidal harmonics $\mathbb{F}_2^m, m = 1,2,3,4,5$ are given by (4.131).

Another expression for (4.259) is

$$\boldsymbol{B}(\boldsymbol{r}) = \frac{\mu_0}{4\pi}\frac{\mathbb{F}_2^1(\rho,\mu,\nu)}{\Lambda-\Lambda'}\sum_{i=1}^{3}\frac{d_i}{\Lambda-\alpha_i^2}\hat{\boldsymbol{x}}_i - \frac{\mu_0}{4\pi}\frac{\mathbb{F}_2^2(\rho,\mu,\nu)}{\Lambda-\Lambda'}\sum_{i=1}^{3}\frac{d_i}{\Lambda'-\alpha_i^2}\hat{\boldsymbol{x}}_i - \frac{15\mu_0}{4\pi}\sum_{\substack{i,j=1\\ i\neq j}}^{3} d_ix_ix_jI_2^{i+j}(\rho)\hat{\boldsymbol{x}}_j + O\left(\frac{1}{\rho^4}\right) \tag{4.260}$$

where

$$\boldsymbol{d} = \frac{\alpha_2^2Q_2x_{03}-\alpha_3^2Q_3x_{02}}{\alpha_2^2+\alpha_3^2}\hat{\boldsymbol{x}}_1 + \frac{\alpha_3^2Q_3x_{01}-\alpha_1^2Q_1x_{03}}{\alpha_1^2+\alpha_3^2}\hat{\boldsymbol{x}}_2 + \frac{\alpha_1^2Q_1x_{02}-\alpha_2^2Q_2x_{01}}{\alpha_1^2+\alpha_2^2}\hat{\boldsymbol{x}}_3. \tag{4.261}$$

We observe here that, in contrast to the spherical case, where the radius of the conductive sphere does not appear in the expression (4.76), the solution (4.259) does depend on the semiaxes $\alpha_1, \alpha_2, \alpha_3$ of the conductive ellipsoid. Another important issue is that although we receive a contribution from the integral term of the Geselowitz formula, which depends on the electric potential, the magnetic field is independent of the conductivity.

Furthermore, at least for the quadrupolic term, the radial sources, which were silent sources for the sphere, are not silent for the ellipsoid. Instead, the silent sources for the quadrupolic solution of the ellipsoidal conductor are to be determined by the vector $\boldsymbol{d}$, given by (4.257) or, in component form, by (4.261).

Indeed, using $\tilde{\boldsymbol{M}}$ we can write the ellipsoidal conductor as

$$\boldsymbol{r}\cdot\tilde{\boldsymbol{M}}^{-1}(\alpha_1)\cdot\boldsymbol{r} = 1 \tag{4.262}$$

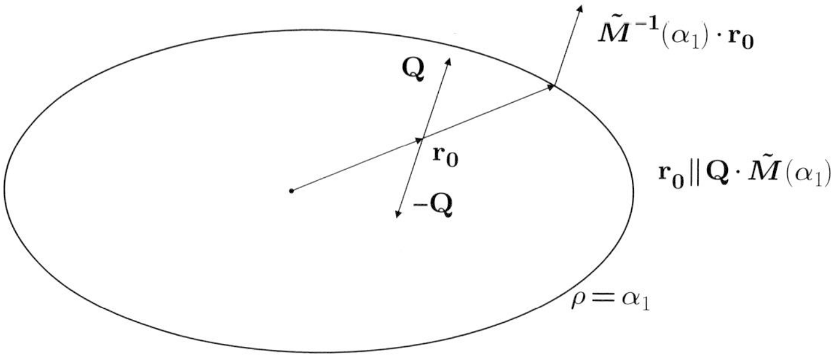


**Fig. 4.6** Ellipsoidal silent sources

from which, by taking the gradient, we obtain the Gaussian map

$$\hat{\boldsymbol{\rho}}(\boldsymbol{r}) = \frac{\tilde{\boldsymbol{M}}^{-1}(\alpha_1)\cdot \boldsymbol{r}}{\|\tilde{\boldsymbol{M}}^{-1}(\alpha_1)\cdot \boldsymbol{r}\|}. \tag{4.263}$$

Hence, the dyadic $\tilde{\boldsymbol{M}}^{-1}(\alpha_1)$ maps position vectors on the ellipsoid to outward vectors normal to the surface of the ellipsoid, and therefore the dyadic $\tilde{\boldsymbol{M}}(\alpha_1)$ maps normal directions on the surface to the corresponding position vectors on the surface. This implies that the vector $\boldsymbol{Q}\cdot\tilde{\boldsymbol{M}}(\alpha_1)$ is parallel to $\boldsymbol{r}_0$ whenever $\boldsymbol{Q}$ is parallel to the Gaussian image of the point where the direction $\boldsymbol{r}_0$ meets the ellipsoidal boundary $\rho = \alpha_1$, as it can be seen in Fig. 4.6. In this case $\boldsymbol{d} = \boldsymbol{0}$ and the dipole becomes silent.

Obviously, for the sphere, $\tilde{\boldsymbol{M}} = \alpha^2 \tilde{\boldsymbol{I}}$, the Gaussian map preserves the orientation of the position vector and hence only radial components of $\boldsymbol{Q}$ are silent.

In order to reduce (4.260) to the spherical case as $(\alpha_1, \alpha_2, \alpha_3) \to (\alpha, \alpha, \alpha)$ we use (4.193)–(4.201) as well as the identities

$$\frac{\mathbb{E}_2^1(\rho,\mu,\nu)}{(\Lambda-\Lambda')(\Lambda-\alpha_m^2)} - \frac{\mathbb{E}_2^2(\rho,\mu,\nu)}{(\Lambda-\Lambda')(\Lambda'-\alpha_m^2)} = r^2 - 3x_m^2 + \alpha_m^2 - \frac{\alpha_1^2+\alpha_2^2+\alpha_3^2}{3} \tag{4.264}$$

$$\frac{I_2^1(\rho) - I_2^2(\rho)}{\Lambda-\Lambda'} = \frac{3}{2} I_3^7(\rho) - \frac{1}{2\rho\sqrt{\rho^2-h_3^2}\sqrt{\rho^2-h_2^2}(\Lambda-\alpha_1^2+\rho^2)(\Lambda'-\alpha_1^2+\rho^2)} \tag{4.265}$$

to obtain the limits

$$\frac{\mathbb{E}_2^1(\rho,\mu,\nu) I_2^1(\rho)}{(\Lambda-\Lambda')(\Lambda-\alpha_m^2)} - \frac{\mathbb{E}_2^2(\rho,\mu,\nu) I_2^2(\rho)}{(\Lambda-\Lambda')(\Lambda'-\alpha_m^2)} \to \frac{r^2 \quad 3x_m^2}{5r^5} \tag{4.266}$$

$$
\begin{aligned}
\left[\frac{\mathbb{F}_2^1(\boldsymbol{r})}{\Lambda-\Lambda'}\tilde{\boldsymbol{\Lambda}}-\frac{\mathbb{F}_2^2(\boldsymbol{r})}{\Lambda-\Lambda'}\tilde{\boldsymbol{\Lambda}}'\right] &\to 5\sum_{m=1}^{3}\frac{r^2-3x_m^2}{5r^5}\hat{\boldsymbol{x}}_m\otimes\hat{\boldsymbol{x}}_m \\
&=\frac{\tilde{\boldsymbol{I}}}{r^3}-\frac{3}{r^5}\sum_{m=1}^{3}x_m^2\hat{\boldsymbol{x}}_m\otimes\hat{\boldsymbol{x}}_m
\end{aligned}
\tag{4.267}
$$

and

$$
\begin{aligned}
\sum_{\substack{i,j=1\\ i\neq j}}^{3}x_ix_jI_2^{i+j}(\rho)\hat{\boldsymbol{x}}_i\otimes\hat{\boldsymbol{x}}_j &\to \frac{1}{5r^5}\sum_{\substack{i,j=1\\ i\neq j}}^{3}x_ix_j\hat{\boldsymbol{x}}_i\otimes\hat{\boldsymbol{x}}_j \\
&=\frac{\boldsymbol{r}\otimes\boldsymbol{r}}{5r^5}-\frac{1}{5r^5}\sum_{i=1}^{3}x_i^2\hat{\boldsymbol{x}}_i\otimes\hat{\boldsymbol{x}}_i
\end{aligned}
\tag{4.268}
$$

which finally give

$$
\boldsymbol{B}(\boldsymbol{r})=\frac{\mu_0}{8\pi}\boldsymbol{Q}\times\boldsymbol{r}_0\cdot\frac{\tilde{\boldsymbol{I}}-3\hat{\boldsymbol{r}}\otimes\hat{\boldsymbol{r}}}{r^3}+O\left(\frac{1}{r^4}\right).
\tag{4.269}
$$

The form (4.269) recovers the quadrupolic term in the multipole expansion of (4.76).

We turn now back to examine the form of the complete eigenfunction expansion of the exterior $\boldsymbol{B}$ field for the ellipsoid [6]. Introducing the notation

$$
A_n^m(\boldsymbol{r}_0)=\frac{\boldsymbol{Q}\cdot\nabla\mathbb{E}_n^m(\boldsymbol{r}_0)}{\sigma\alpha_2\alpha_3\gamma_n^m E_n^{m\prime}(\alpha_1)}
\tag{4.270}
$$

$$
\boldsymbol{B}_n^m(\boldsymbol{r})=\frac{4\pi}{(2n+1)\gamma_n^m}(\nabla\mathbb{F}_n^m(\boldsymbol{r}))E_n^m(\alpha_1)
\tag{4.271}
$$

and

$$
\begin{aligned}
\boldsymbol{C}_{n\kappa}^{m\lambda} &= \oint_{\rho=\alpha_1}E_n^m(\mu)E_n^m(\nu)E_\kappa^\lambda(\mu)E_\kappa^\lambda(\nu)\hat{\boldsymbol{\rho}}(\mu,\nu)ds(\mu,\nu) \\
&=\frac{\alpha_1\alpha_2a_3}{h_1h_2h_3}\sum_{i=1}^{3}\frac{h_i}{\alpha_i}\hat{\boldsymbol{x}}_i\oint_{\rho=\alpha_1}E_n^m(\mu)E_n^m(\nu)E_\kappa^\lambda(\mu)E_\kappa^\lambda(\nu)E_1^i(\mu)E_1^i(\nu)l(\mu,\nu)ds(\mu,\nu)
\end{aligned}
\tag{4.272}
$$

for $n=0,1,2,\ldots,m=1,2,\ldots,2n+1$ and using (4.223) we arrive at the following expansion of the integral term in (4.221)

$$-\oint_{\rho'=\alpha_1} u(\boldsymbol{r}')\hat{\boldsymbol{\rho}}' \times \frac{\boldsymbol{r}-\boldsymbol{r}'}{|\boldsymbol{r}-\boldsymbol{r}'|^3} ds(\boldsymbol{r}') = \sum_{n=0}^{\infty}\sum_{m=1}^{2n+1}\sum_{k=1}^{\infty}\sum_{\lambda=1}^{2n+1} A_\kappa^\lambda(\boldsymbol{r}_0)\boldsymbol{C}_{n\kappa}^{m\lambda} \times \boldsymbol{B}_n^m(\boldsymbol{r}). \tag{4.273}$$

The integral in (4.272) represents the $(nm)$-coefficient of the expansion of the function $E_\kappa^\lambda(\mu)E_\kappa^\lambda(\nu)E_1^i(\mu)E_1^i(\nu)$ in terms of surface ellipsoidal harmonics. Expansion (4.273) is useful, since it "separates" the $\boldsymbol{r}_0$-dependence of the source from the $\boldsymbol{r}$-dependence of the observation. Its importance is due to the fact that the $\boldsymbol{r}_0$-dependence is not explicit in the integral term, but instead, it is implicit within the electric potential u. In the expansion (4.273) this dependence enters apparently via the expressions $\boldsymbol{Q}\cdot\nabla\mathbb{E}_\kappa^\lambda(\boldsymbol{r}_0)$. Similarly, we obtain

$$\boldsymbol{Q}\times\frac{\boldsymbol{r}-\boldsymbol{r}_0}{|\boldsymbol{r}-\boldsymbol{r}_0|^3} = -\boldsymbol{Q}\times\nabla_{\boldsymbol{r}}\frac{1}{|\boldsymbol{r}-\boldsymbol{r}_0|} = -\sum_{n=0}^{\infty}\sum_{m-1}^{2n+1}\frac{4\pi}{2n+1}\frac{1}{\gamma_n^m}\mathbb{E}_n^m(\boldsymbol{r}_0)\boldsymbol{Q}\times\nabla\mathbb{F}_n^m(\boldsymbol{r}) \tag{4.274}$$

Combining, (4.273) and (4.274) we arrive at the following separated (in $\boldsymbol{r}_0$ and $\boldsymbol{r}$) expression

$$\mathbf{B}(\mathbf{r}) = \mu_0\sum_{n=1}^{\infty}\sum_{m=1}^{2n+1}\frac{\nabla\mathbb{F}_n^m(\mathbf{r})}{(2n+1)\gamma_n^m} \times\left[\mathbf{Q}\mathbb{E}_n^m(\mathbf{r}_0) - \frac{1}{\alpha_2\alpha_3}\sum_{\kappa=1}^{\infty}\sum_{\lambda=1}^{2k+1}\mathbf{C}_{n\kappa}^{m\lambda}\frac{E_n^m(\alpha_1)}{E_\kappa^{\lambda\prime}(\alpha_1)}\frac{\mathbf{Q}\cdot\nabla\mathbb{E}_\kappa^\lambda(\mathbf{r}_0)}{\gamma_\kappa^\lambda}\right]. \tag{4.275}$$

In (4.275) we used the fact that the $n=0$ term vanishes, as it should. Indeed, for $n=0$ the leading term in the expansion (4.275) is

$$\boldsymbol{B}_0(\boldsymbol{r}) = \mu_0\frac{\nabla I_0^1(\rho)}{\gamma_0^1} \times\left[\boldsymbol{Q} - \frac{1}{\alpha_2\alpha_3}\sum_{\kappa=1}^{\infty}\sum_{\lambda=1}^{2\kappa+1}\boldsymbol{C}_{0\kappa}^{1\lambda}\frac{1}{E_\kappa^{\lambda\prime}(\alpha_1)}\frac{\boldsymbol{Q}\cdot\nabla\mathbb{E}_\kappa^\lambda(\boldsymbol{r}_0)}{\gamma_\kappa^\lambda}\right] \tag{4.276}$$

where

$$\boldsymbol{C}_{0\kappa}^{1\lambda} = \frac{\alpha_1\alpha_2 a_3}{h_1h_2h_3}\sum_{i=1}^{3}\frac{h_i}{\alpha_i}\hat{\boldsymbol{x}}_i\oint_{\rho'=\alpha_1}E_\kappa^\lambda(\mu')E_\kappa^\lambda(\nu')E_1^i(\mu')E_1^i(\nu')l(\mu',\nu')ds'$$
$$= \frac{\alpha_1\alpha_2 a_3}{h_1h_2h_3}\sum_{i=1}^{3}\frac{h_i}{\alpha_i}\hat{\boldsymbol{x}}_i\gamma_1^i\delta_{\kappa1}\delta_{i\lambda}$$

$$= \frac{4\pi}{3}\alpha_1\alpha_2 a_3 h_1 h_2 h_3 \sum_{i=1}^{3} \frac{\hat{\boldsymbol{x}}_i}{\alpha_i h_i}\delta_{\kappa 1}\delta_{i\lambda}$$

$$= \frac{4\pi}{3}\alpha_1\alpha_2 a_3 h_1 h_2 h_3 \frac{\hat{\boldsymbol{x}}_\lambda}{\alpha_\lambda h_\lambda}\delta_{\kappa 1}. \tag{4.277}$$

Inserting (4.277) in the expression for the $n = 0$ term (4.276) we obtain

$$\begin{aligned}\boldsymbol{B}_0(\boldsymbol{r}) &= \frac{\mu_0}{4\pi}\nabla I_0^1(\rho) \times \left[\boldsymbol{Q} - \sum_{\lambda=1}^{3} \hat{\boldsymbol{x}}_\lambda \frac{h_\lambda}{h_1 h_2 h_3}\boldsymbol{Q}\cdot\nabla\left(\frac{h_1 h_2 h_3}{h_\lambda}x_{0\lambda}\right)\right] \\ &= \frac{\mu_0}{4\pi}\nabla I_0^1(\rho) \times \left[\boldsymbol{Q} - \sum_{\lambda=1}^{3} \boldsymbol{Q}\cdot\hat{\boldsymbol{x}}_\lambda \otimes \hat{\boldsymbol{x}}_\lambda\right] \\ &= \boldsymbol{0}\end{aligned} \tag{4.278}$$

so that

$$\boldsymbol{B}(\boldsymbol{r}) = O\left(\frac{1}{r^3}\right), r \to \infty. \tag{4.279}$$

The coefficients $\boldsymbol{C}_{1\kappa}^{m\lambda}$ with $\kappa \geq 3$ should vanish by orthogonality. This is because the surface ellipsoidal harmonic

$$E_n^m(\mu)E_n^m(\nu)E_1^i(\mu)E_1^i(\nu)$$

lives in the subspace generated by the surface ellipsoidal harmonics of degree less or equal to $(n+1)$. So, by orthogonality

$$\boldsymbol{C}_{n\kappa}^{m\lambda} = 0, \kappa \geqslant n+2 \tag{4.280}$$

because any surface harmonic $E_\kappa^\lambda(\mu)E_\kappa^\lambda(\nu)$ with $k \geqslant n+2$ lives in the orthogonal complement of the aforementioned space.

Then $\boldsymbol{B}$ can be represented by the compact dyadic form

$$\boldsymbol{B}(\boldsymbol{r}) = \mu_0 \sum_{n=1}^{\infty}\sum_{m=1}^{2n+1} \boldsymbol{Q}\cdot\tilde{\boldsymbol{D}}_n^m(\boldsymbol{r}_0) \times \nabla\mathbb{F}_n^m(\boldsymbol{r}) \tag{4.281}$$

where the source dependent dyadic is given by

$$\begin{aligned}\tilde{\boldsymbol{D}}_n^m(\boldsymbol{r}_0) = \frac{1}{(2n+1)\gamma_n^m}&\left[-\mathbb{E}_n^m(\boldsymbol{r}_0)\tilde{\boldsymbol{I}}\right. \\ &\left.+\frac{1}{\alpha_2\alpha_3}\sum_{\kappa=1}^{n+1}\sum_{\lambda=1}^{2\kappa+1}\frac{1}{\gamma_\kappa^\lambda}\frac{E_n^m(\alpha_1)}{E_\kappa^{\lambda\prime}(\alpha_1)}(\nabla\mathbb{E}_\kappa^\lambda(\boldsymbol{r}_0))\otimes\boldsymbol{C}_{n\kappa}^{m\lambda}\right].\end{aligned} \tag{4.282}$$

Expression (4.282) provides the multipole expansion of the magnetic induction field outside the ellipsoid.

We recall that $\boldsymbol{B}$ outside the ellipsoid $\rho = \alpha_1$ is the gradient of a harmonic function $U$, defined by

$$\boldsymbol{B}(\rho, \mu, \nu) = \frac{\mu_0}{4\pi} \nabla U(\rho, \mu, \nu) \tag{4.283}$$

for $\rho > \alpha_1$, which because of (4.279) has the asymptotic behavior

$$U(\boldsymbol{r}) = O\left(\frac{1}{r^2}\right), r \to \infty. \tag{4.284}$$

In order to evaluate the magnetic potential $U$ for the case of the sphere, we had to integrate along a ray, in the direction of $\hat{\boldsymbol{r}}$, from the position $\boldsymbol{r}$ where the potential is evaluated, all the way to infinity where the potential vanishes. In doing so, we actually used only the radial component of $\boldsymbol{B}$ and since $\hat{\boldsymbol{r}}$ was constantly tangent to the path of integrations all the necessary calculations were simplified.

For the case of the ellipsoid though, the radial direction specified by the linear path of integration is not connected to anyone of the ellipsoidal directions $\hat{\boldsymbol{\rho}}$, $\hat{\boldsymbol{\mu}}, \hat{\boldsymbol{\nu}}$ and that makes the calculations cumbersome. This difficulty can be avoided if we choose an appropriate path of integration which is dictated by the intrinsic geometry of the ellipsoidal system.

To this end, we consider the ellipsoidal representation $(\rho, \mu, \nu)$ of the point $\boldsymbol{r}$ where the magnetic field $U$ is to be evaluated. From the point $(\rho, \mu, \nu)$ there passes an ellipsoid specified by the value $\rho$, a hyperboloid of one sheet specified by the value $\mu$ and a hyperboloid of two sheets specified by the value $\nu$. If we fix the values of $\mu$ and $\nu$ and we let the ellipsoidal coordinate vary from $\rho$ to infinity, then we obtain a path $\boldsymbol{C}$ that is generated from the intersection of the two hyperboloids corresponding to the constant values of $\mu$ and $\nu$. This path is a coordinate curve of the ellipsoidal system and its tangent at any point coincides with the ellipsoidal direction $\hat{\boldsymbol{\rho}}$ at the particular point. Since the system is orthogonal, the tangent $\hat{\boldsymbol{\rho}}$ remains normal to $\hat{\boldsymbol{\mu}}$ and $\hat{\boldsymbol{\nu}}$ as we move along the path $\boldsymbol{C}$. Hence, integration along this path for the case of the ellipsoid, corresponds to integration along the ray for the case of the sphere.

Consequently, we evaluate the value of $U$ at the point $\boldsymbol{r} = (\rho, \mu, \nu)$ by integrating along the ellipsoidal coordinate curve

$$\begin{aligned}\boldsymbol{C}(\rho) = \frac{1}{h_1 h_2 h_3} \Big[ & h_1 \rho\mu\nu \hat{\boldsymbol{x}}_1 + h_2 \sqrt{\rho^2 - h_3^2}\sqrt{\mu^2 - h_3^2}\sqrt{h_3^2 - \nu^2}\hat{\boldsymbol{x}}_2 \\ & + h_3 \sqrt{\rho^2 - h_2^2}\sqrt{h_2^2 - \mu^2}\sqrt{h_2^2 - \nu^2}\hat{\boldsymbol{x}}_3 \Big]\end{aligned} \tag{4.285}$$

where $\rho$ varies from the value that corresponds to the observation point all the way to infinity.

Since

$$\nabla = \frac{\hat{\boldsymbol{\rho}}}{h_\rho}\partial_\rho + \frac{\hat{\boldsymbol{\mu}}}{h_\mu}\partial_\mu + \frac{\hat{\boldsymbol{\nu}}}{h_\nu}\partial_\nu \tag{4.286}$$

where $\hat{\boldsymbol{\rho}}, \hat{\boldsymbol{\mu}}, \hat{\boldsymbol{\nu}}$ are the orthogonal unit base vectors and $h_\rho, h_\mu, h_\nu$ are the corresponding Lamé coefficients of the ellipsoidal system, it follows that

$$\frac{\partial}{\partial \rho} = h_\rho \hat{\boldsymbol{\rho}} \cdot \nabla \tag{4.287}$$

and we can represent $U$ as follows

$$\begin{aligned} U(\boldsymbol{r}) &= U(\rho, \mu, \nu) \\ &= -\int_\rho^\infty \frac{\partial}{\partial \rho'} U(\rho', \mu, \nu) d\rho' \\ &= -\int_\rho^\infty h_{\rho'} \hat{\boldsymbol{\rho}}' \cdot \nabla U(\rho', \mu, \nu) d\rho' \\ &= -\frac{4\pi}{\mu_0} \int_\rho^\infty h_{\rho'} \hat{\boldsymbol{\rho}}' \cdot \boldsymbol{B}(\rho', \mu, \nu) d\rho'. \end{aligned} \tag{4.288}$$

Using the identity

$$\nabla \mathbb{F}_n^m(\rho, \mu, \nu) = [\nabla F_n^m(\rho)] E_n^m(\mu) E_n^m(\nu) + F_n^m(\rho)[\nabla E_n^m(\mu) E_n^m(\nu)]$$

$$= \frac{\hat{\boldsymbol{\rho}}}{h_\rho}\left(\frac{\partial}{\partial \rho} F_n^m(\rho)\right) E_n^m(\mu) E_n^m(\nu) + F_n^m(\rho)\left[\frac{\hat{\boldsymbol{\mu}}}{h_\mu}\frac{\partial}{\partial \mu} + \frac{\hat{\boldsymbol{\nu}}}{h_\nu}\frac{\partial}{\partial \nu}\right] E_n^m(\mu) E_n^m(\nu) \tag{4.289}$$

and (4.281) we obtain

$$\hat{\boldsymbol{\rho}}' \cdot (\boldsymbol{Q} \cdot \tilde{\boldsymbol{D}}_n^m) \times \nabla \mathbb{F}_n^m(\rho', \mu, \nu)$$

$$= \hat{\boldsymbol{\rho}}' \cdot (\boldsymbol{Q} \cdot \tilde{\boldsymbol{D}}_n^m) \times \frac{\hat{\boldsymbol{\rho}}'}{h_{\rho'}}\left(\frac{\partial}{\partial \rho'} F_n^m(\rho')\right) E_n^m(\mu) E_n^m(\nu)$$

$$- F_n^m(\rho') \hat{\boldsymbol{\rho}}' \cdot \left[\left(\frac{\hat{\boldsymbol{\mu}}}{h_\mu}\frac{\partial}{\partial \mu} + \frac{\hat{\boldsymbol{\nu}}}{h_\nu}\frac{\partial}{\partial \nu}\right) E_n^m(\mu) E_n^m(\nu)\right] \times (\boldsymbol{Q} \cdot \tilde{\boldsymbol{D}}_n^m)$$

$$
\begin{aligned}
&= -F_n^m(\rho')\hat{\boldsymbol{\rho}}' \times \left[\left(\frac{\hat{\boldsymbol{\mu}}}{h_\mu}\frac{\partial}{\partial\mu} + \frac{\hat{\boldsymbol{\nu}}}{h_\nu}\frac{\partial}{\partial\nu}\right) E_n^m(\mu)E_n^m(\nu)\right] \cdot (\boldsymbol{Q}\cdot\tilde{\boldsymbol{D}}_n^m) \\
&= F_n^m(\rho')\left[\left(\frac{\hat{\boldsymbol{\nu}}}{h_\mu}\frac{\partial}{\partial\mu} - \frac{\hat{\boldsymbol{\mu}}}{h_\nu}\frac{\partial}{\partial\nu}\right) E_n^m(\mu)E_n^m(\nu)\right] \cdot (\boldsymbol{Q}\cdot\tilde{\boldsymbol{D}}_n^m) \qquad (4.290)
\end{aligned}
$$

where we used the fact that the dextral order of the ellipsoidal base is $\hat{\boldsymbol{\rho}}, \hat{\boldsymbol{\nu}}, \hat{\boldsymbol{\mu}}$.

In order to isolate the $\rho'$ dependence in (4.290) we use the expressions (4.95), (4.97), (4.99) and (4.100) to obtain

$$
\begin{aligned}
&\left(\frac{\hat{\boldsymbol{\nu}}}{h_\mu}\frac{\partial}{\partial\mu} - \frac{\hat{\boldsymbol{\mu}}}{h_\nu}\frac{\partial}{\partial\nu}\right) E_n^m(\mu)E_n^m(\nu) \\
&= \frac{1}{h_\nu h_\mu}\sum_{i=1}^{3} x_i\hat{\boldsymbol{x}}_i\left[\frac{\nu E_n^{m\prime}(\mu)E_n^m(\nu)}{\nu^2-\alpha_1^2+\alpha_i^2} - \frac{\mu E_n^m(\mu)E_n^{m\prime}(\nu)}{\mu^2-\alpha_1^2+\alpha_i^2}\right] \\
&= \frac{\sqrt{\mu^2-h_3^2}\sqrt{h_3^2-\nu^2}\sqrt{h_2^2-\mu^2}\sqrt{h_2^2-\nu^2}}{h_1h_2h_3(\mu^2-\nu^2)\sqrt{\rho'^2-\mu^2}\sqrt{\rho'^2-\nu^2}} \\
&\quad\times\sum_{i=1}^{3} h_i\mathbb{E}_1^i(\rho',\mu,\nu)\left[\frac{\nu E_n^{m\prime}(\mu)E_n^m(\nu)}{\nu^2-\alpha_1^2+\alpha_i^2} - \frac{\mu E_n^m(\mu)E_n^{m\prime}(\nu)}{\mu^2-\alpha_1^2+\alpha_i^2}\right]\hat{\boldsymbol{x}}_i \\
&= \sum_{i=1}^{3}\frac{E_1^i(\rho')}{\sqrt{\rho'^2-\mu^2}\sqrt{\rho'^2-\nu^2}} f_{ni}^m(\mu,\nu)\hat{\boldsymbol{x}}_i \qquad (4.291)
\end{aligned}
$$

where

$$
\begin{aligned}
f_{ni}^m(\mu,\nu) =& \frac{E_1^2(\mu)E_1^2(\nu)E_1^3(\mu)E_1^3(\nu)}{h_1h_2h_3(\mu^2-\nu^2)} h_iE_1^i(\mu)E_1^i(\nu) \\
&\times\left[\frac{\nu E_n^{m\prime}(\mu)E_n^m(\nu)}{\nu^2-\alpha_1^2+\alpha_i^2} - \frac{\mu E_n^m(\mu)E_n^{m\prime}(\nu)}{\mu^2-\alpha_1^2+\alpha_i^2}\right] \qquad (4.292)
\end{aligned}
$$

or, because of

$$
\frac{\nu E_1^i(\nu)}{\nu^2-\alpha_1^2+\alpha_i^2} = E_1^{i\prime}(\nu), i=1,2,3 \qquad (4.293)
$$

$$
\frac{\mu E_1^i(\mu)}{\mu^2-\alpha_1^2+\alpha_i^2} = E_1^{i\prime}(\mu), i=1,2,3 \qquad (4.294)
$$

$$f_{ni}^m(\mu,\nu) = \frac{h_i E_2^5(\mu)E_2^5(\nu)}{h_1h_2h_3(\mu^2-\nu^2)}\left[E_1^i(\mu)E_1^{i\prime}(\nu)E_n^{m\prime}(\mu)E_n^m(\nu)\right.$$

$$\left.-E_1^{i\prime}(\mu)E_1^i(\nu)E_n^m(\mu)E_n^{m\prime}(\nu)\right]. \tag{4.295}$$

Finally, we insert (4.291) in (4.290) to obtain the $\rho$-components of the expansion (4.281) which we substitute in (4.288) to arrive at the magnetic potential

$$U(\rho,\mu,\nu) = -4\pi\sum_{n=1}^{\infty}\sum_{m=1}^{2n+1}\sum_{i=1}^{3} f_{ni}^m(\mu,\nu)(\boldsymbol{Q}\cdot\tilde{\boldsymbol{D}}_n^m(\boldsymbol{r}_0)\cdot\hat{\boldsymbol{x}}_i)$$

$$\times\int_\rho^{+\infty}\frac{F_n^m(\rho')E_1^i(\rho')}{\sqrt{\rho'^2-h_3^2}\sqrt{\rho'^2-h_2^2}}d\rho' \tag{4.296}$$

which holds for $\rho > \alpha_1$ with $f_{ni}^m$ given by (4.295), $\tilde{\boldsymbol{D}}_n^m$ by (4.282) and $\boldsymbol{C}_{n\kappa}^{m\lambda}$ by (4.272).

Note that the $\rho$-dependence of $U$ enters through the integral factors of (4.296), the $(\mu,\nu)$-dependence is explicit in $f_{ni}^m$ as they are given by (4.295), while the dependence on the dipolar source occurs in $\boldsymbol{Q}\cdot\tilde{\boldsymbol{D}}_n^m$ as it can be observed in (4.282). Hence (4.296) provides a separable expansion of the magnetic field in ellipsoidal coordinates.

The crucial part in the derivation of (4.296) occurs in (4.288), where the appropriate choice of the contour of integration allowed us to mimic the spherical case and to utilize only the $\rho$-component of the magnetic field to construct $U$. On the other hand, since this procedure can not eliminate the integral part of the Geselowitz formula (4.221) the calculation of the electric potential can not be avoided.

Obviously, if we analyze further the leading term of (4.296) and substitute it in (4.283) we obtain (4.260).

Following similar arguments it is not hard to calculate the vector potential $\boldsymbol{A}$ for the magnetic field, where

$$\boldsymbol{B}(\boldsymbol{r}) = \nabla\times\boldsymbol{A}(\boldsymbol{r}). \tag{4.297}$$

In fact, it is straightforward to see that (4.221) is written as

$$\boldsymbol{B}(\boldsymbol{r}) = \nabla_{\boldsymbol{r}}\times\left[\frac{\mu_0}{4\pi}\frac{\boldsymbol{Q}}{|\boldsymbol{r}-\boldsymbol{r}_0|} - \frac{\mu_0\sigma}{4\pi}\oint_{\rho'=\alpha_1} u(\boldsymbol{r}')\frac{\hat{\boldsymbol{\rho}}'}{|\boldsymbol{r}-\boldsymbol{r}'|}ds(\boldsymbol{r}')\right] \tag{4.298}$$

so that the vector potential is given by

$$\boldsymbol{A}(\boldsymbol{r}) = \frac{\mu_0}{4\pi}\frac{\boldsymbol{Q}}{|\boldsymbol{r}-\boldsymbol{r}_0|} - \frac{\mu_0\sigma}{4\pi}\oint_{\rho'=\alpha_1} u(\boldsymbol{r}')\frac{\hat{\boldsymbol{\rho}}'}{|\boldsymbol{r}-\boldsymbol{r}'|}ds(\boldsymbol{r}') \tag{4.299}$$

or, in separable ellipsoidal coordinates, by

$$\boldsymbol{A}(\boldsymbol{r}) = -\mu_0 \sum_{n=1}^{\infty} \sum_{m=1}^{2n+1} \left( \boldsymbol{Q} \cdot \tilde{\boldsymbol{D}}_n^m(\boldsymbol{r}_0) \right) \mathbb{F}_n^m(\boldsymbol{r}). \tag{4.300}$$

Note that $\boldsymbol{A}$ satisfies the Coulomb gauge

$$\nabla \cdot \boldsymbol{A}(\boldsymbol{r}) = -\mu_0 \sum_{n=1}^{\infty} \sum_{m=1}^{2n+1} \left( \boldsymbol{Q} \cdot \tilde{\boldsymbol{D}}_n^m(\boldsymbol{r}_0) \right) \cdot \nabla \mathbb{F}_n^m(\boldsymbol{r}) = 0. \tag{4.301}$$

Indeed, since $\nabla \times \boldsymbol{B} = \boldsymbol{0}$ outside the ellipsoid and since $|\boldsymbol{r}-\boldsymbol{r}'|^{-1}$ is harmonic for $\boldsymbol{r} \neq \boldsymbol{r}'$, by taking the $\nabla\times$ operator on (4.298) we arrive at

$$\nabla_{\boldsymbol{r}} \nabla_{\boldsymbol{r}} \cdot \left[ \frac{\boldsymbol{Q}}{|\boldsymbol{r}-\boldsymbol{r}_0|} - \sigma \oint_{\rho'=\alpha_1} u(\boldsymbol{r}') \frac{\hat{\boldsymbol{\rho}}'}{|\boldsymbol{r}-\boldsymbol{r}'|} ds(\boldsymbol{r}') \right] = \boldsymbol{0}. \tag{4.302}$$

Equation (4.302) implies that

$$\boldsymbol{Q} \cdot \frac{\boldsymbol{r}-\boldsymbol{r}_0}{|\boldsymbol{r}-\boldsymbol{r}_0|^3} = c + \sigma \oint_{\rho'=\alpha_1} u(\boldsymbol{r}') \hat{\boldsymbol{\rho}}' \cdot \frac{\boldsymbol{r}-\boldsymbol{r}'}{|\boldsymbol{r}-\boldsymbol{r}'|^3} ds(\boldsymbol{r}') \tag{4.303}$$

and taking $r \to \infty$ we obtain $c = 0$. That proves (4.301).
We further present the following invariance property.

Define the dyadics

$$\tilde{\boldsymbol{S}}(\boldsymbol{r}) = \boldsymbol{Q} \otimes \frac{\boldsymbol{r}-\boldsymbol{r}_0}{|\boldsymbol{r}-\boldsymbol{r}_0|^3} \tag{4.304}$$

for the source and

$$\tilde{\boldsymbol{C}}(\boldsymbol{r}) = \sigma \oint_{\rho'=\alpha_1} u(\boldsymbol{r}') \hat{\boldsymbol{\rho}}' \otimes \frac{\boldsymbol{r}-\boldsymbol{r}'}{|\boldsymbol{r}-\boldsymbol{r}'|^3} ds(\boldsymbol{r}') \tag{4.305}$$

for the conductive medium. Let

$$S_S(\boldsymbol{r}) = \boldsymbol{Q} \cdot \frac{\boldsymbol{r}-\boldsymbol{r}_0}{|\boldsymbol{r}-\boldsymbol{r}_0|^3} \tag{4.306}$$

$$\boldsymbol{S}_V(\boldsymbol{r}) = \boldsymbol{Q} \times \frac{\boldsymbol{r}-\boldsymbol{r}_0}{|\boldsymbol{r}-\boldsymbol{r}_0|^3} \tag{4.307}$$

be the scalar and the vector invariants of $\tilde{\boldsymbol{S}}$ respectively, and

$$C_S(\boldsymbol{r}) = \sigma \oint_{\rho'=\alpha_1} u(\boldsymbol{r}') \hat{\boldsymbol{\rho}}' \cdot \frac{\boldsymbol{r}-\boldsymbol{r}'}{|\boldsymbol{r}-\boldsymbol{r}'|^3} ds(\boldsymbol{r}') \tag{4.308}$$

$$\boldsymbol{C}_V(\boldsymbol{r}) = \sigma \oint_{\rho'=\alpha_1} u(\boldsymbol{r}')\hat{\boldsymbol{\rho}}' \times \frac{\boldsymbol{r}-\boldsymbol{r}'}{|\boldsymbol{r}-\boldsymbol{r}'|^3} ds(\boldsymbol{r}') \tag{4.309}$$

the scalar and the vector invariants of $\tilde{\boldsymbol{C}}$, respectively.

Then (4.303) reads as

$$S_S(\boldsymbol{r}) = C_S(\boldsymbol{r}) \tag{4.310}$$

and (4.298) gives

$$\frac{4\pi}{\mu_0}\boldsymbol{B}(\boldsymbol{r}) = \boldsymbol{S}_V(\boldsymbol{r}) - \boldsymbol{C}_V(\boldsymbol{r}). \tag{4.311}$$

In other words, the induction field $\boldsymbol{B}$ outside the conductive medium is the difference between the vector invariant of the source dyadic and the vector invariant of the conductivity dyadic.

We discuss next the case of an inhomogeneous ellipsoidal conductor [13] consisting of an ellipsoid with constant conductivity surrounded by a confocal ellipsoidal shell of different conductivity.

Let

$$S_b : \quad \frac{x_1^2}{b_1^2} + \frac{x_2^2}{b_2^2} + \frac{x_3^2}{b_3^2} = 1 \tag{4.312}$$

be the boundary of the homogeneous ellipsoid with conductivity $\sigma_b$ and let

$$S_a : \quad \frac{x_1^2}{\alpha_1^2} + \frac{x_2^2}{\alpha_2^2} + \frac{x_3^2}{\alpha_3^2} = 1 \tag{4.313}$$

be the outer boundary of the ellipsoidal shell with conductivity $\sigma_\alpha$. Equation (4.312) corresponds to $\rho = b_1$ and (4.313) to $\rho - \alpha_1$. Also by confocality we have that

$$\begin{aligned} h_1^2 &= \alpha_2^2 - \alpha_3^2 = b_2^2 - b_3^2 \\ h_2^2 &= \alpha_1^2 - \alpha_3^2 = b_1^2 - b_3^2 \\ h_3^2 &= \alpha_1^2 - \alpha_2^2 = b_1^2 - b_2^2 \end{aligned} \tag{4.314}$$

where $0 < b_3 < b_2 < b_1$ and $0 < \alpha_3 < \alpha_2 < \alpha_1$ with $b_1 < \alpha_1$. In this case the Geselowitz formula reads

$$\begin{aligned} \boldsymbol{B}(\boldsymbol{r}) = {} & \frac{\mu_0}{4\pi}\boldsymbol{Q} \times \frac{\boldsymbol{r}-\boldsymbol{r}_0}{|\boldsymbol{r}-\boldsymbol{r}_0|^3} - \frac{\mu_0\sigma_\alpha}{4\pi}\oint_{S_\alpha} u_\alpha(\boldsymbol{r}')\hat{\boldsymbol{\rho}}' \times \left(\nabla_{\boldsymbol{r}'}\frac{1}{|\boldsymbol{r}-\boldsymbol{r}'|}\right) ds(\boldsymbol{r}') \\ & + \frac{\mu_0}{4\pi}\oint_{S_b} (\sigma_\alpha u_\alpha(\boldsymbol{r}') - \sigma_b u_b(\boldsymbol{r}'))\hat{\boldsymbol{\rho}}' \times \left(\nabla_{\boldsymbol{r}'}\frac{1}{|\boldsymbol{r}-\boldsymbol{r}'|}\right) ds(\boldsymbol{r}') \end{aligned} \tag{4.315}$$

where the electric potentials $u_\alpha$ and $u_b$ can readily be inferred from (4.214)–(4.220) if we equate the conductivities of the cerebrum, of the cerebrospinal fluid and of the skull. We then obtain

$$u_\alpha(\boldsymbol{r}) = \sum_{n=1}^{\infty} \sum_{m=1}^{2n+1} \frac{\boldsymbol{Q} \cdot \nabla \mathbb{E}_n^m(\boldsymbol{r}_0)}{\gamma_n^m C_n^m} \times \left( I_n^m(\rho) - I_n^m(\alpha_1) + \frac{1}{E_n^m(\alpha_1) E_n^{m\prime}(\alpha_1)\alpha_2\alpha_3} \right) \mathbb{E}_n^m(\rho,\mu,\nu) \tag{4.316}$$

for $\boldsymbol{r} \in V_\alpha$, and

$$u_b(\boldsymbol{r}) = \sum_{n=1}^{\infty} \sum_{m=1}^{2n+1} \frac{\boldsymbol{Q} \cdot \nabla \mathbb{E}_n^m(\boldsymbol{r}_0)}{\gamma_n^m C_n^m} \left[ \frac{C_n^m}{\sigma_b} (I_n^m(b_1) - I_n^m(\rho)) + \left( I_n^m(b_1) - I_n^m(\alpha_1) + \frac{1}{E_n^m(\alpha_1) E_n^{m\prime}(\alpha_1)\alpha_2\alpha_3} \right) \right] \mathbb{E}_n^m(\rho,\mu,\nu) \tag{4.317}$$

for $\boldsymbol{r} \in V_b$, where for each $n = 1, 2, \ldots$ and $m = 1, 2, \ldots, 2n+1$,

$$C_n^m = \sigma_\alpha + (\sigma_b - \sigma_\alpha) \left[ I_n^m(b_1) - I_n^m(\alpha_1) + \frac{1}{E_n^m(\alpha_1) E_n^{m\prime}(\alpha_1)\alpha_2\alpha_3} \right] \times E_n^m(b_1) E_n^{m\prime}(b_1) b_2 b_3. \tag{4.318}$$

As before, the arbitrary additive constant of the corresponding Neumann problem has been set equal to zero. A series of long calculations [13], essentially similar to those performed in the case of the homogeneous conductor, for the evaluation of the integrals in (4.315), leads to the following expression

$$\boldsymbol{B}(\boldsymbol{r}) = \frac{\mu_0}{4\pi} (\boldsymbol{d} - \boldsymbol{d}_b + \boldsymbol{d}_\alpha) \cdot \left[ \frac{\mathbb{F}_2^1(\boldsymbol{r})}{\Lambda - \Lambda'} \tilde{\boldsymbol{\Lambda}} - \frac{\mathbb{F}_2^2(\boldsymbol{r})}{\Lambda - \Lambda'} \tilde{\boldsymbol{\Lambda}}' - 15 \sum_{\substack{i,j=1 \\ i \neq j}}^{3} x_i x_j I_2^{i+j}(\rho) \hat{\boldsymbol{x}}_i \otimes \hat{\boldsymbol{x}}_j \right] + O(\mathrm{el}_3) \tag{4.319}$$

where

$$\boldsymbol{d} = (\boldsymbol{Q} \cdot \tilde{\boldsymbol{M}}(b_1) \times \boldsymbol{r}_0) \cdot \tilde{\boldsymbol{N}}(b_1) \tag{4.320}$$

$$\boldsymbol{d}_b = (\boldsymbol{Q} \cdot \tilde{\boldsymbol{M}}(b_1) \times \boldsymbol{r}_0) \cdot \tilde{\boldsymbol{N}}_c(b_1) \tag{4.321}$$

$$\boldsymbol{d}_\alpha = (\boldsymbol{Q} \cdot \tilde{\boldsymbol{M}}(\alpha_1) \times \boldsymbol{r}_0) \cdot \tilde{\boldsymbol{N}}_c(\alpha_1) \tag{4.322}$$

$$\tilde{\boldsymbol{M}}(b_1) = \sum_{i=1}^{3} b_i^2 \hat{\boldsymbol{x}}_i \otimes \hat{\boldsymbol{x}}_i \tag{4.323}$$

$$\tilde{\boldsymbol{M}}(\alpha_1) = \sum_{i-1}^{3} \alpha_i^2 \hat{\boldsymbol{x}}_i \otimes \hat{\boldsymbol{x}}_i \tag{4.324}$$

$$\tilde{\boldsymbol{N}}(b_1) = \sum_{i=1}^{3} \frac{\hat{\boldsymbol{x}}_i \otimes \hat{\boldsymbol{x}}_i}{b_1^2 + b_2^2 + b_3^2 - b_i^2} \tag{4.325}$$

$$\tilde{\boldsymbol{N}}_c(b_1) = \sum_{i=1}^{3} C^{6-i} \frac{\hat{\boldsymbol{x}}_i \otimes \hat{\boldsymbol{x}}_i}{b_1^2 + b_2^2 + b_3^2 - b_i^2} \tag{4.326}$$

$$\tilde{\boldsymbol{N}}_c(\alpha_1) = \sum_{i=1}^{3} C^{6-i} \frac{\hat{\boldsymbol{x}}_i \otimes \hat{\boldsymbol{x}}_i}{\alpha_1^2 + \alpha_2^2 + \alpha_3^2 - \alpha_i^2} \tag{4.327}$$

and for $i = 1, 2, 3$

$$C^{6-i} = \frac{\sigma_\alpha}{C_2^{6-i}} = \frac{\sigma_\alpha}{\sigma_\alpha + (\sigma_b - \sigma_\alpha)T} \tag{4.328}$$

with

$$T = (I_2^{6-i}(b_1) - I_2^{6-i}(\alpha_1))(b_1^2 + b_2^2 + b_3^2 - b_i^2) + \frac{(b_1^2 + b_2^2 + b_3^2 - b_i^2)b_1 b_2 b_3}{(\alpha_1^2 + \alpha_2^2 + \alpha_3^2 - \alpha_i^2)\alpha_1 \alpha_2 \alpha_3}$$

and $C_2^{6-i}$ given by (4.318).

The constants $\Lambda, \Lambda'$ are the roots of (4.132), and the dyadics $\tilde{\boldsymbol{\Lambda}}, \tilde{\boldsymbol{\Lambda}}'$ are given in (4.230). Comparing (4.319) with (4.259), or (4.260), we observe that the only difference appears to be isolated in the vector $(\boldsymbol{d} - \boldsymbol{d}_b + \boldsymbol{d}_\alpha)$, which involves the physical parameters $\sigma_\alpha$ and $\sigma_b$, the geometrical parameters $\alpha_i, b_i, i = 1, 2, 3$ and the characteristics of the source $\boldsymbol{Q}$ and $\boldsymbol{r}_0$.

In fact, we can recover (4.259) from (4.319) if we set

$$\sigma_b = \sigma_\alpha \tag{4.329}$$

along with the obvious equalities

$$b_i = \alpha_i, i = 1, 2, 3 \tag{4.330}$$

which will give

$$C^{6-i} = 1, i = 1, 2, 3 \tag{4.331}$$

$$\tilde{\boldsymbol{N}}_c(b_1) = \tilde{\boldsymbol{N}}_c(\alpha_1) = \tilde{\boldsymbol{N}}(\alpha_1) \tag{4.332}$$

and finally

$$\boldsymbol{d}_b = \boldsymbol{d}_\alpha = \boldsymbol{d}. \tag{4.333}$$

It is essential to note that the solution (4.319) depends on the conductivity $\sigma_\alpha$ of the shell and the conductivity jump $(\sigma_b - \sigma_\alpha)$ across the interface $\rho = b_1$, which appear in the expression (4.328) for the coefficients $C^{6-i}$. Therefore, shells of different conductivity are "visible" by magnetoencephalographic measurements. Since this property is not preserved in the spherical geometry it is customary to ignore conductivity inhomogeneities in practical applications. Nevertheless, since the brain-head geometry is closer to the ellipsoid than to the sphere, and since the actual shells that surround the brain do have different conductivities, their role should be taken into consideration. For the more realistic model with three ellipsoidal shells we refer to [18].

The question of whether the quadrupolic approximation, in the presence of layered inhomogeneities of the conductivity, has silent dipolar sources remains open. It amounts to search for directions of the moment $\boldsymbol{Q}$ for which

$$\boldsymbol{d} - \boldsymbol{d}_b + \boldsymbol{d}_\alpha = \boldsymbol{0}. \tag{4.334}$$

A similar question can be asked for the three-shell ellipsoidal model [18]. It is very probable that the inhomogeneities in conductivity prevent a dipole source from being silent.

## 4.7 The Inverse MEG Problem

As in most cases in science and modern technology, the main interest for mathematical methods is focused on inverse problems. Inverse problems are very seldom well defined, mainly due to lack of uniqueness and often to lack of stability. Magnetoencephalography is not an exception to this general rule. As we will show, even a complete knowledge of the field outside the head can not recover more than just one scalar function, out of the three, that determine the primary neuronal current [17, 20, 21].

We will discuss here the mathematics involved in the inverse magnetoencephalography problem, which asks for the primary current within the brain that gives rise to the measured magnetic field outside the head. Although there is an extended and continuously growing literature [29, 38] on the inverse MEG problem we will here concentrate on the purely mathematical aspects of the problem as it was developed in [17, 20, 21].

The fact that the knowledge of the exterior magnetic field, generated by a primary current within a bounded conducting medium, *can not* uniquely identify the primary current was known to Helmholtz almost 160 years ago. Nevertheless, a complete quantitative answer as to what exactly can be identified from the interior primary current was given only a couple of years ago, first for the spherical conductor [21] and then for any star-shape conducting medium [17].

Avoiding technical details we will describe next the case of the spherical conductor. Let us assume that the conductor is a sphere with radius one, center at the origin having constant conductivity $\sigma$. A primary current $\boldsymbol{J}^p(\boldsymbol{r}')$ is supported in the interior of the sphere $|\boldsymbol{r}'| < 1$ and let $\boldsymbol{r}$ be any point exterior to the conductor. Then, by Sarvas solution, the magnetic induction field is given by

$$\boldsymbol{B}(\boldsymbol{r}) = \frac{\mu_0}{4\pi}\nabla U(\boldsymbol{r}), |\boldsymbol{r}| > 1 \tag{4.335}$$

where the magnetic potential $U(\boldsymbol{r})$ is obtained by integrating (4.71) over the interior of the sphere

$$U(\boldsymbol{r}) = \int_{|\boldsymbol{r}'|\leq 1} \frac{\boldsymbol{J}^p(\boldsymbol{r}') \times \boldsymbol{r}' \cdot \boldsymbol{r}}{|\boldsymbol{r}-\boldsymbol{r}'|\,[|\boldsymbol{r}||\boldsymbol{r}-\boldsymbol{r}'| + \boldsymbol{r}\cdot(\boldsymbol{r}-\boldsymbol{r}')]} dv(\boldsymbol{r}'). \tag{4.336}$$

As we mentioned before, $U$ is a harmonic function for $|\boldsymbol{r}| > 1$ such that

$$U(\boldsymbol{r}) = O\left(\frac{1}{r^2}\right), r \to \infty \tag{4.337}$$

and (4.336) provides an integral representation of $U$ in terms of the spherical kernel

$$\boldsymbol{K}(\boldsymbol{r};\boldsymbol{r}') = \frac{\boldsymbol{r}}{|\boldsymbol{r}-\boldsymbol{r}'|\,[|\boldsymbol{r}||\boldsymbol{r}-\boldsymbol{r}'| + \boldsymbol{r}\cdot(\boldsymbol{r}-\boldsymbol{r}')]}. \tag{4.338}$$

It is easy to see that the singular support of the kernel $\boldsymbol{K}(\boldsymbol{r};\boldsymbol{r}')$ is the line segment from $\boldsymbol{0}$ to $\boldsymbol{r}'$, i.e. the set

$$S(\boldsymbol{r}') = \{t\hat{\boldsymbol{r}}'|t \in [0, |\boldsymbol{r}'|]\} \tag{4.339}$$

where $\hat{\boldsymbol{r}}' = \boldsymbol{r}'/|\boldsymbol{r}'|$, for every $\boldsymbol{r}'$ with $|\boldsymbol{r}'| < 1$. Hence, an extension of the action of the Laplace's operator to $\mathbb{R}^3$ needs to be established in the sense of distributions. A series of elaborate calculations [21] leads to the weak representation.

$$\int_{\mathbb{R}^3} (\varDelta_{\boldsymbol{r}} U(\boldsymbol{r};\boldsymbol{r}'))\varPhi(\boldsymbol{r}) dv(\boldsymbol{r}) = -\frac{4\pi}{|\boldsymbol{r}'|}\int_0^{|\boldsymbol{r}'|} [(\boldsymbol{J}^p(\boldsymbol{r}') \times \boldsymbol{r}) \cdot (\nabla_{\boldsymbol{r}}\varPhi(\boldsymbol{r}))]|_{\boldsymbol{r}=|\boldsymbol{r}|\hat{\boldsymbol{r}}'} d|\boldsymbol{r}| \tag{4.340}$$

for every test function $\varPhi \in C_0^\infty(\mathbb{R}^3)$, where

$$U(\boldsymbol{r};\boldsymbol{r}') = \frac{\boldsymbol{J}^p(\boldsymbol{r}') \times \boldsymbol{r}' \cdot \boldsymbol{r}}{|\boldsymbol{r}-\boldsymbol{r}'|\,[|\boldsymbol{r}||\boldsymbol{r}-\boldsymbol{r}'| + \boldsymbol{r}\cdot(\boldsymbol{r}-\boldsymbol{r}')]} \tag{4.341}$$

and the integration on the RHS of (4.340) extends over the support $S(\boldsymbol{r}')$ of $\boldsymbol{K}(\boldsymbol{r};\boldsymbol{r}')$.

The weak formulation (4.340) can now be used [21] to prove the integral representation

$$U(\boldsymbol{r}) = -\int_{|\boldsymbol{r}'|\leq 1} \frac{f(\boldsymbol{r}')}{|\boldsymbol{r}-\boldsymbol{r}'|} dv(\boldsymbol{r}'), |\boldsymbol{r}| < 1 \tag{4.342}$$

where the density function $f(\boldsymbol{r}')$ is given by

$$f(\boldsymbol{r}') = \frac{1}{|\boldsymbol{r}'|^2} \int_{|\boldsymbol{r}'|}^{1} [(\nabla_{\boldsymbol{s}} \times \boldsymbol{J}^p(\boldsymbol{s})) \cdot \boldsymbol{s}]|_{\boldsymbol{s}=|\boldsymbol{s}|\hat{\boldsymbol{r}}'} |\boldsymbol{s}| d|\boldsymbol{s}|. \tag{4.343}$$

In contrast to the representation (4.336), which uses the spherical kernel (4.338), the representation (4.342) uses the fundamental solution of the Laplace operator which is not depended on the geometry. Then, the actual spherical character of our problem is confined to the expression of the density function $f$ and not to the kernel. Up to this point there is no indication exactly which part of the current will be silent except of course of its irrotational part. This observation will eliminate the scalar potential $\Phi^p$ of the Helmholtz decomposition

$$\boldsymbol{J}^p(\boldsymbol{r}) = \nabla \Phi^p(\boldsymbol{r}) + \nabla \times \boldsymbol{A}^p(\boldsymbol{r}) \tag{4.344}$$

but the question of whether the whole vector potential $\boldsymbol{A}^p$ is needed remains open at this point.

In order to identify the minimum functional representation of $\boldsymbol{J}^p$, a 1-form decomposition of a compact Riemannian manifold in terms of the Hodge operator has been introduced in [21]. Nevertheless, this decomposition is nothing else but the well known spherical decomposition introduced by Hansen in 1935 as a means of defining vector spherical harmonics. In order to be consistent with the notation in [21] we write the Hansen decomposition as

$$\boldsymbol{J}^p(\boldsymbol{r}) = J_r(\boldsymbol{r})\hat{\boldsymbol{r}} + \hat{\boldsymbol{r}} \times \nabla F(\boldsymbol{r}) - \hat{\boldsymbol{r}} \times (\hat{\boldsymbol{r}} \times \nabla G(\boldsymbol{r})) \tag{4.345}$$

where it is clear that $J_r$ stands for the radial component of $\boldsymbol{J}^p$, while the potentials $F$ and $G$ span the tangential subspace of $\boldsymbol{J}^p$. Indeed, introducing in (4.345) the spherical form of the gradient operator

$$\nabla = \hat{\boldsymbol{r}}\partial_r + \frac{\hat{\boldsymbol{\theta}}}{r}\partial_\theta + \frac{\hat{\boldsymbol{\varphi}}}{r\sin\theta}\partial_\theta \tag{4.346}$$

as well as the spherical analysis of $\boldsymbol{J}^p$

$$\boldsymbol{J}^p = J_r\hat{\boldsymbol{r}} + J_\theta\hat{\boldsymbol{\theta}} + J_\varphi\hat{\boldsymbol{\varphi}} \tag{4.347}$$

we end up with the expressions

$$J_\theta(\boldsymbol{r}) = \frac{1}{r}\left(\partial_\theta G(\boldsymbol{r}) - \frac{1}{\sin\theta}\partial_\varphi F(\boldsymbol{r})\right) \tag{4.348}$$

$$J_\varphi(\boldsymbol{r}) = \frac{1}{r}\left(\partial_\theta F(\boldsymbol{r}) + \frac{1}{\sin\theta}\partial_\varphi G(\boldsymbol{r})\right). \tag{4.349}$$

Straightforward calculations now give the form

$$(\nabla \times \boldsymbol{J}^p(\boldsymbol{r})) \cdot \boldsymbol{r} = \frac{1}{|\boldsymbol{r}|}\mathbb{B}F(\boldsymbol{r}) \tag{4.350}$$

where $\mathbb{B}$ is the Beltrami (surface Laplacian) operator

$$\mathbb{B} = \frac{1}{\sin\theta}\partial_\theta(\sin\theta\partial_\theta) + \frac{1}{\sin^2\theta}\partial_{\varphi\varphi}. \tag{4.351}$$

Inserting (4.350) in (4.343) we obtain the following expression for the density function

$$f(\boldsymbol{r}') = \frac{1}{|\boldsymbol{r}'|^2}\int_{|\boldsymbol{r}'|}^{1} \mathbb{B}F(s,\theta,\varphi)ds, |\boldsymbol{r}'| < 1. \tag{4.352}$$

Consequently, only the potential $F$ is needed to represent the density $f$ and therefore the magnetic potential $U$. The radial component $J_r$ and the potential $G$ are *magnetically silent*. Hence, the best we can do in solving the inverse MEG problem is to reconstruct the part of the current that is given by $\hat{\boldsymbol{r}} \times \nabla F$. Two out of the three scalar functions that specify the vector field $\boldsymbol{J}^p$ belong to the null space of $U$. Although this sounds like a remarkable result it simply states that one scalar function can recover only one scalar function. The other two functions are buried in the representation (4.335) and as it was proved in [17] this is not a consequence of the high symmetry of the spherical geometry. It is true at least for any star-shape conductor.

Then we ask the question: is there an a-priori condition that will identify the hidden functions $J_r$ and G? As it was proved in [21] one such condition is that $\boldsymbol{J}^p$ should have minimum $L^2$-norm inside the sphere. A simple calculus of variations analysis for the minimization of the functional

$$W(\boldsymbol{J}^p) = \int_{|\boldsymbol{r}'|\leq 1} |\boldsymbol{J}^p(\boldsymbol{r})|^2 dv(\boldsymbol{r}) \tag{4.353}$$

leads to the conditions

$$J_r(\boldsymbol{r}) = G(\boldsymbol{r}) = 0, |\boldsymbol{r}| < 1 \tag{4.354}$$

and the minimum value

$$W_0(F) = \int_0^{2\pi}\int_0^{\pi}\int_0^{1}\left[\sin\theta(\partial_\theta F)^2 + \frac{1}{\sin\theta}(\partial_\varphi F)^2\right] dr d\theta d\varphi. \tag{4.355}$$

The final step is to construct the actual inversion process, under the minimizing condition, which will recover $F$ and therefore the $\hat{\boldsymbol{r}} \times \nabla F$ part of the current that we can find.

To this end, we see that

$$\Delta U(\boldsymbol{r}) = \begin{cases} \dfrac{1}{|\boldsymbol{r}|^2} \displaystyle\int_{|\boldsymbol{r}|}^1 \mathbb{B}F(s,\theta,\varphi)ds, & |\boldsymbol{r}| < 1, \\ 0, & |\boldsymbol{r}| > 1. \end{cases} \tag{4.356}$$

Expanding both $F$ and $U$ is surface spherical harmonics we obtain

$$F(r,\theta,\varphi) = \sum_{n=0}^{\infty} \sum_{m=-n}^{n} f_n^m(r) Y_n^m(\hat{\boldsymbol{r}}) \tag{4.357}$$

$$U(r,\theta,\varphi) = \sum_{n=0}^{\infty} \sum_{m=-n}^{n} u_n^m(r) Y_n^m(\hat{\boldsymbol{r}}) \tag{4.358}$$

where $Y_n^m(\hat{\boldsymbol{r}})$ are the normalized surface spherical harmonics.

Then, (4.356) implies the following ordinary integro-differential equations connecting the coefficients $f_n^m$ and $u_n^m$

$$r^2 {u^m_n}''(r) + 2r {u^m_n}'(r) - n(n+1)u_n^m(r) = \begin{cases} -n(n+1) \displaystyle\int_r^1 f_n^m(s)\,\mathrm{d}s, & r < 1 \\ 0, & r > 1, \end{cases} \tag{4.359}$$

where $f_n^m$ should be regular inside the sphere $|\boldsymbol{r}| < 1$ and $u_n^m$ should vanish at least as $r^{-2}$ for $r \to \infty$.

These imply that for $r > 1$

$$u_n^m(r) = c_n^m r^{-(n+1)}, n \geqslant 1. \tag{4.360}$$

For $r < 1$ we use variation of parameters to find a solution of (4.359) in the form

$$u_n^m(r) = A_n^m(r) r^n, r < 1 \tag{4.361}$$

where $A_n^m(r)$ should solve the equation

$$\frac{d}{dr}\left(r^{2n+2}\frac{d}{dr}A_n^m(r)\right) = -n(n+1)r^n \int_r^1 f_n^m(s)ds \tag{4.362}$$

or

$$r^{2n+2}\frac{d}{dr}A_n^m(r) = \frac{d}{dr}A_n^m(r)\bigg|_{r=1} + n(n+1)\int_r^1 t^n \int_t^1 f_n^m(s)dsdt \tag{4.363}$$

which by regularity at $r = 0$ implies that

$$A_n^{m\prime}(r) = -n(n+1)\int_0^1 t^n \int_t^1 f_n^m(s)dsdt. \tag{4.364}$$

On the other hand from (4.360) and (4.361) we obtain

$$\begin{aligned} A_n^{m\prime}(1) &= u_n^{m\prime}(1) - nu_n^m(1) \\ &= \frac{d}{dr}(c_n^m r^{-(n+1)})\Big|_{r=1} - n(c_n^m r^{-(n+1)})|_{r=1} \\ &= -(2n+1)c_n^m. \end{aligned} \tag{4.365}$$

Integrating the RHS of (4.364) by parts and equating the result with the RHS of (4.365) we obtain

$$(2n+1)c_n^m = n\int_0^1 t^{n+1} f_n^m(t)dt \tag{4.366}$$

which reduces the determination of the coefficients $f_n^m$ of $F$ to a moment problem. Hence, even the part of $\boldsymbol{J}^p$ that can be recovered needs the solution of a moment problem in order to recover it. The moment problem (4.366) can be solved completely if the condition of minimizing current is applied.

Indeed, writing the integrand in (4.355) as

$$\sin\theta(\partial_\theta F)^2 + \frac{1}{\sin\theta}(\partial_\varphi F)^2 = [|\nabla F|^2 - (\partial_r F)^2]r^2 \sin\theta \tag{4.367}$$

and using the fact that $F$ is supported in the interior of the unit sphere, which implies that

$$\int_{|\boldsymbol{r}|\leqslant 1} |\nabla F(\boldsymbol{r})|^2 dv(\boldsymbol{r}) = -\int_{|\boldsymbol{r}|\leqslant 1} F(\boldsymbol{r})\Delta F(\boldsymbol{r})dv(\boldsymbol{r}) \tag{4.368}$$

we rewrite (4.355) as

$$W_0(F) = -\int_0^{2\pi}\int_0^{\pi}\int_0^1 [F(\boldsymbol{r})\Delta F(\boldsymbol{r}) + (\partial_r F(\boldsymbol{r}))^2]r^2 \sin\theta dr d\theta d\varphi. \tag{4.369}$$

Writing $W_0(F)$ in this form it is much easier to reduce all the differentiations on the coefficients $f_n^m(r)$ in the expansion (4.357) to multiplications, since the spherical harmonics $Y_n^m$, $m = -n, \ldots, n$ are the eigenfunctions of the Beltrami operator $\mathbb{B}$ corresponding to the eigenvalue $-n(n+1)$, that is

$$\mathbb{B}Y_n^m(\hat{\boldsymbol{r}}) = -n(n+1)Y_n^m(\hat{\boldsymbol{r}}). \tag{4.370}$$

Hence, (4.357) and (4.370) give

$$\begin{aligned}
&F(\boldsymbol{r})\Delta F(\boldsymbol{r}) + (\partial_r F(\boldsymbol{r}))^2 = \\
&\sum_{n,m}\sum_{n',m'} f_n^m(r)\left[\partial_{rr} f_{n'}^{m'}(r) + \frac{2}{r}\partial_r f_{n'}^{m'}(r) - \frac{n'(n'+1)}{r^2} f_{n'}^{m'}(r)\right] Y_n^m(\hat{\boldsymbol{r}})Y_{n'}^{m'}(\hat{\boldsymbol{r}}) \\
&\quad + \sum_{n,m}\sum_{n',m'} (\partial_r f_n^m(r))(\partial_r f_{n'}^{m'}(r)) Y_n^m(\hat{\boldsymbol{r}})Y_{n'}^{m'}(\hat{\boldsymbol{r}}).
\end{aligned} \tag{4.371}$$

Note that if the complex form of the spherical harmonics is used then a complex conjugation will be involved in the corresponding inner product above.

Substituting (4.371) in the integral (4.369) and performing the angular integrations utilizing orthogonality we obtain

$$\begin{aligned}
W_0(F) &= -\sum_{n=0}^{\infty}\sum_{m=-n}^{n}\int_0^1\left[(f_n^{m\prime}(r))^2\right. \\
&\quad \left. + \left(f_n^{m\prime\prime}(r) + \frac{2}{r}f_n^{m\prime}(r) - \frac{n(n+1)}{r^2}f_n^m(r)\right)f_n^m(r)\right] r^2 dr \\
&= -\sum_{n=0}^{\infty}\sum_{m=-n}^{n}\int_0^1 [(r^2 f_n^m(r) f_n^{m\prime}(r))' - n(n+1)(f_n^m(r))^2] dr \\
&= \sum_{n=0}^{\infty}\sum_{m=-n}^{n} n(n+1)\int_0^1 (f_n^m(r))^2 dr
\end{aligned} \tag{4.372}$$

where an integration by parts and the fact that $f_n^m$ vanish on $|\boldsymbol{r}| = 1$ are used just before the last equality.

Since $W_0$ is reduced to a series of positive terms which are independent [because of the independence of the terms in the expansion (4.357)] we can minimize the series (4.372) term by term. Therefore, the problem of minimizing the 3D functional $W_0(F)$ with respect to the function $F(\boldsymbol{r})$ satisfying (4.342), (4.352) is now reduced to a sequence of minimizing problems, where for the $(n,m)$ term we need to minimize the 1D functional

$$w(f_n^m) = \int_0^1 (f_n^m(r))^2 dr \tag{4.373}$$

subjected to the constrain (4.366).

Using the Lagrangian

$$L = (f_n^m(r))^2 + \lambda r^{n+1} f_n^m(r) \tag{4.374}$$

where $\lambda$ is a Lagrange multiplier, the Euler–Lagrange equation gives

$$\frac{\partial L}{\partial f_n^m} = 0 \tag{4.375}$$

from which we immediately obtain

$$f_n^m(r) = -\frac{\lambda}{2} r^{n+1}. \tag{4.376}$$

Finally, choosing $\lambda$ so that (4.366) is satisfied we obtain

$$-\frac{\lambda}{2} = \frac{(2n+1)(2n+3)}{n} c_n^m \tag{4.377}$$

so that the minimizing coefficients are

$$f_n^m(r) = \frac{(2n+1)(2n+3)}{n} c_n^m r^{n+1}. \tag{4.378}$$

Therefore, the minimum current condition, besides securing a unique solution for the inverse problem, it also allows to solve the moment problem and to express the recoverable function $F$ in the form

$$F(\boldsymbol{r}) = \sum_{n=1}^{\infty} \sum_{m=-n}^{n} \frac{(2n+1)(2n+3)}{n} c_n^m r^{n+1} Y_n^m(\hat{\boldsymbol{r}}). \tag{4.379}$$

From (4.360) it follows that $c_0 = 0$, so that for $n = 0$, condition (4.366) holds identically, leaving $f_0^0(r)$ unidentified. Nevertheless, this does not affect the form of the current since $Y_0^0(\hat{\boldsymbol{r}}) = 1$ and therefore

$$\hat{\boldsymbol{r}} \times \nabla f_0^0(r) = \hat{\boldsymbol{r}} \times \hat{\boldsymbol{r}} f_0^{0\prime}(r) = \boldsymbol{0}. \tag{4.380}$$

Hence, without loss of generality we start the expansion in (4.379) with $n = 1$. The coefficients $c_n^m$ in (4.379) are the known coefficients of the expansion

$$U(\boldsymbol{r}) = \sum_{n=1}^{\infty} \sum_{m=-n}^{n} c_n^m \frac{1}{r^{n+1}} Y_n^m(\hat{\boldsymbol{r}}). \tag{4.381}$$

Everything we discussed up to this point, in connection with the inverse MEG problem, concerns a spherical conductor which is the only geometrical model for which a closed form solution for $\boldsymbol{B}$ is known. But what happens with a more general shape where the spherical symmetry is not present? As we will explain in the sequence, the non-uniqueness property of the inverse MEG problem is true at least for star-shape conductors [17].

**Fig. 4.7** Star-shape conductor

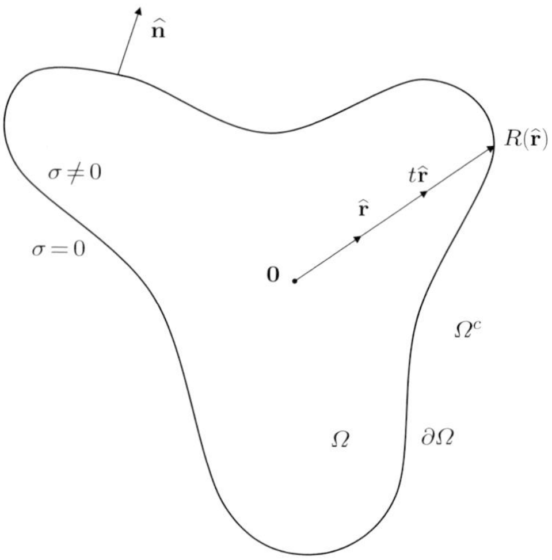


Consider a star-shape conductor (Fig. 4.7)

$$\Omega = \{\boldsymbol{r} \in \mathbb{R}^3 | \boldsymbol{r} = t\hat{\boldsymbol{r}}, t \in [0, R(\hat{\boldsymbol{r}})]\} \tag{4.382}$$

with boundary

$$\partial\Omega = \{R(\hat{\boldsymbol{r}}) | |\hat{\boldsymbol{r}}| = 1\} \tag{4.383}$$

and let $u^-(\boldsymbol{r})$ be the electric potential which solves the Neumann problem

$$\sigma \Delta u(\boldsymbol{r}) = \nabla \cdot \boldsymbol{J}^p(\boldsymbol{r}), \boldsymbol{r} \in \Omega \tag{4.384}$$

$$\partial_n u(\boldsymbol{r}) = 0, \boldsymbol{r} \in \partial\Omega \tag{4.385}$$

where $\boldsymbol{J}^p$ denotes the primary current in the interior of $\Omega$.

Then the magnetic induction $\boldsymbol{B}$, in the exterior to $\Omega$ region $\Omega^c$, is given by

$$\boldsymbol{B}(\boldsymbol{r}) = \frac{\mu_0}{4\pi} \nabla U(\boldsymbol{r}), \boldsymbol{r} \in \Omega^c \tag{4.386}$$

where $U$ is a harmonic function, and also by

$$\begin{aligned} \boldsymbol{B}(\boldsymbol{r}) = & \frac{\mu_0}{4\pi} \int_\Omega \boldsymbol{J}^p(\boldsymbol{r}') \times \frac{\boldsymbol{r} - \boldsymbol{r}'}{|\boldsymbol{r} - \boldsymbol{r}'|^3} dv(\boldsymbol{r}') \\ & - \frac{\mu_0 \sigma}{4\pi} \oint_{\partial\Omega} u(\boldsymbol{r}') \hat{\boldsymbol{n}}' \times \frac{\boldsymbol{r} - \boldsymbol{r}'}{|\boldsymbol{r} - \boldsymbol{r}'|^3} ds(\boldsymbol{r}'), \qquad \boldsymbol{r} \in \Omega^c \end{aligned} \tag{4.387}$$

where the volume integral represents the contribution of the primary current and the surface integral represents the contribution of the induction current.

In order to obtain the magnetic potential $U$ in (4.386) we integrate along the ray from $\boldsymbol{r}$ to infinity and we use (4.387) to obtain

$$
\begin{aligned}
U(\boldsymbol{r}) &= -\int_r^{+\infty} \frac{\partial}{\partial t} U(t\hat{\boldsymbol{r}})dt = -\frac{4\pi}{\mu_0}\int_r^{+\infty} \hat{\boldsymbol{r}} \cdot \boldsymbol{B}(t\hat{\boldsymbol{r}})dt \\
&= \int_\Omega \boldsymbol{J}^P(\boldsymbol{r}') \cdot (\boldsymbol{r}' \times \hat{\boldsymbol{r}}) \left[\int_r^{+\infty} \frac{dt}{|t\hat{\boldsymbol{r}} - \boldsymbol{r}'|^3}\right] dv(\boldsymbol{r}') \\
&\quad - \sigma \oint_{\partial\Omega} u^-(\boldsymbol{r}')\hat{\boldsymbol{n}}' \cdot (\boldsymbol{r}' \times \hat{\boldsymbol{r}}) \left[\int_r^{+\infty} \frac{dt}{|t\hat{\boldsymbol{r}} - \boldsymbol{r}'|^3}\right] ds(\boldsymbol{r}')
\end{aligned} \tag{4.388}
$$

and using (4.69), we finally obtain

$$
U(\boldsymbol{r}) = \boldsymbol{r} \cdot \int_\Omega \frac{\boldsymbol{J}^p(\boldsymbol{r}') \times \boldsymbol{r}'}{F(\boldsymbol{r};\boldsymbol{r}')} dv(\boldsymbol{r}') - \sigma \boldsymbol{r} \cdot \oint_{\partial\Omega} \frac{u(\boldsymbol{r}')\hat{\boldsymbol{n}}' \times \boldsymbol{r}'}{F(\boldsymbol{r};\boldsymbol{r}')} ds(\boldsymbol{r}') \tag{4.389}
$$

where $F(\boldsymbol{r};\boldsymbol{r}')$ is given by (4.70).

The importance of the representation (4.389), which holds for any smooth domain $\Omega$ (even non star-shape), is that it proves that the kernel

$$
\boldsymbol{K}(\boldsymbol{r};\boldsymbol{r}') = \frac{\boldsymbol{r}}{F(\boldsymbol{r};\boldsymbol{r}')} \tag{4.390}
$$

is a characteristic of the Physics of the problem, and it is independent of the Geometry.

That is, a dipole at $\boldsymbol{r}_0$ with moment $\boldsymbol{Q}$, inside the conductor $\Omega$ generates the magnetic potential

$$
U(\boldsymbol{r};\boldsymbol{r}_0) = \boldsymbol{Q} \times \boldsymbol{r}_0 \cdot \boldsymbol{K}(\boldsymbol{r};\boldsymbol{r}_0) - \sigma \oint_{\partial\Omega} u(\boldsymbol{r}';\boldsymbol{r}_0)\hat{\boldsymbol{n}}' \times \boldsymbol{r}' \cdot \boldsymbol{K}(\boldsymbol{r};\boldsymbol{r}')ds(\boldsymbol{r}') \tag{4.391}
$$

where $u(\boldsymbol{r};\boldsymbol{r}_0)$ solves the Neumann problem (4.35), (4.36). Then for a distributed current $\boldsymbol{J}^p$ in $\Omega$, linearity implies (4.389) where

$$
u(\boldsymbol{r}) = \int_\Omega u(\boldsymbol{r};\boldsymbol{r}_0)dv(\boldsymbol{r}_0). \tag{4.392}
$$

Comparing (4.391) with (4.34) we observe that the kernel $\boldsymbol{K}(\boldsymbol{r};\boldsymbol{r}_0)$ for the magnetic potential, plays the role that the field $(\boldsymbol{r} - \boldsymbol{r}_0)|\boldsymbol{r} - \boldsymbol{r}_0|^{-3}$ plays for the magnetic induction, and in that sense (4.389) is the analogue of the *Geselowitz formula for the magnetic potential* $U$.

It is obvious from (4.391) that for the spherical case alone we have that $\hat{\boldsymbol{n}} \times \boldsymbol{r}' = \boldsymbol{0}$, which means that the exterior magnetic potential, and therefore the magnetic induction, "sees" only the dipole at $\boldsymbol{r}_0$ when $\partial\Omega$ is a sphere, whereas in any other case it "sees" both the dipole at $\boldsymbol{r}_0$ and the surface distribution of normal dipoles on the boundary of the conductor.

For the case of a sphere, the contribution of the boundary distribution of dipoles to the magnetic induction is purely tangential at any observation point, and this is the reason why any radial integration, such as the one used in (4.388), is independent of the spherical boundary distribution of dipoles.

We return now to the representation (4.389). Let us assume that we have solved the related EEG problem and therefore the electric potential $u$ is known on the boundary $\partial\Omega$. Hence, if $U$ is the exterior magnetic potential, then the function

$$U_g(\boldsymbol{r}) = U(\boldsymbol{r}) + \sigma\boldsymbol{r} \cdot \oint_{\partial\Omega} \frac{u(\boldsymbol{r}')\hat{\boldsymbol{n}}' \times \boldsymbol{r}'}{F(\boldsymbol{r};\boldsymbol{r}')} ds(\boldsymbol{r}') = \boldsymbol{r} \cdot \int_{\Omega} \frac{\boldsymbol{J}^p(\boldsymbol{r}') \times \boldsymbol{r}'}{F(\boldsymbol{r};\boldsymbol{r}')} dv(\boldsymbol{r}') \tag{4.393}$$

is known, and we want to find what part of $\boldsymbol{J}^p$ can be obtained by inverting (4.393). In fact, following the same analysis as with the sphere we obtain the Green's representation

$$U_g(\boldsymbol{r}) = -\int_{\Omega} \frac{f(\boldsymbol{r}')}{|\boldsymbol{r}-\boldsymbol{r}'|} dv(\boldsymbol{r}'), \boldsymbol{r} \in \Omega^c \tag{4.394}$$

with

$$f(\boldsymbol{r}') = \frac{1}{|\boldsymbol{r}'|^2} \int_{|\boldsymbol{r}'|}^{R(\hat{\boldsymbol{r}}')} [(\nabla_{\boldsymbol{s}} \times \boldsymbol{J}^p(\boldsymbol{s})) \cdot \boldsymbol{s}]|_{\boldsymbol{s}=|\boldsymbol{s}|\hat{\boldsymbol{r}}'} |\boldsymbol{s}| d|\boldsymbol{s}|. \tag{4.395}$$

Then, using the Helmholtz decomposition

$$\boldsymbol{J}^p(\boldsymbol{r}) = \nabla\Phi(\boldsymbol{r}) + \nabla \times \boldsymbol{A}(\boldsymbol{r}) \tag{4.396}$$

with the Coulomb gauge

$$\nabla \cdot \boldsymbol{A}(\boldsymbol{r}) = 0 \tag{4.397}$$

we obtain the expression

$$(\nabla \times \boldsymbol{J}^p(\boldsymbol{r})) \cdot \boldsymbol{r} = -(\Delta\boldsymbol{A}(\boldsymbol{r})) \cdot \boldsymbol{r} \tag{4.398}$$

for the integrand in (4.395). The identity

$$\Delta(\boldsymbol{f} \cdot \boldsymbol{g}) = (\Delta\boldsymbol{f}) \cdot \boldsymbol{g} + \boldsymbol{f} \cdot (\Delta\boldsymbol{g}) + 2\sum_{i=1}^{3} (\nabla f_i) \cdot (\nabla g_i) \tag{4.399}$$

together with (4.397) allow to rewrite (4.398) as

$$(\nabla \times \boldsymbol{J}^p(\boldsymbol{r})) \cdot \boldsymbol{r} = -\Delta(\boldsymbol{r} \cdot \boldsymbol{A}(\boldsymbol{r})). \tag{4.400}$$

Therefore, all we need to express $U$ outside $\Omega$ is the nonharmonic part of the radial component of the solenoidal vector potential of $\boldsymbol{J}^p$, and this is just one scalar function.

Hence, the same order of non-uniqueness that holds true for the sphere also holds true for a star-shape conductor.

## 4.8 Open Mathematical Questions

The number of open computational, technological and mathematical problems in this active area of brain imaging is very large, and certainly lies beyond the knowledge of the present author. Here, we will focus only on a few mathematical problems associated with the understanding of EEG and MEG mechanisms along the spirit of these lectures.

In order to easily identify them we enumerate some of these problems.

**Problem 1.** Suppose the current within the sphere consists of a finite number of dipoles. How can we recover their positions and the tangential components of their moments? If such an algorithm is found, what is its spatial resolution? In other words, what is the minimum separation between two dipoles such that the algorithm "sees" two separate sources?

**Problem 2.** We know that for the spherical model the exterior magnetic potential does not "read" coeccentric shells of different conductivity, but the electric potential does. What is an algorithm that will identify these shells, as well as their conductivities, from surface measurements of the electric potential?

**Problem 3.** For the case of the ellipsoid the confocal shells of different conductivity are "visible" both by the EEG and by the MEG measurements. Which algorithms identify these shells if the electric potential is known?

**Problem 4.** We know that for the sphere, any radial dipole is magnetically silent. For the ellipsoid, we showed that there are directions of the dipole for which the leading quadrupolic term of the magnetic field vanishes, but do there exist directions of the dipole that are silent for the full multipole expansion of $\boldsymbol{B}$? In other words, are there hidden dipoles within an ellipsoidal conductor?

**Problem 5.** In [21] an analytic algorithm is proposed that recovers the part of the current that can be recovered from a knowledge of the exterior magnetic potential for, the case of a sphere. What is the corresponding algorithm for the ellipsoidal model?

**Problem 6.** What part of the current remains silent when both the electric and the magnetic potentials outside the head are known?

**Problem 7.** What differences will we observe, for the direct and the inverse EEG and MEG problems, if we replace the dipole distribution with higher order multipoles at some point? and more generally, if we replace the continuous distribution of dipoles with a finite set of localized higher multipoles?

**Problem 8.** In the inverse problem of identifying a dipole from EEG or from MEG measurements, how sensitive is the location of the dipole and its moment to variations of the principal eccentricities of the ellipsoidal model of the brain? and in particular, what is the error when we replace the average ellipsoidal brain, with semiaxes 6, 6.5 and 9 cm, with an equivalent sphere of the same volume?

**Problem 9.** What part of the current can be recovered if the EEG and/or the MEG data are restricted to a part of the full solid angle?

**Problem 10.** An image system for the representation of the electric potential for the spherical conductor is presented in Sect. 4.3. What is the corresponding image system for the ellipsoidal conductor? Obviously, the answer to this long standing problem has implications far beyond the theory of electromagnetic brain activity.

**Acknowledgements** The present Review-Notes have been prepared in the Department of Applied Mathematics and Theoretical Physics of the University of Cambridge, where the author holds a *Marie Curie Chair of Excellence*, funded by the European Commission under the project BRAIN(EXC 023928). Thanks are due the graduate students Euan Spence and Eleftherios Garyfallidis for their valuable help during the preparation of these notes.

## References

1. Ammari H., Bao G., Fleming J.L. (2002). An inverse source problem for Maxwell's equations in magnetoencephalography. *SIAM Journal of Applied Mathematics*, vol. 62, pp. 1369–1382.
2. Bronzan J.B. (1971). The magnetic scalar potential. *American Journal of Physics,* vol. 39, pp. 1357–1359.
3. Cuffin B.N., Cohen D. (1977). Magnetic fields of a dipole in special volume conductor shapes. *IEEE Transactions on Biomedical Engineering,* vol. 24, pp. 372–381.
4. Dassios G. (2005). On the hidden electromagnetic activity of the brain. In: Fotiadis D., Massalas C. (eds) *Mathematical Methods in Scattering Theory and Biomedical Technology*, Nymfaio, pp. 297–303.
5. Dassios G. (2006). What is recoverable in the inverse magnetoencephalography problem? In: Ammari H., Kang H.(eds). *Inverse Problems, Multi-Scale Analysis and Effective Medium Theory*, American Mathematical Society, Contemporary Mathematics 408, Providence, pp. 181–200.
6. Dassios G. (2007) The magnetic potential for the ellipsoidal MEG problem. *Journal of Computational Mathematics*, vol. 25, pp. 145–156.
7. Dassios G. On the image systems for the electric and the magnetic potential of the brain. In: Alikakos N., DougalisV., (eds). *Modern Mathematical Methods in Sciences and Technology*, Paros, (in press).
8. Dassios G., Kariotou F. (2001). The direct magnetoencephalography problem for a genuine 3-D brain model. In: Fotiadis D., Massalas C., (eds) *Scattering Theory and Biomedical Engineering. Modeling and Applications*, Corfu. World Scientific, London, pp. 289–297.
9. Dassios G., Kariotou F. (2003). On the Geselowitz formula in biomathematics. *Quarterly of Applied Mathematics,* vol. 61, pp. 387–400.
10. Dassios G., Kariotou F. (2003). Magnetoencephalography in ellipsoidal geometry. *Journal of Mathematical Physics*. vol. 44, pp. 220–241.

11. Dassios G., Kariotou F. (2003). The effect on an ellipsoidal shell on the direct EEG problem. In: Fotiadis D., Massalas C., (eds) *Mathematical Methods in Scattering Theory and Biomedical Technology*, Tsepelovo. World Scientific, London, pp.495–503.
12. Dassios G., Kariotou F. (2004). On the exterior magnetic field and silent sources in magnetoencephalography. *Abstract and Applied Analysis,* vol. 2004, pp. 307–314.
13. Dassios G., Kariotou F. (2005). The direct MEG problem in the presence of an ellipsoidal shell inhomogeneity. *Quarterly of Applied Mathematics*, vol. 63, pp. 601–618.
14. Dassios G., Kleinman R. (2000). *Low frequency scattering*. Oxford University Press. Oxford.
15. Dassios G., Kariotou F., Kontzialis K., Kostopoulos V. (2002). Magnetoencephalography for a realistic geometrical model of the brain. *Third European Symposium in Biomedical Engineering and Medical Physics*, Patras, p.22.
16. Dassios G., Giapalaki S., Kariotou F., Kandili A. (2005). Analysis of EEG Images. In: Fotiadis D., Massalas C., (eds). *Mathematical Methods in Scattering Theory and Biomedical Technology*, Nymfaio, pp. 304–311.
17. Dassios G., Fokas A.S., Kariotou F. (2005). On the non-uniqueness of the inverse MEG problem. *Inverse Problems*, vol. 21, pp. L1–L5.
18. Dassios G., Giapalaki S., Kandili A., Kariotou F. (2007). The exterior magnetic field for the multilayer ellipsoidal model of the brain. *Quarterly Journal of Mechanics and Applied Mathematics* vol. 60, pp. 1–25.
19. de Munck J.C. (1988). The potential distribution in a layered anisotropic spheroidal volume conductor. *Journal of Applied Physics*, vol. 2, pp. 464–470.
20. Fokas A.S., Gelfand I.M., Kurylev Y. (1996). Inversion method for magnetoencephalography. *Inverse Problems*, vol. 12, pp. L9–L11.
21. Fokas A.S., Kurylev Y., Marikakis V. (2004). The unique determination of neuronal current in the brain via magnetoencephalography. *Inverse Problems*, vol. 20, pp. 1067–1082.
22. Frank E. (1952). Electric potential produced by two point current sources in a homogeneous conducting sphere. *Journal of Applied Physics*, vol. 23, pp. 1225–1228.
23. Garmier R., Barriot J.P. (2001). Ellipsoidal harmonic expansions of the gravitational potential: Theory and applications. *Celestial Mechanics and Dynamical Astronomy*, vol. 79, pp. 235–275.
24. Geselowitz D.B. (1967). On bioelectric potentials in an inhomogeneous volume conductor. *Biophysical Journal*, vol. 7, pp. 1–11.
25. Geselowitz D.B. (1970). On the magnetic field generated outside an inhomogeneous volume conductor by internal current sources. *IEEE Transactions in Magnetism*, vol. 6, pp. 346–347.
26. Giapalaki S. (2006). *A study of models in Medical Physics through the analysis of mathematical problems in Neurophysiology*. Doctoral Dissertation, Department of Chemical Engineering, University of Patras.
27. Giapalaki S., Kariotou F. (2006). The complete ellipsoidal shell model in EEG Imaging. *Abstract and Applied Analysis*, vol. 2006, pp. 1–18.
28. Gutierrez D., Nehorai A., Preissl H. (2005). Ellipsoidal head model for fetal magnetoencephalography: forward and inverse solutions. *Physics in Medicine and Biology*, vol. 50, pp. 2141–2157.
29. Hämäläinen M., Hari R., Ilmoniemi RJ., Knuutila J., Lounasmaa O. (1993). Magnetoencephalography-theory, instrumentation, and applications to noninvasive studies of the working human brain. *Reviews of Modern Physics*, vol. 65, pp. 413–497.
30. Hobson E.W. (1931). *The theory of spherical and ellipsoidal harmonics*. Cambridge University Press. Cambridge.
31. Jerbi K., Mosher J.C., Baillet S., Leahy R.M. (2002). On MEG forward modeling using multipolar expansions. *Physics in Medicine and Biology*, vol. 47, pp. 523–555.
32. Kamvyssas G., Kariotou F. (2005). Confocal ellipsoidal boundaries in EEG modeling. *Bulletin of the Greek Mathematical Society*, vol. 50, pp. 119–133.

33. Kariotou F. (2001). On the electroencephalography (EEG) problem for the ellipsoidal brain model. Proceedings of the *6th International conference of the HSTAM, Thessaloniki,* pp. 222–226.
34. Kariotou F. (2002). *Mathematical problems of the electromagnetic activity in the neurophysiology of the brain.* Doctoral Dissertation, Department of Chemical Engineering, University of Patras.
35. Kariotou F. (2003). On the mathematics of EEG and MEG in spheroidal geometry. *Bulletin of the Greek Mathematical Society*, vol 47, pp. 117–135.
36. Kariotou F. (2004). Electroencephalography in ellipsoidal geometry. *Journal of Mathematical Analysis and Applications*, vol. 290, pp. 324–342.
37. MacMillan W.D. (1930). *The theory of the potential.* MacGraw-Hill, London.
38. Malmivuo J., Plonsey R. (1995). *Bioelectromagnetism.* Oxford University Press. New York.
39. Nolte G. (2002). Brain current multipole expansion in realistic volume conductors: calculation, interpretation and results. *Recent Research Developments in Biomedical Engineering* vol.1, pp. 171–211.
40. Nolte G. (2003). The magnetic lead field theorem in the quasi-static approximation and its use for MEG forward calculation in realistic volume conductors. *Physics in Medice and Biology*, vol. 48, pp. 3637–3652.
41. Nolte G., Curio G. (1999). Perturbative analytical solutions of the forward problem for realistic volume conductors. *Journal of Applied Physics*, vol. 86, pp. 2800–2811.
42. Nolte G., Dassios G. (2004). Semi-analytic forward calculation for the EEG in multi-shell realistic volume conductors based on the lead field theorem. Poster presentation in BIOMAG 2004, Boston.
43. Nolte G., Dassios G. (2005). Analytic expansion of the EEG lead field for realistic volume conductors. *Physics in Medicine and Biology*, vol. 50, pp. 3807–3823.
44. Nolte G., Fieseler T., Curio G. (2001). Perturbative analytical solutions of the magnetic forward problem for realistic volume conductors. *Journal of Applied Physics*, vol. 89, pp. 2360–2369.
45. Plonsey R., Heppner D.B. (1967). Considerations of quasi-stationarity in electrophysiological systems. *Bulletin of Mathematical Biophysics*, vol. 29, pp. 657–664.
46. Sarvas J. (1987). Basic mathematical and electromagnetic concepts of the biomagnetic inverse problem. *Physics in Medicine and Biology*, vol. 32, pp. 11–22.
47. Snyder W.S., Ford M.R., Warner G.G., Fisher H.L. (1969, 1978). Estimates of absorbed fractions for monoenergetic photon sources uniformly distributed in various organs of a heterogeneous phantom. *Journal of Nuclear Medicine*, Supplement Number 3 (August 1969),vol. 10, Pamphlet No.5, Revised 1978.
48. Whittaker E.T., Watson G.N. (1920). *A course of modern analysis.* Cambridge University Press, Cambridge.
49. Wilson F.N., Bayley R.H. (1950). The electric field of an eccentric dipole in a homogeneous spherical conducting medium. *Circulation*, vol. 1, pp. 84–92.
50. Yeh G.C., Martinek J. (1957). The potential of a general dipole in a homogeneous conducting prolate spheroid. *Annals of the New York Academy of Sciences*, vol. 65, pp. 1003–1006.

# Chapter 5
# Estimation of Velocity Fields and Propagation on Non-Euclidian Domains: Application to the Exploration of Cortical Spatiotemporal Dynamics

**Julien Lefèvre and Sylvain Baillet**

## 5.1 Motivation: Time-Resolved Brain Imaging

Better understanding of the interrelationship between the brain's structural architecture and functional processing is one of the leading questions in today's integrative neuroscience. Non-invasive imaging techniques have revealed as major contributing tools to this endeavor, which obviously requires the cooperation of space and time-resolved experimental evidences. Electromagnetic brain mapping using magneto- and electro-encephalography (M/EEG) source estimation is so far the imaging method with the best trade-off between spatial and temporal resolution ($\sim$1 cm and $<$1 ms respectively, [4, 5]). Combined with individual anatomical information from Magnetic Resonance Imaging (MRI) and statistical inference techniques [35], M/EEG source estimation has now reached considerable maturity and may indeed be considered as a true functional brain imaging technique.

With or without considering the estimation of M/EEG generators as a priority, the analysis of M/EEG data is classically motivated by the detection of salient features in the time course of surface measures either/both at the sensor or/and cortical levels. These features of interest may be extracted from waveform peaks and/or their related time latencies, band-specific oscillatory patterns surging from a time-frequency decomposition of the data or regional activation blobs at the cortical level.

By nature, the extraction of such features usually results from an extremely reductive – though pragmatic – point of view on the spatio-temporal dynamics of brain responses. It is pragmatic because it responds to a need for the reduction in the information mass from the original data. It is reductive

J. Lefèvre and S. Baillet (✉)
Cognitive Neuroscience and Brain Imaging Laboratory, Université Pierre and Marie Curie-Paris 6, CNRS UPR 640 – LENA, Paris, France
e-mail: julien.lefevre@chups.jussieu.fr;sylvain.baillet@chups.jussieu.fr

H. Ammari, *Mathematical Modeling in Biomedical Imaging I*,
Lecture Notes in Mathematics 1983, DOI 10.1007/978-3-642-03444-2_5,

though because most studies report on either/both the localization or/and the dynamical properties of brain events as defined according to the investigator, hence with an uncontrolled level of subjectivity.

This suggests there is a need for innovative methodological solutions that would help detect salient events in the course of brain responses in a given experimental context.

The major issue that must be faced concerns the very large amount of data produced by M/EEG acquisition. At anytime instant, brain activity unfolds at multiple cortical areas. The unique temporal resolution of M/EEG yields a large set of typically 1,000 functional brain images at every second while brain activation unfolds at multiple cortical areas on a complex shape surface object of about 2,500 $\text{cm}^2$.

In this chapter, we suggest to develop new indices related to the appraisal of brain activations in space and time, in direct connection with their anatomical substrate. These indicators are built from the empirical observation of scalp potentials and magnetic fields, but also of distributed brain currents that are literally perceived as *flowing* onto the underlying manifold. Therefore, we have adapted the computation of the optical flow – which is well-known to computer scientists when computed on 2D picture series – to arbitrary surfaces.

Though the motivations stem from neuroscience, we might think of multiple other application fields to this techniques such as, e.g. fluid streams onto surface supports but also realistic models of biological vision from the retina.

It is also interesting to note that this tool may prove as useful in the challenge of relating the experimental evidence of brain activation at the macroscopic/global scale with computational models from system dynamics (see [34] for a wide review and more specifically [27], concerning the issue of wave propagations along the cortical surface).

## 5.2 Velocity Fields and Transport on Riemannian Surfaces

In a large number of applications, spatiotemporal properties of image sequences may be represented and explored using vector fields. The information conveyed by such vectorial representation provides the support for the temporal analysis of evolutionary dynamic patterns and – possibly – their spatial segmentation. Originally, optical flow techniques were developed in computer vision to elucidate the apparent motion of rigid objects in visual scenes via the estimation of a dense vector field. In that respect, Horn and Schunk were pioneers in suggesting a regularized method to this problem [23] (see, e.g. [8] for an updated review of multiple approaches) and further quantitative improvements (see [6], [30] for other specific reviews). It further came out that optical flow usage could be extended to the characterization of complex dynamical patterns. The motion of growing plants [7] and of fluids in meteorologic images were for instance both captured using a div-curl regularized [15, 16] or a mass-conservation law [9] approach to the optical flow model. These techniques have also been investigated until recently as a way of measuring information flow across mass-neural activations as revealed by time-resolved brain imaging modalities [25, 26, 29].

In most previous applications, spatiotemporal processes were analyzed from sequences of 2D flat images, though the underlying physical phenomena may have occurred on non-flat domains – as, e.g. meteorological patterns onto the surface of the earth [15]. Hence, the primary objective of this chapter is to extend the computation of the optical flow to arbitrary surfaces (or submanifolds) of the Euclidian space. Seminal keys in that perspective were recently identified in the study of motion fields on a spherical retina in [24,43] and in [37] for a biological context.

Our secondary objective is to explicit the bounds between some properties of a motion field estimated from optical flow and some physical laws or assumptions involved in the generation of the original data sequences. In that respect, the conservation of intensity along trajectories may be the simplest assumption in the computation of optical flow [23]. Mass conservation laws were also suggested more recently to derive the so-called continuity equation in order to extract velocity fields [9] or the apparent motion [16].

In retrospect, the movement parameters extracted using the optical flow may be turned into prediction tools as in road navigation [21,42] or in weather forecasting [13].

Solving PDEs on arbitrary surfaces is of major interest in applied mathematics. Theoretical and computational developments were only recent though most theoretical aspects of PDEs are more than 50 years-old. A general framework was introduced in [32] to solve variational problems through an implicit representation of manifolds and level-set methods, with applications in medical imaging and computer vision [40]. Alternatively, some authors proposed to solve diffusion PDEs on general triangulations with FEM without embedding the surface into the Euclidian space [14, 18].

In most cited references, the related applications were often restricted to image processing (e.g. denoising in [40]), which may require solving diffusion equations and other parabolic PDEs. In our perspective, we need to mention some earlier works where PDEs are derived from mass conservation laws, for which pure advection equation is a special case [31, 38, 41].

From this brief state-of-the-art background, we know detail how to achieve modeling of optical flow onto arbitrary non-flat surfaces.

### 5.2.1 Vector Fields in Differential Geometry

Let us first recall some necessary background about differential geometry. For more details interested readers may refer to [19].

Let $\mathcal{M}$ be a 2-Riemannian manifold representing an imaging support (e.g. the surface of a planet or the highly circumvoluted brain envelope), parameterized by the local coordinate system $\phi : p \in \mathcal{M} \mapsto (x_1, x_2) \in \mathbb{R}^2$. We introduce a scalar quantity defined in time on a 2D surface (e.g. weather data or time-evolving estimates of brain activation) as a function

$$I : (p, t) \in \mathcal{M} \times \mathbb{R} \longmapsto \mathbb{R}.$$

As for Euclidian spaces, it is possible to define vectors on manifolds and we expose the most intuitive approach to this question.

Considering a curve $\gamma(t)$ defined on $\mathcal{M}$ such as $\gamma(0) = p$, we note that $\gamma'(0)$ does not depend on the local coordinate system. For all curve $\gamma(t)$, the tangent vector $\gamma'(0)$ engenders a tangent space $T_p\mathcal{M}$ at point $p$. The canonical basis of this vectorial space is:

$$\mathbf{e}_\alpha = \gamma'_\alpha(0) := \frac{\partial}{\partial x_\alpha},$$

where $x_\beta\big(\gamma_\alpha(t)\big) = t\delta_{\alpha,\beta}$.

Proceeding identically at any point of the manifold, we define $T\mathcal{M} = \bigcup_p T_p\mathcal{M}$, the tangent bundle of $\mathcal{M}$. Thus a vector field $\mathbf{V}$ is naturally defined as an application

$$\mathbf{V} : \mathcal{M} \longrightarrow T\mathcal{M}.$$

We further proceed by suggesting adaptations to the concepts of angle and distance as defined on a manifold. $\mathcal{M}$ may be equipped with a Riemannian metric. Hence at each point $p$ of $\mathcal{M}$, there exists a positive-definite form:

$$g_p : T_p\mathcal{M} \times T_p\mathcal{M} \longrightarrow \mathbb{R},$$

which is differentiable with respect to $p$. Hereafter, we note $(g_p)_{\alpha,\beta} = g_p\big(\mathbf{e}_\alpha, \mathbf{e}_\beta\big)$. A natural choice for $g_p$ is the restriction of the Euclidian metric to $T_p\mathcal{M}$, which we have adopted for subsequent computations. Next we will only refer to $g_p$ as $g$.

Integrating on a manifold now becomes possible using a *volume form*, i.e. in a few words, a differential 2-form

$$\mathrm{d}\mu_{\mathcal{M}} : T\mathcal{M} \times T\mathcal{M} \longrightarrow \mathbb{R}.$$

The most convenient volume form may be associated to the metric $g$ via: $\sqrt{\det(g_{\alpha,\beta})}\mathrm{d}x_1\mathrm{d}x_2$.

### 5.2.2 Optical Flow on a Riemannian Manifold

As in classical computational approaches to optical flow, we now assume that the activity of a point moving on a curve $p(t)$ in $\mathcal{M}$ is constant with time. The condition

$$\frac{d}{dt}\Big[I\big(p(t),t\big)\Big] = 0$$

yields

$$\partial_t I + D_{p(t)} I(\dot{\mathbf{p}}) = 0, \tag{5.1}$$

where $D_p I$ is the differential of $I$ at point $p$, that is the tangent linear application:

$$D_p I : T\mathcal{M} \longrightarrow \mathbb{R}.$$

$\dot{\mathbf{p}} = \mathbf{V} = (V^1, V^2)$ stands for the unknown motion field we wish to compute. Though being mathematically rigorous, the concept of differential is not intuitive when it comes to manipulate vector fields. In this regard, we adopt an alternative approach to [12] (in the context of Maxwell equations) where differential forms are preferred to vector fields. We will come back to this point at the quantization step (Sect. 5.3). Therefore, we express the linear application $D_p I$ as a scalar product and thus introduce $\nabla_{\mathcal{M}} I$, the gradient of $I$, defined as the vector field satisfying at each point $p$:

$$\forall \mathbf{V} \in T_p\mathcal{M}, g(\nabla_{\mathcal{M}} I, \mathbf{V}) = D_p I\big(\mathbf{V}\big).$$

Equation (5.1) may thereby be transformed into an optical-flow type of equation:

$$\partial_t I + g(\mathbf{V}, \nabla_{\mathcal{M}} I) = 0. \tag{5.2}$$

We note that (5.2) takes the same form as general conservation laws defined on manifolds in [38]. Here, only the component of the flow $\mathbf{V}$ in the direction of the gradient is accessible to estimation. This corresponds to the well-known *aperture problem* [23], which requires additional constraints on the flow to yield a unique solution.

### 5.2.3 Regularization

The previous approach classically reduces to minimizing an energy functional such as in [23]:

$$\mathcal{E}(\mathbf{V}) = \int_{\mathcal{M}} \left[\frac{\partial I}{\partial t} + g(\mathbf{V}, \nabla_{\mathcal{M}} I)\right]^2 \mathrm{d}\mu_{\mathcal{M}} + \lambda \int_{\mathcal{M}} \mathcal{C}(\mathbf{V}) \mathrm{d}\mu_{\mathcal{M}}. \tag{5.3}$$

The first term is a measure of fit of the optical flow model to the data, while the second one acts as a spatial regularizer of the flow estimate. The scalar parameter $\lambda$ tunes the respective contribution of these two terms in the net energy cost $\mathcal{E}(\mathbf{V})$. Here we rewrite the smoothness term from [23], which can be expressed as a Frobenius norm:

$$\mathcal{C}(\mathbf{V}) = \mathrm{Tr}({}^t\nabla\mathbf{V}.\nabla\mathbf{V}), \tag{5.4}$$

where

$$(\nabla\mathbf{V})_\alpha^\beta = \partial_\alpha V^\beta + \sum_\gamma \Gamma_{\alpha\gamma}^\beta V^\gamma$$

is the covariant derivative of $\mathbf{V}$, a generalization of vectorial gradient. $\partial_\alpha V^\beta$ is the classical Euclidian expression of the gradient, and $\sum_\gamma \Gamma_{\alpha\gamma}^\beta V^\gamma$ reflects local deformations of the tangent space basis since the Christoffel symbols $\Gamma_{\alpha\gamma}^\beta$ are the coordinates of $\partial_\beta \mathbf{e}_\alpha$ along $\mathbf{e}_\gamma$. This rather complex expression ensures the tensoriality property of $\mathbf{V}$, i.e. invariance with parametrization changes.

This constraint will tend to generate a regularized vector field with small spatial derivatives, that is, a field with weak local variations. Such a regularization scheme may be problematic in situations where spatial discontinuities occur in the image sequences. In the case of a moving object on a static background for example, the severe velocity discontinuities about the object contours are eventually blurred in the regularized flow field (see [45] for a taxonomy of other possible terms).

### *5.2.4 Variational Formulation*

Variational formulation of 2D-optical flow equation was first proposed by Schnörr in [39]. The advantage of such formulation is twofold: (1) theoretically, it ensures the problem is well-posed; that is there exists a unique solution in a specific and convenient function space, e.g. a Sobolev space [39], or a space of functions with bounded variations [2]; (2) numerically, it allows to solve the problem on discrete irregular surface tessellations and to yield discrete solutions belonging to the chosen function space. Obviously, a possible restriction may occur when dealing with non-quadratic regularizing terms where iterative methods come to replace matrix inversions.

We now derive a variational formulation in the case of Horn & Schunk isotropic smoothness priors, but the corresponding general framework remains the same for alternatives such as, e.g. Nagel's anisotropic image-driven regularization approach [33].

Considering $\mathcal{M}$, we need to define a working space of vector fields $\Gamma^1(\mathcal{M})$ on which functional $\mathcal{E}(\mathbf{V})$ will be minimized. Let us first denote the Sobolev space $H^1(\mathcal{M})$ defined in [20] as the completion of $C^1(\mathcal{M})$ (the space of differentiable functions on the manifold) with respect to $\| \, . \, \|_{H^1}$ derived from the following scalar product:

$$< u, v >_{H^1} = \int_{\mathcal{M}} uv \, \mathrm{d}\mu_{\mathcal{M}} + \int_{\mathcal{M}} g(\nabla u, \nabla v) \, \mathrm{d}\mu_{\mathcal{M}}.$$

We chose a space of vector fields in which the coordinates of each element are located in a classical Sobolev space:

$$\Gamma^1(\mathcal{M}) = \Big\{ \mathbf{V} : \mathcal{M} \to T\mathcal{M} \;/\; \mathbf{V} = \sum_{\alpha=1}^{2} V^\alpha \mathbf{e}_\alpha, \; V^\alpha \in H^1(\mathcal{M}) \Big\}, \tag{5.5}$$

with the following scalar product:

$$< \mathbf{U}, \mathbf{V} >_{\Gamma^1(\mathcal{M})} = \int_{\mathcal{M}} g(\mathbf{U}, \mathbf{V}) \, \mathrm{d}\mu_{\mathcal{M}} + \int_{\mathcal{M}} \mathrm{Tr}({}^t\nabla \mathbf{U} \nabla \mathbf{V}) \, \mathrm{d}\mu_{\mathcal{M}}.$$

$\mathcal{E}(\mathbf{V})$ may be simplified from (5.3) as a combination of the following constant, linear and bilinear forms, respectively:

$$\begin{aligned}
K(t) &= \int_{\mathcal{M}} (\partial_t I)^2 \mathrm{d}\mu_{\mathcal{M}}, \\
f(\mathbf{U}) &= -\int_{\mathcal{M}} g(\mathbf{U}, \nabla_{\mathcal{M}} I) \partial_t I \, \mathrm{d}\mu_{\mathcal{M}}, \\
a_1(\mathbf{U}, \mathbf{V}) &= \int_{\mathcal{M}} g(\mathbf{U}, \nabla_{\mathcal{M}} I) g(\mathbf{V}, \nabla_{\mathcal{M}} I) \mathrm{d}\mu_{\mathcal{M}} \\
a_2(\mathbf{U}, \mathbf{V}) &= \int_{\mathcal{M}} \mathrm{Tr}({}^t\nabla \mathbf{U} \nabla \mathbf{V}) \, \mathrm{d}\mu_{\mathcal{M}} \\
a(\mathbf{U}, \mathbf{V}) &= a_1(\mathbf{U}, \mathbf{V}) + \lambda a_2(\mathbf{U}, \mathbf{V})
\end{aligned}$$

Minimizing $\mathcal{E}(\mathbf{V})$ on $\Gamma^1(\mathcal{M})$ is then equivalent to the problem:

$$\min_{\mathbf{V} \in \Gamma^1(\mathcal{M})} \big( a(\mathbf{V}, \mathbf{V}) - 2f(\mathbf{V}) + K(t) \big). \tag{5.6}$$

**Theorem 5.1.** *There is an unique solution* $\mathbf{V}$ *to the previous problem. Moreover, this solution satisfies:*

$$a(\mathbf{V}, \mathbf{U}) = f(\mathbf{U}), \forall \; \mathbf{U} \in \Gamma^1(\mathcal{M}). \tag{5.7}$$

*Proof.* Lax–Milgram theorem ensures uniqueness of the solution with the following assumptions [1]:

1. $a$ and $f$ are continuous forms.

2. $\Gamma^1(\mathcal{M})$ is complete, the bilinear form $a(.,.)$ is symmetric and coercive (elliptic), that is, there exists a constant $C$ such that:

$$\forall\, \mathbf{V} \in \Gamma^1(\mathcal{M}), a(\mathbf{V},\mathbf{V}) \geq C \parallel \mathbf{V} \parallel^2_{\Gamma^1(\mathcal{M})} .$$

Continuity of $f$ and $a$ are straightforward. Completeness of $\Gamma^1(\mathcal{M})$ is ensured because any Cauchy sequence has components in $H^1(\mathcal{M})$ which are also Cauchy sequences since $\parallel . \parallel_{H^1}$ is bounded by $\parallel . \parallel_{\Gamma^1(\mathcal{M})}$. Coercivity of $a$ is demonstrated thanks to the following proposition.

**Proposition 5.1.** *$a$ is coercive.*

*Proof.* As in [39], we suppose by absurd that $a$ is not coercive. So one can find a sequence $\mathbf{U}_n$ in $\Gamma^1(\mathcal{M})$ with $||\mathbf{U}_n|| = 1$ and $a(\mathbf{U}_n,\mathbf{U}_n) \to 0$.

Since $\mathbf{U}_n$ is bounded, $U^1_n$ and $U^2_n$ are also bounded therefore by Rellich–Kondrakov theorem [19], we extract two subsequences which converge in $L^2(\mathcal{M})$, note the limits $U^1_\infty$ and $U^2_\infty$ and keep the notation $U^1_n$, $U^2_n$ for the two subsequences.

Next since $a(\mathbf{U}_n,\mathbf{U}_n) \to 0$ we have:

$$\int_{\mathcal{M}} \mathrm{Tr}({}^t\nabla \mathbf{U}_n \nabla \mathbf{U}_n)\, \mathrm{d}\mu_{\mathcal{M}} \longrightarrow 0;$$

which can be expressed as:

$$\sum_{\alpha\beta} ||\partial_\alpha U^\beta_n + \Gamma^\beta_{\alpha,1} U^1_n + \Gamma^\beta_{\alpha,2} U^2_n||^2_{L^2} \longrightarrow 0.$$

By a classical theorem in measure theory, we can say that $\partial_\alpha U^\beta_n + \Gamma^\beta_{\alpha,1} U^1_n + \Gamma^\beta_{\alpha,2} U^2_n$ converges almost everywhere to zero. Thus since $U^1_n$ and $U^2_n$ converge, $\partial_\alpha U^\beta_n$ converges almost everywhere. Finally $U^1_n$ and $U^2_n$ converge in $H^1(\mathcal{M})$. From $\mathbf{U}_n$ we have therefore extracted a subsequence which converges to $\mathbf{U}_\infty$.

So with the previous results $\nabla \mathbf{U}_\infty = 0$, but this implies that $\mathbf{U}_\infty = 0$ since the domain is non Euclidian. This clearly contradicts the assumption that $||\mathbf{U}_n|| = 1$ and $||\mathbf{U}_\infty|| = 1$ taking the limit. □

A significant difference with [39] is that coercivity and therefore well-posedness do not require any assumption about linear independency of the two components of the gradient $\nabla_{\mathcal{M}} I$.

### *5.2.5 Vectorial Heat Equation*

In this short section, we study the heat equation for vector fields on a manifold, which will be necessary for defining regular vector fields in Sect. 5.4.

Considering the 2D Euclidian case and the parabolic PDE on a domain $\Omega$ with an initial condition $\mathbf{V}_0$ at time $t = 0$:

$$\frac{\partial \mathbf{V}}{\partial t} = \mathbf{\Delta V} \ , \ \mathbf{V} = 0 \text{ on } \partial\Omega, \tag{5.8}$$

where

$$\mathbf{\Delta V} = \nabla\big(\text{div } \mathbf{V}\big) - \mathbf{rot}\big(\mathbf{rot}\mathbf{V}\big) = \big(\Delta V_1, \Delta V2\big).$$

Equation (5.8) can be transformed in the weak problem, introducing test vectors $\mathbf{W}$ with $\mathbf{W} = 0$ on $\partial\Omega$:

$$\forall \mathbf{W}, \ \int_\Omega \frac{\partial \mathbf{V}}{\partial t} \cdot \mathbf{W} = \int_\Omega \mathbf{\Delta V} \cdot \mathbf{W} = -\int_\Omega \text{Tr}\big({}^t\nabla\mathbf{V}\nabla\mathbf{W}\big).$$

By analogy, the previous variational formulation may be transposed to manifolds and using notations from Sect. 5.2.4 we obtain:

$$\forall \mathbf{W}, \ \frac{\partial}{\partial t} \int_\mathcal{M} g(\mathbf{V}, \mathbf{W}) = -a_2(\mathbf{V}, \mathbf{W}). \tag{5.9}$$

We note that this vectorial diffusion is similar to regularization in the previous variational formulation, as previously underlined in the Euclidian case [23].

### 5.2.6 *Advection on Surfaces*

The optical flow equation (5.2) may be viewed as the estimation of a motion field thereby yielding some evolution. Conversely given any vector field, (5.2) with initial condition $I(p, 0) = I_0(p)$ is a first-order PDE belonging to the family of hyperbolic problems and refers to *pure advection.* It derives from more general conservation laws which can be expressed through a PDE defined on the conserved quantity $I$:

$$\frac{\partial I}{\partial t} + \text{div}\big(\mathbf{f}(I)\big) = 0, \tag{5.10}$$

where $f : \mathbb{R} \to T\mathcal{M}$ is the flux function [38]. For example, the pure advection equation:

$$\frac{\partial I}{\partial t} + g(\nabla_\mathcal{M} I, \mathbf{V}) = 0 \tag{5.11}$$

corresponds to $\mathbf{f}(I) = I\mathbf{V}$ with non-divergential conditions $\text{div}(\mathbf{V}) = 0$.

However it is important to distinguish the optical flow and advection equations in mechanic terms of respectively Lagrangian and Eulerian descriptions, attached to a moving point or to fixed positions in space respectively. Moving from one to the other is generally non-trivial and is the objective of this section.

From this class of hyperbolic problems, direct variational formulations may be derived but suffer from critical limitations such as instable schemes or oscillatory behaviors [28]. Least-squares techniques (see, e.g. [22]) allow to get rid of numerical instabilities. They have been applied in [36] for pure advection equation in a space-time integrated framework.

Here we apply such techniques for transport or advection equations on closed surfaces, with a velocity field $\mathbf{V}$ depending on both space and time. Finding a solution of (5.11) with respect to $I$ is equivalent to minimizing the following functional

$$\mathcal{F}(I) = \int_{[0,T]} \int_{\mathcal{M}} \left[ \frac{\partial I}{\partial t} + g(\nabla_{\mathcal{M}} I, \mathbf{V}) \right]^2 \mathrm{d}\mu_{\mathcal{M}}, \tag{5.12}$$

on the function space $H^1(\mathcal{M}, \mathbf{V})$:

$$H^1(\mathcal{M}, \mathbf{V}) = \left\{ I \in L^2(\mathcal{M}) / g(\nabla_{\mathcal{M}} I, V) \in L^2(\mathcal{M}) \right\}. \tag{5.13}$$

This problem is well-posed provided that a curved Poincaré inequality is fulfilled, which ensures the coercivity of the positive bilinear form:

$$\Phi(I, J) = \int_{[0,T]} \int_{\mathcal{M}} \left[ \frac{\partial I}{\partial t} + g(\nabla_{\mathcal{M}} I, \mathbf{V}) \right] \left[ \frac{\partial J}{\partial t} + g(\nabla_{\mathcal{M}} J, \mathbf{V}) \right] \mathrm{d}\mu_{\mathcal{M}}.$$

Finally Lax–Milgram theorem ensures that the functional (5.12) has a unique minimum $I$ which satisfies:

$$\forall\, J \in H^1(\mathcal{M}, \mathbf{V}) \quad \Phi(I, J) = 0. \tag{5.14}$$

It is also proven that the problem (5.11) with a Dirichlet condition ($I(p, 0)$ fixed) is well-posed if the following inequality holds for a certain constant $C'$:

$$\int_{\mathcal{M}} I^2 \mathrm{d}\mu_{\mathcal{M}} \le C' \int_{\mathcal{M}} \left( g(\nabla_{\mathcal{M}} I, \mathbf{V}) \right)^2 \mathrm{d}\mu_{\mathcal{M}}. \tag{5.15}$$

This inequality is fulfilled for time-dependent velocity fields whose divergence is bounded [3]. This result can be generalized when the domain is a surface in an easier way than for optical flow demonstration.

## 5.3 Discretization with the Finite Element Method

After proving the well-posedness of the regularized optical flow problem and advection equation on a manifold $\mathcal{M}$, we derive computational methods from the variational formulations, which are defined on a tessellation $\widehat{\mathcal{M}}$ approximating the manifold. The tessellation consists of $N$ nodes and $T$ triangles.

### 5.3.1 Optical Flow Equation

The FEM aims at approximating continuous (and possibly more regular) functions by n-degree polynomials on each elementary triangle of the tessellation. In the scalar case, we may define $N$ functions:

$$w_i : p \in \mathcal{M} \longmapsto \mathbb{R}$$

which are continuous piecewise affine, with the property to be equal to 1 at node $i$ and 0 at all other triangle nodes. They are the *basis functions* for the approximation on the vectorial functional space of finite dimension $N$.

In the vectorial case, we want to define a similar approximation method. As suggested in Sect. 5.2.2, vector fields may be considered through two different mathematical viewpoints. In [12], the question is raised and an analogy to finite elements is found for differential forms, that is the Whitney elements. These latter have been mainly used in electromagnetism so far but also in very recent developments in advanced visualization applications [44]. We have not chosen this representation though because it is slightly more computationally demanding than our alternative since the dimensionality of the problem would be augmented by $2T$ and the velocity fields obtained at each triangle would not exactly "live" in the tangent bundle of $\widehat{\mathcal{M}}$.

We rather consider the vector space of continuous piecewise affine vector fields on $\widehat{\mathcal{M}}$ which belong to the tangent space at each node of the tessellation. A convenient basis is:

$$\mathbf{W}_\alpha(i) = w_i \mathbf{e}_\alpha(i) \text{ for } 1 \leq i \leq N,\ \alpha \in \{1,2\},$$

where $\mathbf{e}_\alpha(i)$ is some basis of the tangent space at node $i$.

The variational formulation in (5.7) yields a classical linear system:

$$\forall j\ \forall \beta \in \{1,2\}, \sum_{i=1}^{N} \sum_{\alpha=1}^{2} a\big(\mathbf{W}_\alpha(i), \mathbf{W}_\beta(j)\big) V_\alpha(i) = f\big(\mathbf{W}_\beta(j)\big), \tag{5.16}$$

where $V_\alpha(i)$ are the components of the velocity field $\mathbf{V}$ in the basis $\mathbf{W}_\alpha(i)$. Note that $a\big(\mathbf{W}_\alpha(i), \mathbf{W}_\beta(j)\big)$ and $f\big(\mathbf{W}_\beta(j)\big)$ can be explicitly computed with first-order finite elements by estimating the integrals on each triangle $T$ of the tessellation and summing out the corresponding contributions. $\nabla_\mathcal{M} I$ is obtained in practice on each triangle $T = [i,j,k]$ from the linear interpolation:

$$\nabla_\mathcal{M} I \approx I(i)\nabla_T w_i + I(j)\nabla_T w_j + I(k)\nabla_T w_k.,$$

with

$$\nabla_T w_i = \frac{\mathbf{h_i}}{\| \mathbf{h_i} \|^2}, \tag{5.17}$$

where $\mathbf{h_i}$ is the height of triangle $T$ from vertex $i$.

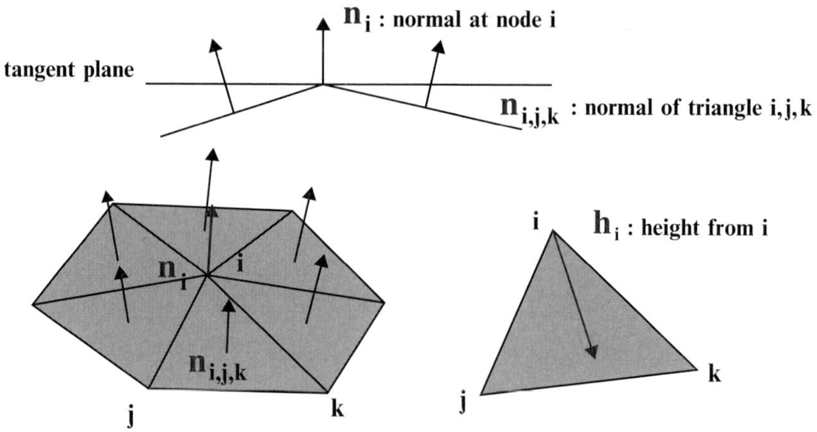


**Fig. 5.1** Illustration of local computations and associated definitions from FEM on a surface mesh (5.17)

Temporal derivatives are discretized with finite differences

$$\partial_t I(i) \approx \frac{I_{t_{k+1}}(i) - I_{t_k}(i)}{t_{k+1} - t_k}.$$

Let us define $P_{\mathbf{n}}(i)$ as the projection operator onto the local tangent plane, which is obtained at node $i$ by estimating the local normal $\mathbf{n}$ as the sum of normals of each triangle containing $i$ (see Fig. 5.1). For each $i$, $\mathbf{e}_\alpha(i)$ is chosen as a basis of the kernel of $P_{\mathbf{n}}(i)$. The Christoffel symbols $\Gamma_{ij}^k$ vanish in our discretization with first-order finite elements since the $\mathbf{e}_\alpha(i)$ are variationless on each triangle.

Finally, we need to compute integrals of the basis functions:

$$\int_T w_i w_j = \begin{cases} \frac{\mathcal{A}(T)}{6} & \text{if } i = j \\ \frac{\mathcal{A}(T)}{12} & \text{if } i \neq j \end{cases}$$

Hence we obtain the following discretized expressions:

$$a_1\big(\mathbf{W}_\alpha(i), \mathbf{W}_\beta(j)\big) = \sum_{T \ni i,j} \big(\nabla_{\mathcal{M}} I \cdot \mathbf{e}_\alpha(i)\big)\big(\nabla_{\mathcal{M}} I \cdot \mathbf{e}_\beta(j)\big) \int_T w_i w_j; \quad (5.18)$$

$$a_2\big(\mathbf{W}_\alpha(i), \mathbf{W}_\beta(j)\big) = \sum_{T \ni i,j} \big(\mathbf{e}_\alpha(i) \cdot \mathbf{e}_\beta(j)\big)\Big(\frac{\mathbf{h_i}}{\| \mathbf{h_i} \|^2} \cdot \frac{\mathbf{h_j}}{\| \mathbf{h_j} \|^2}\Big)\mathcal{A}(T); \quad (5.19)$$

$$f\big(\mathbf{W}_\alpha(i)\big) = -\frac{1}{12} \sum_{T \ni i} \big(\mathbf{e}_\alpha(i) \cdot \nabla_{\mathcal{M}} I\big)\big(2\partial_t I(i) + \partial_t I(j) + \partial_t I(k)\big). \quad (5.20)$$

### 5.3.2 Pure Advection Equation

Our approach to discretization uses a time-marching technique, which was first proposed in [10]. From the variational formulation, we know that an approximate solution $I(p,t)$ satisfies the spatiotemporal constraints on each temporal interval $[t_k, t_{k+1}] \subset [0,T]$, that is for all $J$:

$$\int_{t_k}^{t_{k+1}} \Phi(I,J)\mathrm{dt} = 0. \tag{5.21}$$

Following [10], we introduce two functions for linear temporal interpolation:

$$a(t) = \frac{t_{k+1} - t}{t_{k+1} - t_k}, \; b(t) = \frac{t - t_k}{t_{k+1} - t_k}.$$

Hence $I(p,t)$ is decomposed on the basis functions for $t \in [t_k, t_{k+1}]$:

$$I(p,t) = \sum_{i=1}^{N} w_i(p)\big[I_{i,k}a(t) + I_{i,k+1}b(t)\big]. \tag{5.22}$$

With a proper choice of test functions $\big(b(t)w_j(p)\big)$, which are null at $t = t_k$, (5.21) yields a linear system:

$$[M]_k[I]_{k+1} + [N]_k[I]_k = 0, \tag{5.23}$$

with:

$$[M]_k = \left(\int_{t_k}^{t_{k+1}} b(t)^2 \Phi(w_i, w_j)\mathrm{dt}\right)_{i,j}, \; [N]_k = \left(\int_{t_k}^{t_{k+1}} a(t)b(t)\Phi(w_i, w_j)\mathrm{dt}\right)_{i,j},$$

and $[I]_k$ is the vector of coordinates $I_{i,k}$ at time $t_k$.

From a computational point of view, $[M]_k$ and $[N]_k$ may be simply expressed from three kind of integrals, related to $\phi_1$, $\phi_2$ and $\phi_3$:

$$\int_{\mathcal{M}} w_i w_j \mathrm{d}\mu_{\mathcal{M}}, \; \int_{\mathcal{M}} w_i g(\nabla_{\mathcal{M}} w_j, \mathbf{V})\mathrm{d}\mu_{\mathcal{M}}$$

$$\int_{\mathcal{M}} g(\nabla_{\mathcal{M}} w_i, \mathbf{V}) g(\nabla_{\mathcal{M}} w_j, \mathbf{V})\mathrm{d}\mu_{\mathcal{M}}.$$

These three integrals may now be discretized on each triangle of the tesselation in the same way as for (5.18), (5.19) and (5.20).

### 5.3.3 Vectorial Diffusion

Using FEM discretization, (5.9) yields the differential equation:

$$[G]\frac{\partial[V]}{\partial t} + [A_2][V] = 0, \tag{5.24}$$

where $[G]$ and $[A_2]$ are the matrices associated to the bilinear forms $\int_{\mathcal{M}} g(\mathbf{V}, \mathbf{W})$ and $a_2$; $[V]$ are the coordinates of $\mathbf{V}$.

The temporal approximation of this equation is achieved with an implicit upwind scheme which is unconditionally stable [1]:

$$[G]\Big([V]_{k+1} - [V]_k\Big) + [A_2][V]_{k+1} = 0.$$

Thus each iteration of the diffusion process readily implies a matrix inversion:

$$[V]_{k+1} = \Big([G] + [A_2]\Big)^{-1}[G][V]_k. \tag{5.25}$$

## 5.4 Simulations

We address the quantitative and qualitative evaluation of optical flow and advection equations with simple and illustrative simulations on a selection of surfaces. More precisely, we propose to link these two equations in order to assess the quality of the new methodology introduced here.

Knowledge of the true displacement field is a major issue when evaluating the accuracy of the optical flow. On an Euclidian surface, the velocity field is the same for each point on a translating object. However it has no sense anymore for a skew surface since two different points do not share the same tangent space.

Hence we proceeded in four complementary steps: (A) a constant vector field in time was designed which enabled (B) to advect a scalar field in a second step; (C) we were therefore able to compute the optical flow of this evolving scalar field and to compare it with the original velocity fields. Step (B) allows to measure the quality and accuracy of advection processes whereas step (C) is the appraisal of optical flow estimation.

Specific quantitative indices will be defined further below as they strongly depend on the objects – scalar or vector fields – to be evaluated.

### 5.4.1 Definition of Simple Vector Fields

We first define a time-constant vector field on a surface. We use the vectorial heat equation from Sect. 5.2.5 and stop the diffusion process when the

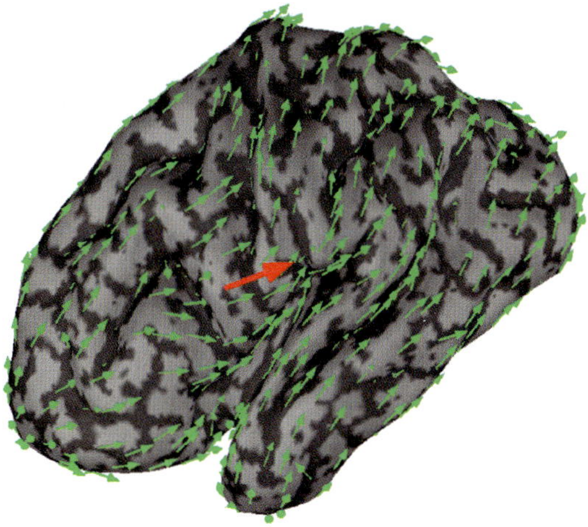

**Fig. 5.2** A smooth vector field is readily obtained by diffusing the vector shown in *red* using (5.8), yielding the vector field in *green* once magnitude normalization was applied

vector fields is considered as being sufficiently smooth. Figure 5.2 shows a regularized and normalized vector field where initial conditions are shown in red. For illustrating purposes, we used a mesh of a left cerebral hemisphere with 62,406 triangles and 31,189 vertices, which were slightly smoothed for display purposes. The black (resp. grey) stripes correspond to regions where the scalar curvature is strong (resp. weak).

### 5.4.2 Evaluation of the Advection Process

We now evaluate the accuracy of the discretization approach of the advection equation introduced in Sect. 5.3. In that purpose, we introduce several quantitative measurements:

(1) Global intensity:

$$\mathcal{I}(t) = \int_{\mathcal{M}} I(p,t)\mathrm{d}\mu_{\mathcal{M}}, \tag{5.26}$$

which is theoretically supposed to be constant with time when originating from an advection process (mass conservation law).

(2) Maximum of intensity:

$$I_M(t) = \max_{p\in\mathcal{M}} I(p,t). \tag{5.27}$$

(3) $\mathrm{L}_2$ norm:

$$||I||_{L_2} = \sqrt{\int_M I(x,t)^2\mathrm{d}\mu_{\mathcal{M}}}. \tag{5.28}$$

First a simple example is illustrated and consisted of the transport of a Gaussian activity pattern on a sphere by a velocity field defined – using cylindrical coordinates – as $\mathbf{e}_\phi = (-\sin\theta\sin\phi, \sin\theta\cos\phi, 0)$. This allows to test precisely for the accuracy of the FEM discretization since the solution of advection equation is analytical in this case: it is a rotation of the intensity at the initial time. The spherical surface envelope was composed of 2,562 vertices and 5,120 triangular faces.

As shown Fig. 5.3 (bottom row) and as predicted, the global intensity is properly conserved along the time. However, the maximum of the numerical solution decreases gradually from time step 1 to time step 100, which logically induces an increase in the spatial extension of the patch area. Further, the relative error in $L_2$ norm between the predicted and theoretical intensity distributions may be evaluated and shown also Fig. 5.3 (bottom row). This reveals a slow increase of this error from 0% to 30% at time step 100.

We also detail another simulation in a less canonical case where the theoretical advected patterns are not known a priori (Fig. 5.4). Advection of

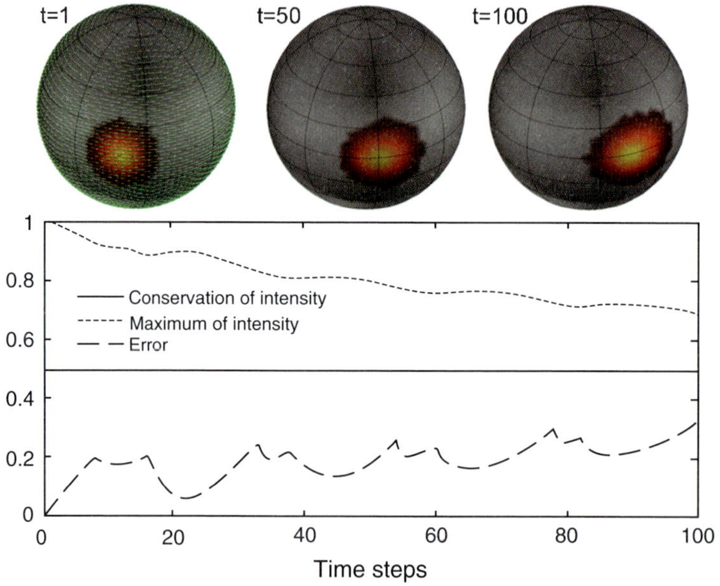


**Fig. 5.3** Advection of a Gaussian shaped activity pattern on a sphere. *Top row:* a Gaussian patch is transported onto a spherical surface under a velocity field (*shown in green, left*). *Center and right:* the patch is being transported and snapshots are shown at time steps 50 and 100. *Bottom row:* variations in time of three evaluation indices are plotted. Global intensity $\mathcal{I}$ is constant across time. The $L_2$ norm of the difference of analytical and numerical solutions (*dashed line*) increases whereas the maximum of intensity (*dotted line*) decreases

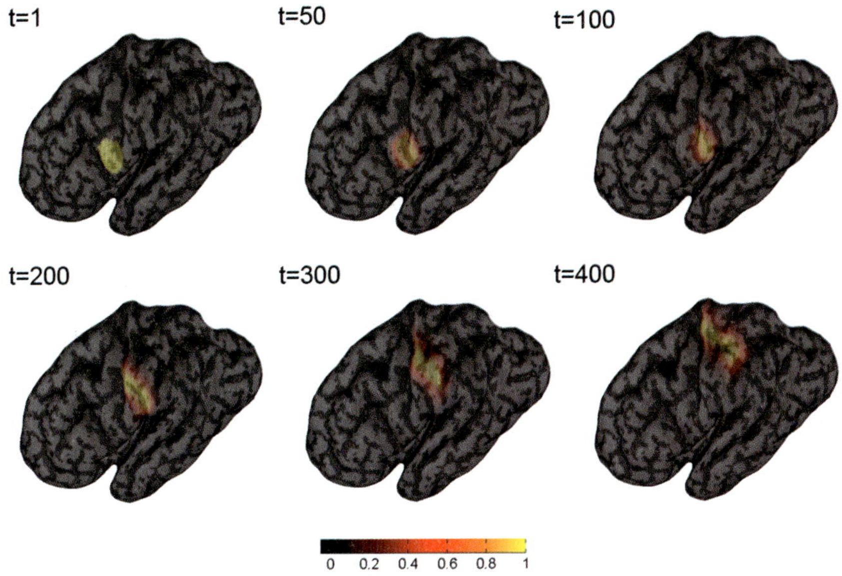


**Fig. 5.4** Advection of a constant patch on the cortical mesh: $t = 1$ shows the initial state of the intensity patch that will be propagated through the velocity field of Fig. 5.2. $t = 50$, 100, 200, 300, 400 show the resulting transported intensity distribution at corresponding times

a constant patch was performed via the velocity field from Fig. 5.2 on the smoothed cortical surface. The corresponding displacement of patches is qualitatively fair though conservation of global intensity is not observed anymore on Fig. 5.5. Besides, we observe on Fig. 5.4 and quantitatively on Fig. 5.5 that the spatial extension of the advected patch tends to be smoothed with time.

This phenomenon results from a diffusion effect which is well-known in the numerical discretization of hyperbolic equations. Several strategies could be investigated to limit this diffusivity such as: higher-order FEM, regularization schemes using $L_1$ norm or other conservative models [11], but this would reach far beyond the scope of this study.

### *5.4.3 Evaluation of the Optical Flow*

We have previously described how to advect scalar quantities in a velocity field. Likewise, given the evolution of some scalar intensity, we aimed at retrieving the velocity field which had engendered this evolution. In that

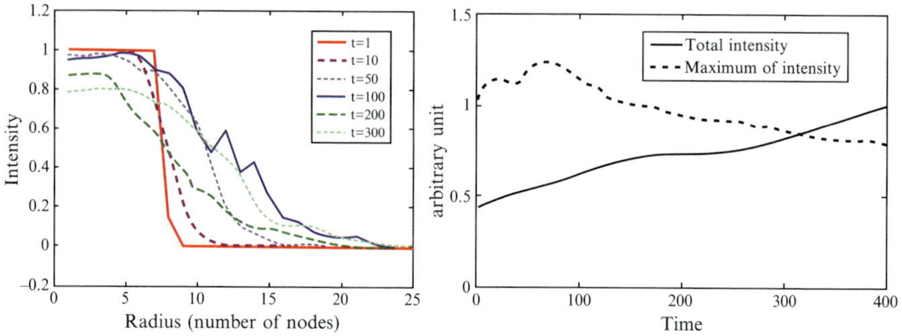


**Fig. 5.5** *Left:* different sections of intensity patches at time $t = 1$, 50, 100, 200, 300. The corresponding surface areas are evaluated from the number of nodes from the center to the outside. *Right:* variations of the global intensity and the maximum of intensity as defined in (5.26) and (5.27)

purpose, we estimated the optical flow of the sequence of intensities obtained from the advection in Sect. 5.4.2, and compared it to the original velocity field from Sect. 5.4.1.

Most optical flow techniques refer the angular error (AE) between the true velocity field $\mathbf{V}$ and the estimated optical flow $\widehat{\mathbf{V}}$ rather than a RMS or $L_2$ norm error in order to prevent from errors at high speed to be amplified [6]. We have adapted the angular error index originally introduced in [6] following:

$$AE(\mathbf{V}, \widehat{\mathbf{V}}) = \arccos\left(\frac{g(\mathbf{V}, \widehat{\mathbf{V}})}{||\mathbf{V}||.||\widehat{\mathbf{V}}||}\right). \tag{5.29}$$

AE provides a simple evaluation index as well as a quantitative criterion to adjust the regularizing parameter $\lambda$ in (5.3). In some preliminary work, we have set $\lambda = 0.1$ [29].

From a qualitative point a view, Fig. 5.6 shows the estimated optical flow superimposed to the moving intensity patterns at four time instants. Vectors whose norm was larger than $0.1\times$ the maximum of the norm in the vector field are displayed. We observe effects from the diffusion caused by the regularization term in (5.3), since the displacement field does not strictly vanish outside the three Gaussian patches. However the global optical flow is qualitatively directed along the proper orientations, following the displacement of the patch along the surface. We note that the average angular error on Fig. 5.7 remains constant across time and seems to be independent from the local surface curvature.

We therefore conclude that the discretization approach we have taken is satisfactory for a wide range of applications and could be readily enhanced using the technical improvements we have mentioned above.

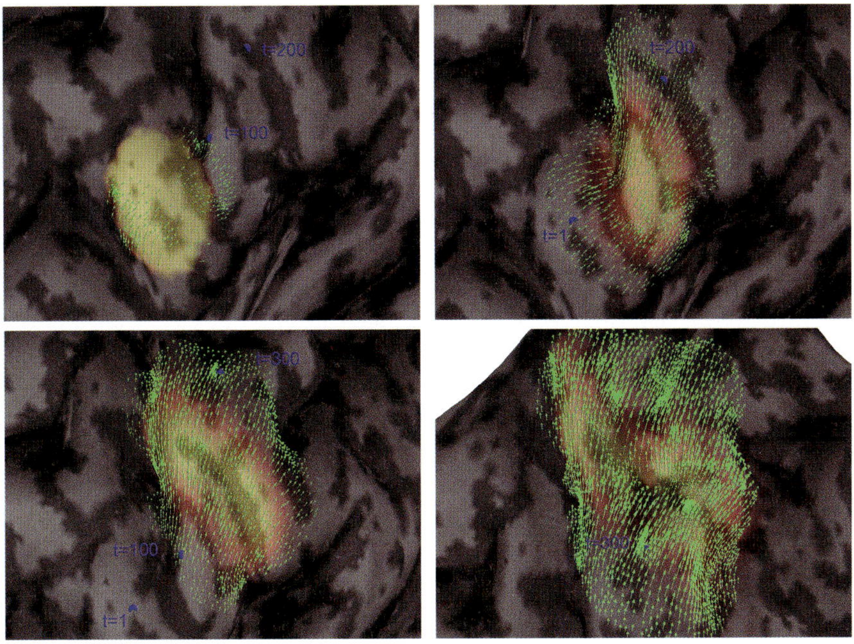


**Fig. 5.6** Optical flow (in *green*) computed at steps 1, 100, 200 and 400 from the displacement of intensity patches on the rabbit mesh. Centroid of patches at several time instants are plotted in *blue*

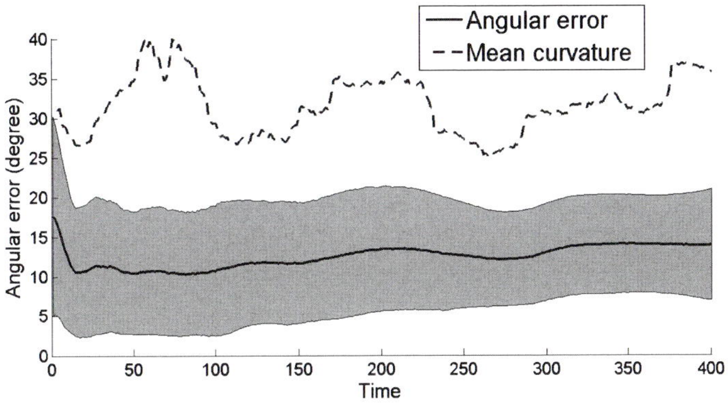


**Fig. 5.7** Mean angular error ($AE$) between the original velocity field and its optical flow estimate (in *black*, with the sample standard deviation error in shade of *gray*). The local surface curvature by the centroid of the moving patch is superimposed in arbitrary unit using a gray *dashed line*

## 5.5 Application to Time-Resolved Brain Imaging

Here we used our framework for optical flow estimation on data from a simple experimental paradigm, where elementary propagation patterns are likely to occur onto the brain surface. The experimental paradigm dealt with visual stimulation consisting of expanding checkerboard rings at 5 Hz presented to the subjects for 2.5 s (see Fig. 5.8). The cortical activities are obtained after brain signals were band-passed around the stimulation frequency (3–7 Hz) and processed using a regularized minimum-norm estimator (BrainStorm software: http://neuroimage.usc.edu/brainstorm), to obtain an estimate of the neural currents distributed on the individual cortical surface of the subject [4].

The corresponding displacement fields were computed from (5.16) on a triangulated mesh of the cortical envelope consisting of 10,000 nodes. For illustration purposes, we have restricted our investigations to a large part of the left visual cortex as shown Fig. 5.8. We then performed the advection

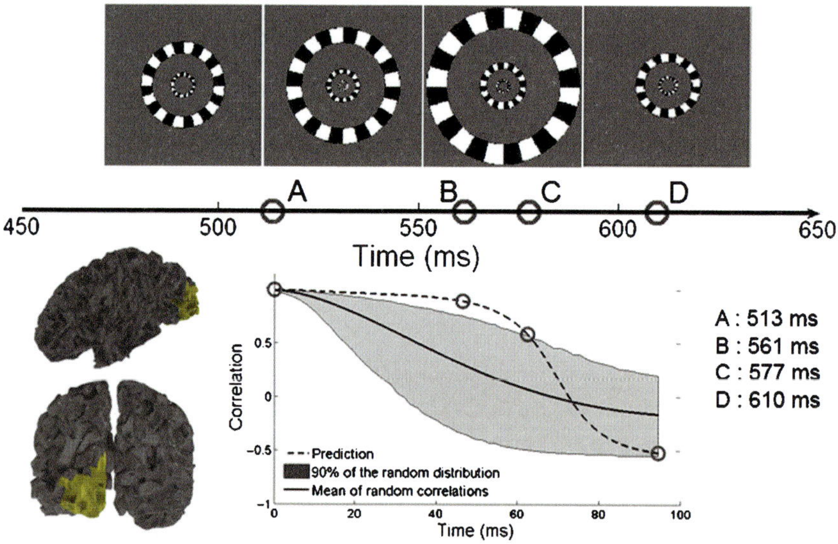


**Fig. 5.8** *First row:* visual stimulus used during the MEG experiment. This latter consisted of expanding checkerboard rings at the frequency of 5 Hz. Each cycle was composed by the four successive images shown, each being displayed for 50 ms [17]. The time line below indicates the succession of images after stimulation started at 0 ms. Two views of the subject's cortical surface are shown. The region of interest (ROI) consisted of a large part of the left visual cortex (colored in *yellow*). The correlation coefficients (*dashed line*) between brain activations in the ROI and the predicted ones through advection with the displacement field obtained from A (513 ms) to C (610 ms) are also shown. Random fluctuations for correlations were obtained from 314 displacement fields computed during the prestimulus period and applied to the same initial condition. The sample distribution of such random correlations is shown by the *gray area* representing 90% of the distribution

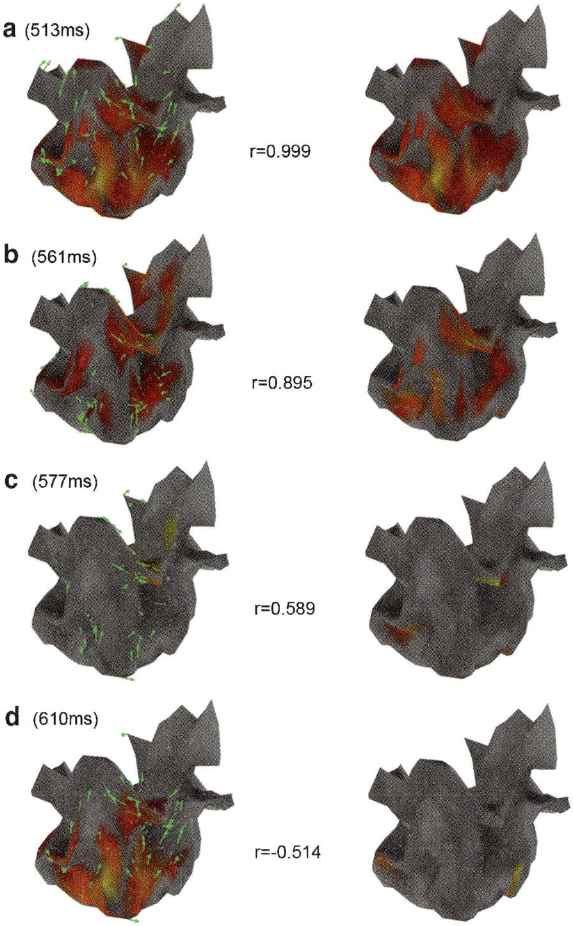


**Fig. 5.9** *Left:* real cortical activities on the left visual areas and corresponding displacement fields in *green*. *Right:* activities modeled through unsteady advection with the displacement field of estimated activities of the left

of neural activity from 513 to 610 ms using the previously computed vector field. Figure 5.9 compares the real neural activities vs. the ones "predicted" from advection. For visualization purposes, activities have been thresholded at 30% of the maximum activity (likewise for the optical flow: at 10% of the maximal vector magnitude).

Finally, the graph on Fig. 5.8 evaluates the relevance of our methodology to identify propagation patterns from time-resolved imaging of brain activations. The dashed line indicates the Pearson correlation coefficient $C(t)$ between the real data $I(p,t)$ and the predictions $\widehat{I}(p,t)$ during 97 ms, whereas the black line represents an index of random fluctuations. In order to obtain this latter index, displacement fields $\widetilde{\mathbf{V}}(p, t_k)$ with $t_k \in [-0.5, 0]s$ were computed during the prestimulus baseline period (314 samples) during which the activity of the brain would be considered as irrelevant to the subsequent brain response during visual stimulation. Next, the neural activity at 513 ms was used as the initial condition of 314 successive advection problems whose vector fields were $\widetilde{\mathbf{V}}(p, t-t_k)$ with $k \in [|1, 314|]$. This yielded $k$ advected activities $\widetilde{I}_k(p,t)$ with $\widetilde{I}_k(p,0) = \widehat{I}(p, 0.513) = I(p, 0.513)$, which could be compared to real data in terms of the same correlation index. Hence we obtained 314 coefficients $\widetilde{C}_k(t)$ for each $t \in 510 + [0, 97\,\text{ms}]$, from which the 90% confidence interval of the associated distribution could be deduced (gray area on Fig. 5.8).

Results show that correlation between the advection model and real data reached above the 90% level during more than 40 ms before returning to the level of random fluctuations 64 ms after initialization. This suggests that the evolution of MEG activations may be described in terms of propagation during a relatively long time period regarding the time scale of the brain in action. These results are in line with the spatiotemporal modeling of MEG proposed in [27].

## 5.6 Conclusion

This chapter was motivated by two point of views on time-resolved imaging of brain activity. The first viewpoint is essentially pragmatic: there is a need to extract some relevant information in the course of brain responses with minimum subjectivity and arbitrariness. The second motivation is essentially driven by the necessity to bridge predictions from computational models of the working brain with experimental evidence.

Though this contribution has essentially focussed on the technical aspects of a new set of tools derived from the optical flow, we think these later open exciting perspectives in response to the early motivations of this chapter. The robust definition of a vector field on the intricate cortical surface brings a dynamical index of supplementary interest to the classical measure of instantaneous activation levels in terms of amplitudes. We have shown how the surface vector field may be utilized in the technical framework of advection to infer on the stability of brain activity.

## References

1. G. Allaire, *Analyse numérique et optimisation*, Editions de l'Ecole Polytechnique, 2005.
2. G. Aubert, R. Deriche, and P. Kornprobst, Computing optical flow via variational techniques, SIAM J. Appl. Math., 1999, 60(1), 156–182.
3. P. Azerad and G. Pousin, Inégalite de Poincaré courbe pour le traitement variationnel de l'équation de transport, CR Acad. Sci. Paris, 1996, 721–727.

4. S. Baillet, J.C. Mosher, and R.M. Leahy, Electromagnetic brain mapping, IEEE Signal Processing Magazine, November, 2001.
5. J.M. Barrie, W.J. Freeman, and M.D. Lenhart, Spatiotemporal Analysis of Prepyriform, Visual, Auditory, and Somesthetic Surface EEGs in Trained Rabbits, Journal of Neurophysiology, 1996, 76, 520–539.
6. J.L. Barron, D.J. Fleet, and S.S. Beauchemin, Performance of optical flow techniques, International Journal of Computer Vision, 1994, 12, 43–77.
7. J.L. Barron and A. Liptay, Measuring 3-D plant growth using optical flow, Bioimaging, 1997, 5, 82–86.
8. S.S. Beauchemin and J.L. Barron, The computation of optical flow, ACM Computing Surveys, 1995, 27(3), 433–467.
9. D. Bereziat, I. Herlin, and L. Younes, A Generalized Optical Flow Constraint and its Physical Interpretation, Proc. Conf. Computer Vision and Pattern Recognition, 2000, 02, 2487.
10. O. Besson and G. De Montmollin, Space-time integrated least squares: a time-marching approach, International Journal for Numerical Methods in Fluids, 2004, 44(5), 525–543.
11. D.L. Book, J.P. Boris, and K. Hain, Flux-Corrected Transport. II – Generalizations of the method, Journal of Computational Physics, 1975, 18, 248–283.
12. A. Bossavit, Whitney forms: a class of finite elements for three-dimensional computations in electromagnetism, IEE Proceedings, 1988, 135, Part A, 493–500, 8, November.
13. N.E.H. Bowler, C.E. Pierce, and A. Seed, Development of a precipitation nowcasting algorithm based upon optical flow techniques, Journal of Hydrology, 2004, 288, 74–91.
14. U. Clarenz, M. Rumpf, and A. Telea, Finite elements on point based surfaces, Proc. of Symp. on Point-Based Graphics, 2004, 201–211.
15. T. Corpetti, D. Heitz, G. Arroyo, E. Mémin, and A. Santa-Cruz, Fluid experimental flow estimation based on an optical-flow scheme, Experiments in fluids, 2006, 40, 80–97.
16. T. Corpetti, E. Mémin, and P. Pérez, Dense Estimation of Fluid Flows, IEEE Transactions on Pattern Analysis and Machine Intelligence, 2002, 24 (3), 365–380.
17. D. Cosmelli, O. David, J.P. Lachaux, J. Martinerie, L. Garnero, B. Renault, and F. Varela, Waves of consciousness: ongoing cortical patterns during binocular rivalry, Neuroimage, 2004, 23, 128–140.
18. U. Diewald, T. Preusser, and M. Rumpf, Anisotropic Diffusion in Vector Field Visualization on Euclidean Domains and Surfaces, IEEE Transactions on Visualization and Computer Graphics, 2000, 6, 139–149.
19. M.P. Do Carmo, *Riemannian Geometry*, Birkhuser, 1993.
20. O. Druet, E. Hebey, and F. Robert, *Blow-up Theory for Elliptic PDEs in Riemannian Geometry*, Princeton University Press, Princeton, N.J., 2004.
21. A. Giachetti, M. Campani, and V. Torre, The use of optical flow for road navigation, IEEE Trans. Robotics and Automation, 1998, 14, 34–48.
22. J.L. Guermond, A finite element technique for solving first order PDE's in L1, SIAM J. Numer. Anal., 2004, 42(2), 714–737.
23. B.K.P. Horn and B.G. Schunck, Determining optical flow, Artificial Intelligence, 1981, 17, 185–204.
24. A. Imiya, H. Sugaya, A. Torii, and Y. Mochizuki, Variational Analysis of Spherical Images, Proc. Computer Analysis of Images and Patterns, 2005.
25. T. Inouye, K. Shinosaki, A. Iyama, Y. Matsumoto, S. Toi, and T. Ishthara, Potential flow of frontal midline theta activity during a mental task in the human electroencephalogram, Neurosci. Lett., 1994, 169, 145–148.
26. T. Inouye, K. Shinosaki, S. Toi, Y. Matsumoto, and N. Hosaka, Potential flow of alpha-activity in the human electroencephalogram, Neurosci. Lett., 1995, 187, 29–32.
27. V.K. Jirsa, K.J. Jantzen, A. Fuchs, and J.A.S. Kelso, Spatiotemporal forward solution of the EEG and MEG using networkmodeling, IEEE Trans. Medical Imaging, 2002, 21, 493–504.
28. C. Johnson, *Numerical Solution of Partial Differential Equations by the Finite Element Method*, Cambridge University Press, New York, 1987.

29. J. Lefèvre, G. Obozinski, and S. Baillet, Imaging Brain Activation Streams from Optical Flow Computation on 2-Riemannian Manifold, Proc. Information Processing in Medical Imaging, 2007, 470–481.
30. H. Liu, T. Hong, M. Herman, T. Camus, and R. Chellapa, Accuracy vs. efficiency tradeoff in optical flow algorithms, Comput. Vision Image Understand., 1998, 72, 271–286.
31. L.M. Lui, Y. Wang, and T.F. Chan, Solving PDEs on Manifold using Global Conformal Parameterization, Proc. Variational, Geometric, and Level Set Methods in Computer Vision, 2005, 307–319.
32. F. Mémoli, G. Sapiro, and S. Osher, Solving variational problems and partial differential equations mapping into general target manifolds, J. Comput. Phys., 2004, 195, 263–292.
33. H.H. Nagel, On the estimation of optical flow: relations between different approaches and some new results., Artificial Intelligence, 1987, 33, 299–324.
34. P.L. Nunez, Toward a quantitative description of large-scale neocortical dynamic function and EEG, Behavioral and Brain Sciences, 2000, 23, 371–398.
35. D. Pantazis, T.E. Nichols, S. Baillet, and R.M. Leahy, A comparison of random field theory and permutation methods for the statistical analysis of MEG data., Neuroimage, 2005, 25, 383–394.
36. P. Perrochet and P. Azérad, Space-time integrated least-squares: Solving a pure advection equation with a pure diffusion operator., J.Computer.Phys., 1995, 117, 183–193.
37. W. Rekik, D. Bereziat, and S. Dubuisson, Optical flow computation and visualization in spherical context. Application on 3D+ t bio-cellular sequences., Engineering in Medicine and Biology Society, 2006. EMBS'06. 28th Annual International Conference of the IEEE, 2006, 1645–1648.
38. J.A. Rossmanith, D.S. Bale, and R.J. LeVeque, A wave propagation algorithm for hyperbolic systems on curved manifolds, J. Comput. Phys., 2004, 199, 631–662.
39. C. Schnörr, Determining optical flow for irregular domains by minimizing quadratic functionals of a certain class., Int. J. Computer Vision, 1991, 6(1), 25–38.
40. N. Sochen, R. Deriche, and L. Lopez Perez, The Beltrami Flow over Implicit Manifolds, Proc. of the Ninth IEEE International Conference on Computer Vision, 2003, 832.
41. A. Spira and R. Kimmel, Geometric curve flows on parametric manifolds, J. Comput. Phys., 2007, 223, 235–249.
42. Z. Sun, G. Bebis, and R. Miller, On-Road Vehicle Detection: A Review, IEEE Trans. Pattern Anal. Mach. Intell., 2006, 28, 694–711.
43. A. Torii, A. Imiya, H. Sugaya, and Y. Mochizuki, Optical Flow Computation for Compound Eyes: Variational Analysis of Omni-Directional Views, Proc. Brain, Vision and Artificial Intelligence, 2005.
44. K. Wang, Weiwei, Y. Tong, M. Desbrun, and P. Schroder, Edge subdivision schemes and the construction of smooth vector fields, ACM Trans. Graph., 2006, 25, 1041–1048.
45. J. Weickert and C. Schnörr, A theoretical framework for convex regularizers in PDE-based computation of image motion, International Journal of Computer Vision, 2001, 45(3), 245–264.

# Index

# Lecture Notes in Mathematics

For information about earlier volumes
please contact your bookseller or Springer
LNM Online archive: springerlink.com

Vol. 1801: J. Azéma, M. Émery, M. Ledoux, M. Yor (Eds.), Séminaire de Probabilités XXXVI (2003)

Vol. 1802: V. Capasso, E. Merzbach, B. G. Ivanoff, M. Dozzi, R. Dalang, T. Mountford, Topics in Spatial Stochastic Processes. Martina Franca, Italy 2001. Editor: E. Merzbach (2003)

Vol. 1803: G. Dolzmann, Variational Methods for Crystalline Microstructure – Analysis and Computation (2003)

Vol. 1804: I. Cherednik, Ya. Markov, R. Howe, G. Lusztig, Iwahori-Hecke Algebras and their Representation Theory. Martina Franca, Italy 1999. Editors: V. Baldoni, D. Barbasch (2003)

Vol. 1805: F. Cao, Geometric Curve Evolution and Image Processing (2003)

Vol. 1806: H. Broer, I. Hoveijn. G. Lunther, G. Vegter, Bifurcations in Hamiltonian Systems. Computing Singularities by Gröbner Bases (2003)

Vol. 1807: V. D. Milman, G. Schechtman (Eds.), Geometric Aspects of Functional Analysis. Israel Seminar 2000-2002 (2003)

Vol. 1808: W. Schindler, Measures with Symmetry Properties (2003)

Vol. 1809: O. Steinbach, Stability Estimates for Hybrid Coupled Domain Decomposition Methods (2003)

Vol. 1810: J. Wengenroth, Derived Functors in Functional Analysis (2003)

Vol. 1811: J. Stevens, Deformations of Singularities (2003)

Vol. 1812: L. Ambrosio, K. Deckelnick, G. Dziuk, M. Mimura, V. A. Solonnikov, H. M. Soner, Mathematical Aspects of Evolving Interfaces. Madeira, Funchal, Portugal 2000. Editors: P. Colli, J. F. Rodrigues (2003)

Vol. 1813: L. Ambrosio, L. A. Caffarelli, Y. Brenier, G. Buttazzo, C. Villani, Optimal Transportation and its Applications. Martina Franca, Italy 2001. Editors: L. A. Caffarelli, S. Salsa (2003)

Vol. 1814: P. Bank, F. Baudoin, H. Föllmer, L.C.G. Rogers, M. Soner, N. Touzi, Paris-Princeton Lectures on Mathematical Finance 2002 (2003)

Vol. 1815: A. M. Vershik (Ed.), Asymptotic Combinatorics with Applications to Mathematical Physics. St. Petersburg, Russia 2001 (2003)

Vol. 1816: S. Albeverio, W. Schachermayer, M. Talagrand, Lectures on Probability Theory and Statistics. Ecole d'Eté de Probabilités de Saint-Flour XXX-2000. Editor: P. Bernard (2003)

Vol. 1817: E. Koelink, W. Van Assche (Eds.), Orthogonal Polynomials and Special Functions. Leuven 2002 (2003)

Vol. 1818: M. Bildhauer, Convex Variational Problems with Linear, nearly Linear and/or Anisotropic Growth Conditions (2003)

Vol. 1819: D. Masser, Yu. V. Nesterenko, H. P. Schlickewei, W. M. Schmidt, M. Waldschmidt, Diophantine Approximation. Cetraro, Italy 2000. Editors: F. Amoroso, U. Zannier (2003)

Vol. 1820: F. Hiai, H. Kosaki, Means of Hilbert Space Operators (2003)

Vol. 1821: S. Teufel, Adiabatic Perturbation Theory in Quantum Dynamics (2003)

Vol. 1822: S.-N. Chow, R. Conti, R. Johnson, J. Mallet-Paret, R. Nussbaum, Dynamical Systems. Cetraro, Italy 2000. Editors: J. W. Macki, P. Zecca (2003)

Vol. 1823: A. M. Anile, W. Allegretto, C. Ringhofer, Mathematical Problems in Semiconductor Physics. Cetraro, Italy 1998. Editor: A. M. Anile (2003)

Vol. 1824: J. A. Navarro González, J. B. Sancho de Salas, $\mathscr{C}^\infty$ – Differentiable Spaces (2003)

Vol. 1825: J. H. Bramble, A. Cohen, W. Dahmen, Multiscale Problems and Methods in Numerical Simulations, Martina Franca, Italy 2001. Editor: C. Canuto (2003)

Vol. 1826: K. Dohmen, Improved Bonferroni Inequalities via Abstract Tubes. Inequalities and Identities of Inclusion-Exclusion Type. VIII, 113 p, 2003.

Vol. 1827: K. M. Pilgrim, Combinations of Complex Dynamical Systems. IX, 118 p, 2003.

Vol. 1828: D. J. Green, Gröbner Bases and the Computation of Group Cohomology. XII, 138 p, 2003.

Vol. 1829: E. Altman, B. Gaujal, A. Hordijk, Discrete-Event Control of Stochastic Networks: Multimodularity and Regularity. XIV, 313 p, 2003.

Vol. 1830: M. I. Gil', Operator Functions and Localization of Spectra. XIV, 256 p, 2003.

Vol. 1831: A. Connes, J. Cuntz, E. Guentner, N. Higson, J. E. Kaminker, Noncommutative Geometry, Martina Franca, Italy 2002. Editors: S. Doplicher, L. Longo (2004)

Vol. 1832: J. Azéma, M. Émery, M. Ledoux, M. Yor (Eds.), Séminaire de Probabilités XXXVII (2003)

Vol. 1833: D.-Q. Jiang, M. Qian, M.-P. Qian, Mathematical Theory of Nonequilibrium Steady States. On the Frontier of Probability and Dynamical Systems. IX, 280 p, 2004.

Vol. 1834: Yo. Yomdin, G. Comte, Tame Geometry with Application in Smooth Analysis. VIII, 186 p, 2004.

Vol. 1835: O.T. Izhboldin, B. Kahn, N.A. Karpenko, A. Vishik, Geometric Methods in the Algebraic Theory of Quadratic Forms. Summer School, Lens, 2000. Editor: J.-P. Tignol (2004)

Vol. 1836: C. Năstăsescu, F. Van Oystaeyen, Methods of Graded Rings. XIII, 304 p, 2004.

Vol. 1837: S. Tavaré, O. Zeitouni, Lectures on Probability Theory and Statistics. Ecole d'Eté de Probabilités de Saint-Flour XXXI-2001. Editor: J. Picard (2004)

Vol. 1838: A.J. Ganesh, N.W. O'Connell, D.J. Wischik, Big Queues. XII, 254 p, 2004.

Vol. 1839: R. Gohm, Noncommutative Stationary Processes. VIII, 170 p, 2004.

Vol. 1840: B. Tsirelson, W. Werner, Lectures on Probability Theory and Statistics. Ecole d'Eté de Probabilités de Saint-Flour XXXII-2002. Editor: J. Picard (2004)
Vol. 1841: W. Reichel, Uniqueness Theorems for Variational Problems by the Method of Transformation Groups (2004)
Vol. 1842: T. Johnsen, A. L. Knutsen, $K_3$ Projective Models in Scrolls (2004)
Vol. 1843: B. Jefferies, Spectral Properties of Noncommuting Operators (2004)
Vol. 1844: K.F. Siburg, The Principle of Least Action in Geometry and Dynamics (2004)
Vol. 1845: Min Ho Lee, Mixed Automorphic Forms, Torus Bundles, and Jacobi Forms (2004)
Vol. 1846: H. Ammari, H. Kang, Reconstruction of Small Inhomogeneities from Boundary Measurements (2004)
Vol. 1847: T.R. Bielecki, T. Björk, M. Jeanblanc, M. Rutkowski, J.A. Scheinkman, W. Xiong, Paris-Princeton Lectures on Mathematical Finance 2003 (2004)
Vol. 1848: M. Abate, J. E. Fornaess, X. Huang, J. P. Rosay, A. Tumanov, Real Methods in Complex and CR Geometry, Martina Franca, Italy 2002. Editors: D. Zaitsev, G. Zampieri (2004)
Vol. 1849: Martin L. Brown, Heegner Modules and Elliptic Curves (2004)
Vol. 1850: V. D. Milman, G. Schechtman (Eds.), Geometric Aspects of Functional Analysis. Israel Seminar 2002-2003 (2004)
Vol. 1851: O. Catoni, Statistical Learning Theory and Stochastic Optimization (2004)
Vol. 1852: A.S. Kechris, B.D. Miller, Topics in Orbit Equivalence (2004)
Vol. 1853: Ch. Favre, M. Jonsson, The Valuative Tree (2004)
Vol. 1854: O. Saeki, Topology of Singular Fibers of Differential Maps (2004)
Vol. 1855: G. Da Prato, P.C. Kunstmann, I. Lasiecka, A. Lunardi, R. Schnaubelt, L. Weis, Functional Analytic Methods for Evolution Equations. Editors: M. Iannelli, R. Nagel, S. Piazzera (2004)
Vol. 1856: K. Back, T.R. Bielecki, C. Hipp, S. Peng, W. Schachermayer, Stochastic Methods in Finance, Bressanone/Brixen, Italy, 2003. Editors: M. Fritelli, W. Runggaldier (2004)
Vol. 1857: M. Émery, M. Ledoux, M. Yor (Eds.), Sémi naire de Probabilités XXXVIII (2005)
Vol. 1858: A.S. Cherny, H.-J. Engelbert, Singular Stochastic Differential Equations (2005)
Vol. 1859: E. Letellier, Fourier Transforms of Invariant Functions on Finite Reductive Lie Algebras (2005)
Vol. 1860: A. Borisyuk, G.B. Ermentrout, A. Friedman, D. Terman, Tutorials in Mathematical Biosciences I. Mathematical Neurosciences (2005)
Vol. 1861: G. Benettin, J. Henrard, S. Kuksin, Hamiltonian Dynamics – Theory and Applications, Cetraro, Italy, 1999. Editor: A. Giorgilli (2005)
Vol. 1862: B. Helffer, F. Nier, Hypoelliptic Estimates and Spectral Theory for Fokker-Planck Operators and Witten Laplacians (2005)
Vol. 1863: H. Führ, Abstract Harmonic Analysis of Continuous Wavelet Transforms (2005)
Vol. 1864: K. Efstathiou, Metamorphoses of Hamiltonian Systems with Symmetries (2005)
Vol. 1865: D. Applebaum, B.V. R. Bhat, J. Kustermans, J. M. Lindsay, Quantum Independent Increment Processes I. From Classical Probability to Quantum Stochastic Calculus. Editors: M. Schürmann, U. Franz (2005)
Vol. 1866: O.E. Barndorff-Nielsen, U. Franz, R. Gohm, B. Kümmerer, S. Thorbjønsen, Quantum Independent Increment Processes II. Structure of Quantum Lévy Processes, Classical Probability, and Physics. Editors: M. Schürmann, U. Franz, (2005)
Vol. 1867: J. Sneyd (Ed.), Tutorials in Mathematical Biosciences II. Mathematical Modeling of Calcium Dynamics and Signal Transduction. (2005)
Vol. 1868: J. Jorgenson, S. Lang, $Pos_n(R)$ and Eisenstein Series. (2005)
Vol. 1869: A. Dembo, T. Funaki, Lectures on Probability Theory and Statistics. Ecole d'Eté de Probabilités de Saint-Flour XXXIII-2003. Editor: J. Picard (2005)
Vol. 1870: V.I. Gurariy, W. Lusky, Geometry of Müntz Spaces and Related Questions. (2005)
Vol. 1871: P. Constantin, G. Gallavotti, A.V. Kazhikhov, Y. Meyer, S. Ukai, Mathematical Foundation of Turbulent Viscous Flows, Martina Franca, Italy, 2003. Editors: M. Cannone, T. Miyakawa (2006)
Vol. 1872: A. Friedman (Ed.), Tutorials in Mathematical Biosciences III. Cell Cycle, Proliferation, and Cancer (2006)
Vol. 1873: R. Mansuy, M. Yor, Random Times and Enlargements of Filtrations in a Brownian Setting (2006)
Vol. 1874: M. Yor, M. Émery (Eds.), In Memoriam Paul-André Meyer - Séminaire de Probabilités XXXIX (2006)
Vol. 1875: J. Pitman, Combinatorial Stochastic Processes. Ecole d'Eté de Probabilités de Saint-Flour XXXII-2002. Editor: J. Picard (2006)
Vol. 1876: H. Herrlich, Axiom of Choice (2006)
Vol. 1877: J. Steuding, Value Distributions of $L$-Functions (2007)
Vol. 1878: R. Cerf, The Wulff Crystal in Ising and Percolation Models, Ecole d'Eté de Probabilités de Saint-Flour XXXIV-2004. Editor: Jean Picard (2006)
Vol. 1879: G. Slade, The Lace Expansion and its Applications, Ecole d'Eté de Probabilités de Saint-Flour XXXIV-2004. Editor: Jean Picard (2006)
Vol. 1880: S. Attal, A. Joye, C.-A. Pillet, Open Quantum Systems I, The Hamiltonian Approach (2006)
Vol. 1881: S. Attal, A. Joye, C.-A. Pillet, Open Quantum Systems II, The Markovian Approach (2006)
Vol. 1882: S. Attal, A. Joye, C.-A. Pillet, Open Quantum Systems III, Recent Developments (2006)
Vol. 1883: W. Van Assche, F. Marcellàn (Eds.), Orthogonal Polynomials and Special Functions, Computation and Application (2006)
Vol. 1884: N. Hayashi, E.I. Kaikina, P.I. Naumkin, I.A. Shishmarev, Asymptotics for Dissipative Nonlinear Equations (2006)
Vol. 1885: A. Telcs, The Art of Random Walks (2006)
Vol. 1886: S. Takamura, Splitting Deformations of Degenerations of Complex Curves (2006)
Vol. 1887: K. Habermann, L. Habermann, Introduction to Symplectic Dirac Operators (2006)
Vol. 1888: J. van der Hoeven, Transseries and Real Differential Algebra (2006)
Vol. 1889: G. Osipenko, Dynamical Systems, Graphs, and Algorithms (2006)
Vol. 1890: M. Bunge, J. Funk, Singular Coverings of Toposes (2006)
Vol. 1891: J.B. Friedlander, D.R. Heath-Brown, H. Iwaniec, J. Kaczorowski, Analytic Number Theory, Cetraro, Italy, 2002. Editors: A. Perelli, C. Viola (2006)
Vol. 1892: A. Baddeley, I. Bárány, R. Schneider, W. Weil, Stochastic Geometry, Martina Franca, Italy, 2004. Editor: W. Weil (2007)

Vol. 1893: H. Hanßmann, Local and Semi-Local Bifurcations in Hamiltonian Dynamical Systems, Results and Examples (2007)
Vol. 1894: C.W. Groetsch, Stable Approximate Evaluation of Unbounded Operators (2007)
Vol. 1895: L. Molnár, Selected Preserver Problems on Algebraic Structures of Linear Operators and on Function Spaces (2007)
Vol. 1896: P. Massart, Concentration Inequalities and Model Selection, Ecole d'Été de Probabilités de Saint-Flour XXXIII-2003. Editor: J. Picard (2007)
Vol. 1897: R. Doney, Fluctuation Theory for Lévy Processes, Ecole d'Été de Probabilités de Saint-Flour XXXV-2005. Editor: J. Picard (2007)
Vol. 1898: H.R. Beyer, Beyond Partial Differential Equations, On linear and Quasi-Linear Abstract Hyperbolic Evolution Equations (2007)
Vol. 1899: Séminaire de Probabilités XL. Editors: C. Donati-Martin, M. Émery, A. Rouault, C. Stricker (2007)
Vol. 1900: E. Bolthausen, A. Bovier (Eds.), Spin Glasses (2007)
Vol. 1901: O. Wittenberg, Intersections de deux quadriques et pinceaux de courbes de genre 1, Intersections of Two Quadrics and Pencils of Curves of Genus 1 (2007)
Vol. 1902: A. Isaev, Lectures on the Automorphism Groups of Kobayashi-Hyperbolic Manifolds (2007)
Vol. 1903: G. Kresin, V. Maz'ya, Sharp Real-Part Theorems (2007)
Vol. 1904: P. Giesl, Construction of Global Lyapunov Functions Using Radial Basis Functions (2007)
Vol. 1905: C. Prévôt, M. Röckner, A Concise Course on Stochastic Partial Differential Equations (2007)
Vol. 1906: T. Schuster, The Method of Approximate Inverse: Theory and Applications (2007)
Vol. 1907: M. Rasmussen, Attractivity and Bifurcation for Nonautonomous Dynamical Systems (2007)
Vol. 1908: T.J. Lyons, M. Caruana, T. Lévy, Differential Equations Driven by Rough Paths, Ecole d'Été de Probabilités de Saint-Flour XXXIV-2004 (2007)
Vol. 1909: H. Akiyoshi, M. Sakuma, M. Wada, Y. Yamashita, Punctured Torus Groups and 2-Bridge Knot Groups (I) (2007)
Vol. 1910: V.D. Milman, G. Schechtman (Eds.), Geometric Aspects of Functional Analysis. Israel Seminar 2004-2005 (2007)
Vol. 1911: A. Bressan, D. Serre, M. Williams, K. Zumbrun, Hyperbolic Systems of Balance Laws. Cetraro, Italy 2003. Editor: P. Marcati (2007)
Vol. 1912: V. Berinde, Iterative Approximation of Fixed Points (2007)
Vol. 1913: J.E. Marsden, G. Misiołek, J.-P. Ortega, M. Perlmutter, T.S. Ratiu, Hamiltonian Reduction by Stages (2007)
Vol. 1914: G. Kutyniok, Affine Density in Wavelet Analysis (2007)
Vol. 1915: T. Bıyıkoğlu, J. Leydold, P.F. Stadler, Laplacian Eigenvectors of Graphs. Perron-Frobenius and Faber-Krahn Type Theorems (2007)
Vol. 1916: C. Villani, F. Rezakhanlou, Entropy Methods for the Boltzmann Equation. Editors: F. Golse, S. Olla (2008)
Vol. 1917: I. Veselić, Existence and Regularity Properties of the Integrated Density of States of Random Schrödinger (2008)
Vol. 1918: B. Roberts, R. Schmidt, Local Newforms for GSp(4) (2007)
Vol. 1919: R.A. Carmona, I. Ekeland, A. Kohatsu-Higa, J.-M. Lasry, P.-L. Lions, H. Pham, E. Taflin, Paris-Princeton Lectures on Mathematical Finance 2004. Editors: R.A. Carmona, E. Çinlar, I. Ekeland, E. Jouini, J.A. Scheinkman, N. Touzi (2007)
Vol. 1920: S.N. Evans, Probability and Real Trees. Ecole d'Été de Probabilités de Saint-Flour XXXV-2005 (2008)
Vol. 1921: J.P. Tian, Evolution Algebras and their Applications (2008)
Vol. 1922: A. Friedman (Ed.), Tutorials in Mathematical BioSciences IV. Evolution and Ecology (2008)
Vol. 1923: J.P.N. Bishwal, Parameter Estimation in Stochastic Differential Equations (2008)
Vol. 1924: M. Wilson, Littlewood-Paley Theory and Exponential-Square Integrability (2008)
Vol. 1925: M. du Sautoy, L. Woodward, Zeta Functions of Groups and Rings (2008)
Vol. 1926: L. Barreira, V. Claudia, Stability of Nonautonomous Differential Equations (2008)
Vol. 1927: L. Ambrosio, L. Caffarelli, M.G. Crandall, L.C. Evans, N. Fusco, Calculus of Variations and Non-Linear Partial Differential Equations. Cetraro, Italy 2005. Editors: B. Dacorogna, P. Marcellini (2008)
Vol. 1928: J. Jonsson, Simplicial Complexes of Graphs (2008)
Vol. 1929: Y. Mishura, Stochastic Calculus for Fractional Brownian Motion and Related Processes (2008)
Vol. 1930: J.M. Urbano, The Method of Intrinsic Scaling. A Systematic Approach to Regularity for Degenerate and Singular PDEs (2008)
Vol. 1931: M. Cowling, E. Frenkel, M. Kashiwara, A. Valette, D.A. Vogan, Jr., N.R. Wallach, Representation Theory and Complex Analysis. Venice, Italy 2004. Editors: E.C. Tarabusi, A. D'Agnolo, M. Picardello (2008)
Vol. 1932: A.A. Agrachev, A.S. Morse, E.D. Sontag, H.J. Sussmann, V.I. Utkin, Nonlinear and Optimal Control Theory. Cetraro, Italy 2004. Editors: P. Nistri, G. Stefani (2008)
Vol. 1933: M. Petkovic, Point Estimation of Root Finding Methods (2008)
Vol. 1934: C. Donati-Martin, M. Émery, A. Rouault, C. Stricker (Eds.), Séminaire de Probabilités XLI (2008)
Vol. 1935: A. Unterberger, Alternative Pseudodifferential Analysis (2008)
Vol. 1936: P. Magal, S. Ruan (Eds.), Structured Population Models in Biology and Epidemiology (2008)
Vol. 1937: G. Capriz, P. Giovine, P.M. Mariano (Eds.), Mathematical Models of Granular Matter (2008)
Vol. 1938: D. Auroux, F. Catanese, M. Manetti, P. Seidel, B. Siebert, I. Smith, G. Tian, Symplectic 4-Manifolds and Algebraic Surfaces. Cetraro, Italy 2003. Editors: F. Catanese, G. Tian (2008)
Vol. 1939: D. Boffi, F. Brezzi, L. Demkowicz, R.G. Durán, R.S. Falk, M. Fortin, Mixed Finite Elements, Compatibility Conditions, and Applications. Cetraro, Italy 2006. Editors: D. Boffi, L. Gastaldi (2008)
Vol. 1940: J. Banasiak, V. Capasso, M.A.J. Chaplain, M. Lachowicz, J. Miękisz, Multiscale Problems in the Life Sciences. From Microscopic to Macroscopic. Będlewo, Poland 2006. Editors: V. Capasso, M. Lachowicz (2008)
Vol. 1941: S.M.J. Haran, Arithmetical Investigations. Representation Theory, Orthogonal Polynomials, and Quantum Interpolations (2008)

Vol. 1942: S. Albeverio, F. Flandoli, Y.G. Sinai, SPDE in Hydrodynamic. Recent Progress and Prospects. Cetraro, Italy 2005. Editors: G. Da Prato, M. Röckner (2008)
Vol. 1943: L.L. Bonilla (Ed.), Inverse Problems and Imaging. Martina Franca, Italy 2002 (2008)
Vol. 1944: A. Di Bartolo, G. Falcone, P. Plaumann, K. Strambach, Algebraic Groups and Lie Groups with Few Factors (2008)
Vol. 1945: F. Brauer, P. van den Driessche, J. Wu (Eds.), Mathematical Epidemiology (2008)
Vol. 1946: G. Allaire, A. Arnold, P. Degond, T.Y. Hou, Quantum Transport. Modelling, Analysis and Asymptotics. Cetraro, Italy 2006. Editors: N.B. Abdallah, G. Frosali (2008)
Vol. 1947: D. Abramovich, M. Mariño, M. Thaddeus, R. Vakil, Enumerative Invariants in Algebraic Geometry and String Theory. Cetraro, Italy 2005. Editors: K. Behrend, M. Manetti (2008)
Vol. 1948: F. Cao, J-L. Lisani, J-M. Morel, P. Musé, F. Sur, A Theory of Shape Identification (2008)
Vol. 1949: H.G. Feichtinger, B. Helffer, M.P. Lamoureux, N. Lerner, J. Toft, Pseudo-Differential Operators. Quantization and Signals. Cetraro, Italy 2006. Editors: L. Rodino, M.W. Wong (2008)
Vol. 1950: M. Bramson, Stability of Queueing Networks, Ecole d'Eté de Probabilités de Saint-Flour XXXVI-2006 (2008)
Vol. 1951: A. Moltó, J. Orihuela, S. Troyanski, M. Valdivia, A Non Linear Transfer Technique for Renorming (2009)
Vol. 1952: R. Mikhailov, I.B.S. Passi, Lower Central and Dimension Series of Groups (2009)
Vol. 1953: K. Arwini, C.T.J. Dodson, Information Geometry (2008)
Vol. 1954: P. Biane, L. Bouten, F. Cipriani, N. Konno, N. Privault, Q. Xu, Quantum Potential Theory. Editors: U. Franz, M. Schuermann (2008)
Vol. 1955: M. Bernot, V. Caselles, J.-M. Morel, Optimal Transportation Networks (2008)
Vol. 1956: C.H. Chu, Matrix Convolution Operators on Groups (2008)
Vol. 1957: A. Guionnet, On Random Matrices: Macroscopic Asymptotics, Ecole d'Eté de Probabilités de Saint-Flour XXXVI-2006 (2009)
Vol. 1958: M.C. Olsson, Compactifying Moduli Spaces for Abelian Varieties (2008)
Vol. 1959: Y. Nakkajima, A. Shiho, Weight Filtrations on Log Crystalline Cohomologies of Families of Open Smooth Varieties (2008)
Vol. 1960: J. Lipman, M. Hashimoto, Foundations of Grothendieck Duality for Diagrams of Schemes (2009)
Vol. 1961: G. Buttazzo, A. Pratelli, S. Solimini, E. Stepanov, Optimal Urban Networks via Mass Transportation (2009)
Vol. 1962: R. Dalang, D. Khoshnevisan, C. Mueller, D. Nualart, Y. Xiao, A Minicourse on Stochastic Partial Differential Equations (2009)
Vol. 1963: W. Siegert, Local Lyapunov Exponents (2009)
Vol. 1964: W. Roth, Operator-valued Measures and Integrals for Cone-valued Functions and Integrals for Cone-valued Functions (2009)
Vol. 1965: C. Chidume, Geometric Properties of Banach Spaces and Nonlinear Iterations (2009)
Vol. 1966: D. Deng, Y. Han, Harmonic Analysis on Spaces of Homogeneous Type (2009)
Vol. 1967: B. Fresse, Modules over Operads and Functors (2009)
Vol. 1968: R. Weissauer, Endoscopy for GSP(4) and the Cohomology of Siegel Modular Threefolds (2009)
Vol. 1969: B. Roynette, M. Yor, Penalising Brownian Paths (2009)
Vol. 1970: M. Biskup, A. Bovier, F. den Hollander, D. Ioffe, F. Martinelli, K. Netočný, F. Toninelli, Methods of Contemporary Mathematical Statistical Physics. Editor: R. Kotecký (2009)
Vol. 1971: L. Saint-Raymond, Hydrodynamic Limits of the Boltzmann Equation (2009)
Vol. 1972: T. Mochizuki, Donaldson Type Invariants for Algebraic Surfaces (2009)
Vol. 1973: M.A. Berger, L.H. Kauffmann, B. Khesin, H.K. Moffatt, R.L. Ricca, De W. Sumners, Lectures on Topological Fluid Mechanics. Cetraro, Italy 2001. Editor: R.L. Ricca (2009)
Vol. 1974: F. den Hollander, Random Polymers: École d'Été de Probabilités de Saint-Flour XXXVII – 2007 (2009)
Vol. 1975: J.C. Rohde, Cyclic Coverings, Calabi-Yau Manifolds and Complex Multiplication (2009)
Vol. 1976: N. Ginoux, The Dirac Spectrum (2009)
Vol. 1977: M.J. Gursky, E. Lanconelli, A. Malchiodi, G. Tarantello, X.-J. Wang, P.C. Yang, Geometric Analysis and PDEs. Cetraro, Italy 2001. Editors: A. Ambrosetti, S.-Y.A. Chang, A. Malchiodi (2009)
Vol. 1978: M. Qian, J.-S. Xie, S. Zhu, Smooth Ergodic Theory for Endomorphisms (2009)
Vol. 1979: C. Donati-Martin, M. Émery, A. Rouault, C. Stricker (Eds.), Śeminaire de Probablitiés XLII (2009)
Vol. 1980: P. Graczyk, A. Stos (Eds.), Potential Analysis of Stable Processes and its Extensions (2009)
Vol. 1981: M. Chlouveraki, Blocks and Families for Cyclotomic Hecke Algebras (2009)
Vol. 1982: N. Privault, Stochastic Analysis in Discrete and Continuous Settings. With Normal Martingales (2009)
Vol. 1983: H. Ammari (Ed.), Mathematical Modeling in Biomedical Imaging I. Electrical and Ultrasound Tomographies, Anomaly Detection, and Brain Imaging (2009)

## Recent Reprints and New Editions

Vol. 1702: J. Ma, J. Yong, Forward-Backward Stochastic Differential Equations and their Applications. 1999 – Corr. 3rd printing (2007)
Vol. 830: J.A. Green, Polynomial Representations of $GL_n$, with an Appendix on Schensted Correspondence and Littelmann Paths by K. Erdmann, J.A. Green and M. Schoker 1980 – 2nd corr. and augmented edition (2007)
Vol. 1693: S. Simons, From Hahn-Banach to Monotonicity (Minimax and Monotonicity 1998) – 2nd exp. edition (2008)
Vol. 470: R.E. Bowen, Equilibrium States and the Ergodic Theory of Anosov Diffeomorphisms. With a preface by D. Ruelle. Edited by J.-R. Chazottes. 1975 – 2nd rev. edition (2008)
Vol. 523: S.A. Albeverio, R.J. Høegh-Krohn, S. Mazzucchi, Mathematical Theory of Feynman Path Integral. 1976 – 2nd corr. and enlarged edition (2008)
Vol. 1764: A. Cannas da Silva, Lectures on Symplectic Geometry 2001 – Corr. 2nd printing (2008)

Edited by J.-M. Morel, F. Takens, B. Teissier, P.K. Maini

**Editorial Policy** (for Multi-Author Publications: Summer Schools/Intensive Courses)

1. Lecture Notes aim to report new developments in all areas of mathematics and their applications - quickly, informally and at a high level. Mathematical texts analysing new developments in modelling and numerical simulation are welcome. Manuscripts should be reasonably self-contained and rounded off. Thus they may, and often will, present not only results of the author but also related work by other people. They should provide sufficient motivation, examples and applications. There should also be an introduction making the text comprehensible to a wider audience. This clearly distinguishes Lecture Notes from journal articles or technical reports which normally are very concise. Articles intended for a journal but too long to be accepted by most journals, usually do not have this "lecture notes" character.
2. In general SUMMER SCHOOLS and other similar INTENSIVE COURSES are held to present mathematical topics that are close to the frontiers of recent research to an audience at the beginning or intermediate graduate level, who may want to continue with this area of work, for a thesis or later. This makes demands on the didactic aspects of the presentation. Because the subjects of such schools are advanced, there often exists no textbook, and so ideally, the publication resulting from such a school could be a first approximation to such a textbook. Usually several authors are involved in the writing, so it is not always simple to obtain a unified approach to the presentation.

   For prospective publication in LNM, the resulting manuscript should not be just a collection of course notes, each of which has been developed by an individual author with little or no co-ordination with the others, and with little or no common concept. The subject matter should dictate the structure of the book, and the authorship of each part or chapter should take secondary importance. Of course the choice of authors is crucial to the quality of the material at the school and in the book, and the intention here is not to belittle their impact, but simply to say that the book should be planned to be written by these authors jointly, and not just assembled as a result of what these authors happen to submit.

   This represents considerable preparatory work (as it is imperative to ensure that the authors know these criteria before they invest work on a manuscript), and also considerable editing work afterwards, to get the book into final shape. Still it is the form that holds the most promise of a successful book that will be used by its intended audience, rather than yet another volume of proceedings for the library shelf.
3. Manuscripts should be submitted either online at www.editorialmanager.com/lnm/ to Springer's mathematics editorial, or to one of the series editors. Volume editors are expected to arrange for the refereeing, to the usual scientific standards, of the individual contributions. If the resulting reports can be forwarded to us (series editors or Springer) this is very helpful. If no reports are forwarded or if other questions remain unclear in respect of homogeneity etc, the series editors may wish to consult external referees for an overall evaluation of the volume. A final decision to publish can be made only on the basis of the complete manuscript; however a preliminary decision can be based on a pre-final or incomplete manuscript. The strict minimum amount of material that will be considered should include a detailed outline describing the planned contents of each chapter.

   Volume editors and authors should be aware that incomplete or insufficiently close to final manuscripts almost always result in longer evaluation times. They should also be aware that parallel submission of their manuscript to another publisher while under consideration for LNM will in general lead to immediate rejection.

4. Manuscripts should in general be submitted in English. Final manuscripts should contain at least 100 pages of mathematical text and should always include
   - a general table of contents;
   - an informative introduction, with adequate motivation and perhaps some historical remarks: it should be accessible to a reader not intimately familiar with the topic treated;
   - a global subject index: as a rule this is genuinely helpful for the reader.

   Lecture Notes volumes are, as a rule, printed digitally from the authors' files. We strongly recommend that all contributions in a volume be written in the same LaTeX version, preferably LaTeX2e. To ensure best results, authors are asked to use the LaTeX2e style files available from Springer's web-server at

   ftp://ftp.springer.de/pub/tex/latex/svmonot1/ (for monographs) and
   ftp://ftp.springer.de/pub/tex/latex/svmultt1/ (for summer schools/tutorials).

   Additional technical instructions are available on request from: lnm@springer.com.
5. Careful preparation of the manuscripts will help keep production time short besides ensuring satisfactory appearance of the finished book in print and online. After acceptance of the manuscript authors will be asked to prepare the final LaTeX source files and also the corresponding dvi-, pdf- or zipped ps-file. The LaTeX source files are essential for producing the full-text online version of the book. For the existing online volumes of LNM see: http://www.springerlink.com/openurl.asp?genre=journal&issn=0075-8434.

   The actual production of a Lecture Notes volume takes approximately 12 weeks.
6. Volume editors receive a total of 50 free copies of their volume to be shared with the authors, but no royalties. They and the authors are entitled to a discount of 33.3% on the price of Springer books purchased for their personal use, if ordering directly from Springer.
7. Commitment to publish is made by letter of intent rather than by signing a formal contract. Springer-Verlag secures the copyright for each volume. Authors are free to reuse material contained in their LNM volumes in later publications: a brief written (or e-mail) request for formal permission is sufficient.

**Addresses:**

Professor J.-M. Morel, CMLA,
École Normale Supérieure de Cachan,
61 Avenue du Président Wilson,
94235 Cachan Cedex, France
E-mail: Jean-Michel.Morel@cmla.ens-cachan.fr

Professor F. Takens, Mathematisch Instituut,
Rijksuniversiteit Groningen, Postbus 800,
9700 AV Groningen, The Netherlands
E-mail: F.Takens@rug.nl

Professor B. Teissier,
Institut Mathématique de Jussieu,
UMR 7586 du CNRS,
Équipe "Géométrie et Dynamique",
175 rue du Chevaleret,
75013 Paris, France
E-mail: teissier@math.jussieu.fr

*For the "Mathematical Biosciences Subseries" of LNM:*

Professor P.K. Maini, Center for Mathematical Biology,
Mathematical Institute, 24-29 St Giles,
Oxford OX1 3LP, UK
E-mail: maini@maths.ox.ac.uk

Springer, Mathematics Editorial I, Tiergartenstr. 17,
69121 Heidelberg, Germany,
Tel.: +49 (6221) 487-8259
Fax: +49 (6221) 4876-8259
E-mail: lnm@springer.com

Printing: Ten Brink, Meppel, The Netherlands
Binding: Stürtz, Würzburg, Germany